Intelligent Manufacturing

Chao'an Lai

Intelligent Manufacturing

 Springer

Chao'an Lai
South China University of Technology
Guangzhou, Guangdong, China

ISBN 978-981-19-0169-0   ISBN 978-981-19-0167-6 (eBook)
https://doi.org/10.1007/978-981-19-0167-6

This Springer imprint is published by the registered company Springer Nature Singapore Pte Ltd.
The registered company address is: 152 Beach Road, #21-01/04 Gateway East, Singapore 189721,
Singapore

Preface

In the report of the 19th National Congress, General Secretary Xi Jinping has addressed the importance of "accelerating the construction of manufacturing power, accelerating the development of advanced manufacturing industry, and promoting the deep integration of the Internet, big data, artificial intelligence technology, and the real economy." In response, this book explores the new mode of integration and development of the new generation of information technology and manufacturing industry, including the Internet, big data, and artificial intelligence technology. It promotes transforming and upgrading from "manufacturing" to "intelligent manufacturing." Intelligent manufacturing has been called the "fourth industrial revolution". This book shows the path and method of intelligent manufacturing system planning, design and implementation, through the deep integration of the Internet, big data, artificial intelligence and manufacturing process, to promote the transformation and upgrading of enterprises. This book consists of six chapters, shows the implementation of intelligent manufacturing process with 12 benchmarking enterprises, discusses the planning, implementation and control of intelligent manufacturing system technology and method of theory, analysis the five hierarchies of intelligent manufacturing system, the five stages of life cycle, five kinds of intelligent depth, as well as a comprehensive and structured implementation method, cultivate the reader's vocational ability to develop intelligent solutions and implementation based on complex, uncertain environment needs. This book will be interesting and useful to a wide readership in the various fields of management, information science, and engineering science.

I want to thank Ms. Lai Sirun for polishing the text. Thanks to Prof. Sun Yanming, Vice President of Guangzhou University, and Prof. Song Tiebo of South China University of Technology, for their guidance in academic research, and the Guangdong Province Soft Science Project (2019A101002006) for its support.

Guangzhou, China

Chao'an Lai

Contents

Chapter 1
Background, Basic Concepts and Methods

1.1 Inspiration Case: Huawei's Comprehensive Cloud

Huawei's digital transformation is a long-term process. Since 2002, Huawei started its transformation by introducing IBM's Integrated Product Development (IPD) and then introduced Siemens' manufacturing technology. When the company's sales revenue reached 20 billion yuan, Ren Zhengfei, president of the company, raised a question: If the company could achieve 50 billion yuan, would senior executives be able to manage such a scale? The answer is no, but IBM, the industry giant, has the ability, so Huawei takes it as a benchmark to learn and change. The IPD R&D process reform guided by IBM ensures the reliability of Huawei's entire R&D process. After reviewing the R&D process, Huawei believes that IBM's IPD may be a relatively slow and cumbersome R&D management system. Still, it conforms to Huawei's "stable" characteristics, so it has adhered to the implementation but constantly changed. The company has been thinking about adapting to the wave of Internet change, promoting the integration of CT (Communication Technology) and IT (Information Technology), and the adjustment of IPD. Hence, the IPD operated by Huawei now is entirely different from the IPD implemented more than a decade ago.

Above is just one example of Huawei's overall digital revolution, with only a starting point and no endpoint. According to the 2017 Global Connectivity Index (GCI) research report released by Huawei, the development of the global digital economy is accelerating. The correlation analysis between GCI indicators found that the cloud is one of the key engines. Based on the concept that others cannot restrain "key technologies", the company's overall cloud and overall optimization are ultimately decided. A digital Huawei will be built in 3–5 years.

The original version of this chapter was revised: Author provided correction in Fig. 1.13 have been incorporated. The correction to this chapter is available at
https://doi.org/10.1007/978-981-19-0167-6_7

C. Lai, *Intelligent Manufacturing*,
https://doi.org/10.1007/978-981-19-0167-6_1

Huawei needs to first digitize key business processes, then help companies from various industries complete digital transformation. Digitalization within an enterprise means opening up the business processes, from marketing to R&D, production, service, finance, and human resources. Huawei's current application systems, including information systems in production and office, currently have more than 600 information systems. This is similar to a workshop with more than 600 chimneys in the state of isolated islands. The flow resistance between the chimneys consumes much energy, so the transition from "chimney" to "pipe" must be promoted to form a continuous flow of information from design to production. Ren Zhengfei pointed out that IT reform should have a focal point. The speed and quality of the information flow equal profit. Every time you add a process, you subtract two; there are two review points removed for each additional review point. Huawei has also been making information-based changes, but never as determined as the comprehensive cloud.

The company has always emphasized making changes with a global perspective and experience. The transformation revolves around these two sentences: first, "more grain, increase the fertility of the land"; Second, "To get rich, first build roads." The first sentence means increasing sales revenue and laying the groundwork for the next sales team. The second sentence refers to the ability of the whole IT system to be improved. In the next five years, the company's sales revenue will probably exceed 1 trillion yuan. To build this system, the concept of a global resource pool is required first. For example, after the 5G R&D project is determined, resources will be organized in dozens of R&D centers worldwide to develop and lead the formulation of international standards for 5G carrier network construction. After completing the project, resources will be released back to the resource pool. The second is a unified data platform. A unified data platform should achieve complete and whole-process data homology and build a database. By creating such a base to support a variety of applications on the upper layer, but also the various applications call the data from the same data source, there will be no data deviation. This is an end-to-end intelligent operation based on the R&D cloud, terminal cloud, logistics cloud, manufacturing cloud, collaborative office cloud, etc. All applications are on the cloud, a unified data platform.

R&D cloud is the first Huawei cloud. This is used to support software development and system CAD/CAE. The R&D cloud is the spillover of the software development capability that Huawei has accumulated over the past 30 years. It started in 2008, and the cloud's first goal was to keep software code from getting leaked, because this kind of code leakage occurred every year. Before R&D cloud implementation, for visitors coming to the computer area, Huawei confidential areas are labeled as blue, red, or yellow, all mobile phones of visitors can't contain a camera, all people into and out of the region need to go through "inhuman" body search, but even so, there have been many code leaks, such as R&D department personnel getting to be self-employed, in the production of similar products. After the cloud implementation, all the code cannot be copied out. The second purpose of cloud development is to improve efficiency. In iterative development mode, the number of software iterations per day is a magnitude previously unimaginable, which requires high efficiency in the compilation process. If each department built its own server, it would be inefficient because a single server would not have the flexibility and computing power of the cloud. Therefore, Huawei created the R&D cloud based on security and efficiency and also built the test cloud, design cloud, and simulation cloud for the same reason. At present, in the office of more than 80,000 R&D personnel of Huawei, all R&D personnel have only one screen and no-host machine. They are all connected to the cloud and developed on the desktop cloud.

The second representative Huawei cloud is a terminal cloud. At first, the electronic mall was unwilling to put on the cloud because The Cloud had just started. The mall will do sales promotion at 10 a.m. on Monday, Wednesday, and Friday. At this time, there will be panic buying. There is often much concurrent shopping, which first has a great impact on the network and then on the back-end system. The traditional way is capacity expansion, which is challenging to achieve elastic growth. The concurrent effect is easy to cause a system shutdown. The terminal cloud can take advantage of the unique scalability and agility of cloud to withstand the shock of uncertainty. So later, the terminal business of the mall was put on the cloud platform, which was the Huawei Terminal cloud. At present, Huawei is starting to carry out intelligent and customized work. For example, the mobile phones ordered in the mall can be marked with buyers' names. In the future, the company should personalize it on a large scale to realize ultra-flexible manufacturing.

The third Huawei cloud is the logistics cloud. In logistics, the application of big data and AI (artificial intelligence) is of great size. The first is to use big data and AI algorithms to optimize the entire logistics route. Huawei delivered 3.7 million orders and 70,000 logistics routes in 2017. Using data to optimize the routes is crucial. Before the cloud, Huawei used only internal data. No external data was introduced, such as port congestion location, unrest events, weather conditions, etc. So the prediction effect of the algorithm was inferior. After submitting external data to optimize the algorithm, and through the decentralized collection and centralized analysis of cloud platform data, logistics cloud supports the annual delivery of 3.7 million orders, supports the whole process of transportation goods management, and realizes intelligent operation. In the logistics department, there is a big screen showing the status of each order of goods. It clearly shows which port the goods will arrive at, predicting the risk of accidents based on the big data of logistics and how many days the accidents will affect the delivery time. The most immediate advantage of this is that, in addition to less air transportation, more sea transportation will reduce the cost. Shipping alone saved more than $17 million in 2017.

The fourth Huawei cloud is a collaborative office cloud. At present, Huawei can completely solve all office problems and realize the connection of all elements with only one mobile phone. The first is the connection between employees. When employees from different countries communicate, they can use the translation function of the collaborative office cloud to achieve direct communication. The second is the connection of businesses. All Huawei businesses correspond to several apps (small applications). The software WeLink (Huawei collaborative office platform) integrates email, messages, meetings, knowledge, videos, to-do approval and other office scenes, which solves the connection of all businesses without needing a full-screen APP. Huawei has operations all over the world, and the cost of communicating across borders is rising. At present, through WeLink video conferencing, Huawei employees, customers, and suppliers can access the video conferencing via mobile phones, tablets, PCS, and other terminals by clicking on a link, thus creating a real-time "face-to-face" communication experience. The third is connected knowledge. All new knowledge supports listening and watching, which is similar to listening applications such as Himalaya and Zhihu.

The fourth is connection equipment. Huawei has a large amount of equipment all over the world. Many laboratories have idle equipment, and some places have no equipment. Now you can access it through the office's WIFI. You can install Radio Frequency Identification (RFID) tags on each device, add location information and send it back so that all the information about the device can be seen from inside Huawei. There are only two or three former asset managers left. This is the internal ecosystem of Huawei connectivity.

The last kind of Huawei Cloud is manufacturing clouds. Huawei puts much emphasis on quality. A comprehensive platform of manufacturing clouds supports the whole end-to-end quality. Benchmarking analysis of German and Japanese enterprises plays a vital role in forming Huawei's "big quality concept." Germany's "standard first, construction of quality management system that doesn't depend on people" and Japan's "lean production theory as the core, bad quality is a waste, reduce waste and execute a continuous improvement cycle," these two different theoretical systems help Huawei construct "zero defect" culture and customer-oriented quality loop. Drawing on and applying German process flow and industrial software, the Japanese lean production mode and quality management method of "do one, deliver one, check one" without breakpoint is embedded into it, achieving a smooth flow of 98% throughout efficiency. There are MES (Manufacturing Execution System), ERP (Enterprise Resource Planning), and other systems on the cloud. A large screen in the Songshan Lake R&D center displays real-time production capacity, including quality and data, of Huawei's five global supply centers, 15 OEM and ODM plants. All the data sent back from the supplier is shown above. For factories in foreign countries, for example, the new mobile phone Mate10/Pro production line is far away in Latin America instead of Dongguan base in Shenzhen. At this time, how to import all product-related capabilities such as process and raw materials remotely is a crucial problem. Through data homology, Huawei can supervise all production line data and realize allocation and remote control of front-line resources.

After Huawei went into the cloud, it first helped its equipment operators transform to the digital world and open up new markets. Huawei helps operators develop video businesses. At the same time, the cloud services will enable industry digitization, providing computing, storage, networking, enterprise communications, enterprise connectivity. IoT services can help corporate and government customers participating in the $15 trillion industry digitization market over the next decade.

The second is to build a new digital transformation ecosystem based on vertical industry alliances. Huawei will continue to increase the construction and investment in industry alliances, business alliances, open-source communities, developer platforms, and other fields to achieve a win–win situation. In the next five years, Huawei plans to take the lead in realizing ROADS (real-time, on-demand, all-online, DIY, and social sharing) experience in business fields such as R&D, sales service, and supply. The ROADS experience becomes the core experience of the end-user requirements criteria. The first measure of user experience is the accuracy of the supply transaction process. One of the main problems with supply is uncertainty. The supply to be managed is at both ends. The first is the supplier, which is also the raw material source. The second is the client. The front-line project manager requires the manufacturing department to prepare the goods according to the forecast sales volume, but the actual customer demand fluctuates wildly. The simple requirement of "accurate rate of delivery" cannot be met by traditional forecasting methods, so it need to turn to analysis based on big data. In 2017, storage prices soared, but Huawei failed to anticipate this risk, so it was very passive in cost competition. The second indicator of user experience is the level of Internet governance. Huawei attaches more importance to network security and has built a security capability center inside the company to strictly guarantee the security of products and solutions launched by Huawei. Increase investment in research on new threat technologies and defense solutions, threat intelligence system building, and security emergency response to better serve users and ensure customer network security.

Open ROADS Community advocates an "Open" attitude to cooperate with all parties, brainpower and implement the development and interests of the communications industry. Huawei has committed to four openings, namely open to all sectors, available to eco-partners, open to industry organizations, and open cloud lab for innovation incubation validation.

In China National Petroleum Corporation, Huawei has built the largest data center in Asia, which can analyze and process the production process of oil and gas exploration, storage and transportation, refining and marketing, and the massive data generated. As early as 2013, Sinopec started the construction of smart factories. The intelligent factory of Sinopec will be jointly built by Huawei and The Pacific Century Company. At present, the utilization rate of advanced control and automatic collection rate of production data of the four pilot enterprises have reached more than 90%, the automatic monitoring rate of discharge pollution sources has reached 100%, and the labor productivity has also increased by more than 10%. By 2018, data collection of 1870 oil wells in the northwest oil field has been fully covered. Automatic data acquisition and video monitoring of process parameters of more than 600 critical wells and 125 stations have been realized, and unattended storage of 19 stations has been achieved. The number of oil and gas wells in the northwest oil field has increased by more than 600 compared with 2011, and the total number of external labors has decreased by more than 1000.

Thinking Exercise

1. What stage has Huawei's intelligent manufacturing gone through, and is it entering? What are the main achievements of Huawei's work at each stage?
2. What business changes has Huawei made? What enabling techniques are used? Which is the fundamental driver, business or technology?
3. How to understand integrated product Development (IPD) integrating product development, component sourcing, and production start-up? What are the steps in the product development process?

4. Try to analyze the hierarchical nature of the products and services Huawei provides for the oil industry.
5. What is the significance of product technology standardization for Huawei's intelligent manufacturing and company growth?

1.2   The Introduction

What is the background of the era of social change in which intelligent manufacturing (IM) comes into being? Where is it driven? What connotation, goal, and evaluation measure should it have? What reference architecture models can adopt the implementation? What criteria should be followed in the implementation of IM? The analysis of the above problems is very important for the practitioners of IM to establish a correct system viewpoint and grasp the core model ideas. The following course will answer these questions.

1.3   Background

1.3.1   China's Manufacturing Industry Environment

China's modern industry originated from the Anqing Ordnance Institute, founded by Zeng Guofan in 1861. In 1862, the first Marine steam engine designed and manufactured by Chinese people was born, which meant that the Chinese industry entered modern times. After 70 years of development since the founding of the People's Republic of China and 40 years of reform and opening-up, China has ranked first in the world in terms of the output of 220 of 500 major industrial products, making it the world's largest manufacturing country. A comprehensive, independent, and complete industrial system supported China's status as a significant country in the world.

Since the reform and opening-up, China's manufacturing industry has achieved success mainly on a large scale and at a low cost and developed rapidly from the low-end manufacturing industry. However, after the turning point of the financial crisis in 2008, the two premises of the previous manufacturing development model have changed. First, with the general shortening of the product life cycle, as shown in Fig. 1.1, the variety of products and materials caused by personalized customization has increased. It weakens the advantage of economies of scale. Suppose an enterprise does not have the voice and dominate the market and customer contact and does not benefit from mass production. In that case, it will put tremendous pressure on profits. Second, the advantages of demographic dividends are gradually disappearing, and the costs of various factors are rising. As a result, the benefits of low cost in China have been significantly reduced. Changes in these two premises mean that traditional manufacturing models will fail.

In sharp contrast to the overcapacity of the low-end manufacturing industry, China's high-end manufacturing industry is seriously lacking. For example, the C919, a domestic medium aircraft, is not expected to enter service until 2020. The domestic

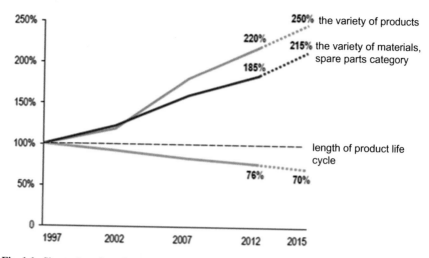

Fig. 1.1 Shortening of product life cycle and increasing of product and material types

Fig. 1.2 Four industrial revolutions

engine is not likely to be delivered until 2025. By contrast, the same Boeing 737 was in service 50 years ago; the production of Boeing's 2017 order was scheduled to be seven years later. China's ICT (Information Communications Technology) core components industry is far away from the world's first echelon. For example, Taiwan Semiconductor Manufacturing CO. (TSMC) has been trial-producing 7 nm chips. Our country lacks "Equipment for manufacturing equipment," specifically including industrial software, high precision NC machine tools and production of chips, and overloading high-speed precision industrial robots and reducers. General Secretary Xi Jinping stressed that "core technologies are the most important in the country.

We need to be determined, persistent, and focused on accelerating breakthroughs in core technologies in the information field."

After the world financial crisis in 2008, developed countries implemented the strategy of "reindustrialization" one after another and competed to raise the revival of the manufacturing industry to a strategic height. Southeast Asian countries are also actively undertaking the transfer and distribution of the global manufacturing industry. Since 2009, the US has issued the National Manufacturing Innovation Network and Advanced Manufacturing 2.0. Germany issued "A strategic proposal for implementing industrial 4.0" in 2013. A series of policies issued by the United States, Britain, France, Germany, and Japan are intended to make high-end manufacturing flow back to developed countries and prevent the transformation and upgrading of China's manufacturing industry.

Xi said that in the past, what we had to solve was the problem of "to have it or not." However, now we have to solve the problem of "excellence or not?". Solving the question requires structural reforms, supply-side reform, and upgrading of IM.

At present, China's manufacturing industry is facing a severely competitive environment. China's manufacturing industry is in urgent need of transformation and upgrading. Its path has three directions: seeking new cost advantages, developing and building the ability to differentiate, and realizing business model innovation. These three directions are also the goals of IM.

The five development concepts of "innovation, coordination, green, openness, and sharing" put forward at the 18th National Congress have led to profound changes in China. It pointed out the direction for the development of the manufacturing industry and the long-term planning of IM.

1. Innovation is the driving force of an enterprise's development and the core of a nation's progress. There are three forms of innovation: first, the original innovation, such as the four great inventions of China; Second, comprehensive innovation, also known as integrated innovation, is formed by the integration of different disciplines and technologies, such as the C919 mid-sized passenger plane. Third, digestion, absorption, and innovation. A typical example is China's high-speed rail technology, through exchanging markets for technology, the introduction of digestion and absorption of Germany's Siemens, Japan's Kawasaki heavy industries, France's Alstom, and Canada's Bombardier technology, through networked collaboration, finally realizes the innovation. In our country, there is a lack of original innovation. Originality will produce patent barriers, technical barriers, and standard barriers, which increase the cost of subsequent entrants and make subsequent entrants always in the state of passive tracking. Therefore, it is urgent to promote the original innovation of our country. Whether it is made in China 2025 or other fields of scientific research, there is an urgent need for original innovation, which runs through all our work.

2. Coordinated Development Focuses on solving the problem of unbalanced and uncoordinated development. We will focus on coordinated development among regions, integrated military-civilian development, the Guangdong-Hong Kong-Macao Greater Bay Area, the Beijing-Tianjin-Hebei Region, and the Yangtze

River Economic Belt. We will accelerate the rational distribution of industries, the upstream and downstream linkage mechanism. Also, promote world-class advanced manufacturing's synergy and cluster (agglomeration) effects through integration and cooperation, realizing transformation and upgrading through IM. It will also be a powerful economic engine for our country.

3. Green development pays attention to solving the contradiction between people and nature. General Secretary Xi said, "We want both clear water and green mountains, and we also want gold and silver mountains." "Clear water and green mountains are gold and silver mountains." The contradiction between "clear water and green mountains" and "gold and silver mountains" needs to be solved through the transformation and upgrading of IM. Only when the contradiction is solved can innovation be achieved. Low-carbon energy development is a long-term trend and a new direction of international technology competition. We should promote green and low-carbon development and transform the economic structure in the process of eliminating the "three high industries," reduce the proportion of the three high industries in the economy, and increase the balance of green and low-carbon initiatives in the economic structure. This is about green manufacturing, which is one of the goals of advanced manufacturing.

4. Open development pays attention to addressing the problem of the internal and external linkage. China's manufacturing cost advantage has been dramatically reduced. The country can promote its position in the global value chain by constructing "One Belt and One Road," upgrading its IM capacity, and exporting China's industrial capacity and high-end IM service capacity overseas. The "One Belt and One Road" route covers more than 60 countries, covering 4.4 billion people, accounting for 63% of the global population. Its economy of 21 trillion dollars accounts for one-third of the global GDP, with huge market space in infrastructure construction, logistics equipment and services, energy applications, and other fields.

5. Shared development focuses on addressing issues of social equity and justice. We will pursue shared development so that all our people will have a greater sense of gain from shared development. At present, there are four main modes of capacity sharing in China's manufacturing industry, namely intermediary sharing platform, mass innovation sharing platform, service sharing platform, and collaborative sharing platform. Intermediary sharing platforms usually do not have manufacturing resources but only provide docking services for supply and demand parties, such as ZhuBajie.com and Alibaba's Tao Factory. Mass innovation sharing platforms are generally open platforms built by large manufacturing enterprises, such as the COSMOPlat industrial Internet platform of Haier, the Midea Cloud platform of Midea Group, and the Aerospace Cloud Network of Aerospace Group. Service sharing platforms are usually built by industrial technology-based enterprises, such as "Foxconn Cloud" and the I5 platform of Shenyang Machine Tool Factory. The Collaborative sharing platform allows multiple enterprises to jointly use cloud services and various production resources to achieve collaborative production, such as "Business help" (http://wayboo.org.cn/). In 2017, the manufacturing capacity sharing

market size was about 412 billion yuan, a 25% increase over the previous year. The number of enterprises providing services through capacity-sharing platforms exceeded 200,000.

As one of the "new drivers" of economic development, the sharing economy has a bright future. Its essence is to integrate idle resources, invigorate the stock economy, reduce waste, and avoid exploiting new resources. China is a global leader in sharing economy and mobile payment. At the same time, we should be soberly aware that success in these areas is mainly due to China's huge market and high penetration rate of mobile phones. It is a success driven by scenarios and models rather than technological innovation. The focus of industrial development will undoubtedly change from mode innovation to technological innovation in the future.

[Case: "Mold Cloud Design Platform" of Qingdao Haier Mold Company]
To solve the contradiction between the enterprise's ability to receive orders and the enterprise's personnel cost, Qingdao Haier mold Company started to build a "mold cloud design platform" in 2013, to change the closed situation that the previous work was all completed by the enterprise employees. Now, hundreds of certified and credit-approved external corporate engineers are populating the platform. These engineers, known as "cloud resources," will receive orders remotely according to technical requirements, price, time limit, and other information. After the task is delivered and accepted, the salary will be paid online in real-time. The whole work will be connected and coordinated through the network. In this way, through the sharing and collaboration mode of "crowdsourcing," technical personnel with various skills in society can take advantage of their time resources and intellectual resources to undertake the work matching their skills and realize the value of their "surplus wisdom." To solve the contradiction of opaque information of supply and demand and mismatching issue, avoid overcapacity, and find a suitable factory for orders, a factory in the future is bound to break its organization "walls". It will give their orders, equipment, materials, personnel, and other information via social and open sharing platforms to realize the open collaboration of intra-enterprise and inter-enterprise, building a network of social and open sharing ecosystems.

1.3.2 New Industrial Revolution

As shown in Fig. 1.2, steam engines emerged during the first Industrial Revolution, which used machinery to convert heat into power. It was an era of "steel + steam" as power. During the second industrial Revolution, electric motors and production lines appeared, providing immensely stable lighting energy and large-scale mechanized production with continuous electric energy, and society entered the electrical era. During the third industrial Revolution, programmable logic controllers, computers, and the Internet appeared. The software was added into product elements and systems as a manufacturing factor, and human beings have entered the information age.

Every industrial revolution is a long process. The core of the fourth Industrial Revolution is IM. The Boeing 777, the world's first fully digital aircraft designed and built in 1991, is an essential milestone in IM. In the process of its development, more than about 1700 engineers in the global use of 8 large Computer, 3200 sets of CAD (Computer-Aided Design) workstations, more than 20,000 connected to the PC (Personal Computer), more than 800 sets of related software, formation of 14 BOM (Bill of Material), the realization of PDM (Product Data Management) for configuration management, workflow management and engineering change, rather than just conventional file management, dramatically improves the efficiency. The 777 has been both a technical and commercial success. The first 777 is of better quality than the Boeing 747, which has already built 400. Its development cycle is only four and a half years, significantly shorter than the Boeing 767's 12 years.

Manufacturing is the main focus of "Internet Plus." At the same time, IM is the core technology of the new round of industrial revolution and the main focus of Made in China 2025. The new generation of information technology and intelligent technology is accelerating the in-depth integration of the manufacturing industry, namely the in-depth integration of "intelligent + manufacturing", which is the upgraded version of the integration of information and industrialization. The fusion process of "intelligent + manufacturing" is shown in Fig. 1.3. It does not affect the manufacturing industry from a single technological aspect but brings profound changes to the manufacturing industry from research and development design, production and manufacturing, industrial form, and business model. This profound intellectual transformation has the potential to be a hallmark of the fourth industrial revolution. All countries worldwide have launched a fierce competition for this, and intelligent manufacturing has become a strategic commanding point to lead the future development of the world's manufacturing industry.

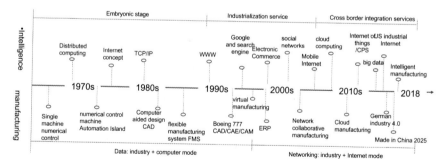

Fig. 1.3 Deep fusion process of "intelligent + manufacturing"

1.3.3 Next-Generation Manufacturing Model

Over the past decades, many different manufacturing modes (or paradigms) have been formed in the practice evolution, including lean manufacturing, flexible manufacturing, concurrent engineering, green manufacturing, agile manufacturing, digital manufacturing, computer integrated manufacturing, distributed networked manufacturing, virtual manufacturing, cloud manufacturing, IM, etc.

Lean Production (LP) encompasses the concepts of just-in-time (JIT), Theory of Constraints (TOC), agile manufacturing and also complements the Six Sigma Constraints with the aim of reducing defects.

Flexible Manufacturing (FM) refers to the system's responsiveness to adapt to changes in the external environment and internal disturbances (such as machine failure) and maintain stable production, as well as the mode of multi-variety and small-batch production capacity that adapts to personalized customization. A Flexible Manufacturing System (FMS) is an automatic Manufacturing System composed of several numerical control equipment, material transportation, and storage devices, and a computer control system, which can be adjusted rapidly according to the changes of Manufacturing tasks and product varieties.

Concurrent Engineering (CE) refers to the requirements of product development staff from the start, considering the whole product life cycle from concept formation to product scrap's stage, factors such as user and function demand, production, assembly, quality, cost, maintenance, recycling, and the environment. The core of the model is parallel design, etc., and stressed that all departments work together, through the establishment of policymakers in different stages of the effective mechanism of information exchange and communication between the possible problems in the later stages were found in the early stages of design, and resolved, Thus, it can improve the manufacturability, maintainability, recycling, and regeneration of the products, minimize the design repetition and shorten the product development and manufacturing cycle. The core of the model is parallel design. The concept of environment-oriented design and environment-oriented Manufacturing For Environment contained in concurrent engineering is consistent with Green Manufacturing.

Green Manufacturing (GM) is a manufacturing model that comprehensively considers the environmental impact and resource benefits and reflects the sustainable development strategy of humankind. Its goal is to minimize the adverse effects on the environment, maximize the utilization of resources, and coordinate and optimize the economic and social benefits of enterprises throughout the product life cycle from design, manufacturing, packaging, transportation, and use to scrap treatment.

Agile Manufacturing (AM) refers to the Manufacturing enterprise in the face of market demand and new opportunities, using modern means of communication, through the rapid configuration of various resources, including technology, management, personnel resources, and different companies and organizations, to form virtual enterprise and dynamic alliance, in an effective and coordinated response to market demand and opportunity, realize the agility of manufacturing mode. When market opportunities disappear, or tasks are completed, virtual enterprises disintegrate.

Digital Manufacturing (DM) refers to on the basis of 3D digital modeling technology, and in the numerical control processing, CAD/CAM/CAE, rapid prototyping, support technology such as information management, under the support of the implementation of product design and the function of simulation and prototype manufacturing, then quickly produce products from user requirements.

Computer Integrated Manufacturing (CIM) refers to the organic integration of various isolated automatic subsystems dispersed in the product design and manufacturing process through computer technology to form an Integrated Manufacturing mode suitable for multi-varieties, small batch production, and improve the overall efficiency. At present, CIM has been changed into "modern integrated manufacturing" in China, and "integration" has a broader content.

Networked Manufacturing (NM) is the combination of agile Manufacturing, Collaborative Manufacturing mode, which is based on network technology, based on the digital, flexible, effective, and dynamic alliance of mutual benefit, effectively realize the reorganization of research, design, production, and sales resources of the supply chain, to improve the rapid response and market competition ability of the enterprise. NM can also be called distributed network manufacturing and networked collaborative manufacturing.

Virtual Manufacturing (VM) refers to the use of Virtual Reality technology, simulation technology, computer technology in manufacturing activities to find possible problems in manufacturing before the actual production products, to reduce costs, shorten the product development cycle, increase the competitiveness of products. Therefore, it is a crucial way to realize concurrent engineering.

Cloud manufacturing (CM) refers to the "manufacturing as a service" concept, based on taking including Cloud computing, manufacturing technology, and the emerging IoT technology, modern information technology, support manufacturing in a wide range of network resources environment, provide dynamic, easy extension and often virtualized resources, to achieve low cost and the globalization of networked manufacturing service model.

The above manufacturing modes have played a positive historical role in guiding the technological upgrading of the manufacturing industry. However, at the same time, the different emphasis of the manufacturing paradigm has caused many troubles for the transformation and upgrading practice of manufacturing enterprises. In the face of emerging new demands, new technologies, new ideas, and new models, it is necessary to summarize and extract a comprehensive and new manufacturing model.

From a global perspective, the evolution stage of the manufacturing industry is roughly the same as the four industrial revolution stages, which can be divided into the mechanized manufacturing stage, electrical manufacturing stage, automatic manufacturing stage, and intelligent manufacturing stage. According to the development level of manufacturing technology, production organization, and management concept, the development process of manufacturing mode is summarized into eight stages: manual workshop production, machine production, mass production, low-cost mass production, high-quality production, networked manufacturing, service-oriented manufacturing, and Intelligent Manufacturing. Roughly divided, intelligent manufacture on the development and implementation in time sequence logic can be divided into three phases: digital, networked, intelligent.

According to the NIST (National Institute of Standards and Technology)'s understanding. **Intelligent Manufacturing (IM)** is the next generation of manufacturing model to solve the following problems: more diverse customized services, smaller production batches, and unpredictable supply chain changes and disruptions. In a word, the goal is to solve the uncertainty caused by the complexity of customized production. Uncertainty is the fundamental characteristic of the decision-making process. As users tend to personalized, diversified, and constantly changing products, the product configuration leads to the complexity of the supply chain, manufacturing process, and project management, thus bringing high uncertainty. Certainty is what the industry seeks. The essence of IM is to reduce uncertainty.

As shown in Fig. 1.4, after the historical process from manual customization to mass manufacturing to mass customization, the production mode in the future will be the coexistence of single-variety mass production, multi-variety small-batch production, and personalized customization. Customization has a long history, with companies such as Germany's Mercedes-Benz and Britain's Rolls-Royce taking pride in customizing cars for the royal family. Due to the need to develop special automobile molds and process equipment for each particular customized vehicle, the high cost of molds, tools, and other hardware leads to the high cost of personalized, customized vehicles. So the industry began to research, if in the form of software instead of hardware in response to a variety of personalized needs, through the software in use rather than hardware such as die change to complete the personalized product design,

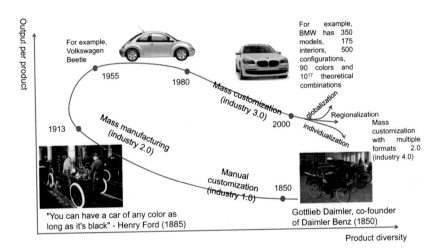

Fig. 1.4 Evolution of manufacturing patterns from the perspective of personalized needs

process, and manufacture, using a large number of industrial software to deal with the uncertainty of the complex product individuation, can lower cost to realize the personalized custom, which is the nature of IM.

Therefore, uncertainty is the most challenging problem for IM to solve, and software to control the automatic orderly flow of data and solve the uncertainty caused by personalized orders for complex products are the essential characteristics of IM. IM is to regularize the operation of the physical world by mapping from the physical world to model, from model to algorithm, from algorithm to code, from code to software, then using software to optimize the operation of the physical world. *Manufacturing intelligence* is a "three-body intelligence" composed of the physical entity, conscious human body, and digital virtual body. The software can be used to assign value, power, and wisdom to R&D and manufacturing activities. At present, a highly configured car has nearly 100 embedded systems and nearly 10 million lines of software code. As you can see, software definitions are now actually occurring across the manufacturing industry, especially in complex product systems.

[Case: Foxconn's camera and Huazhong CNC chromatogram]
With the help of Tencent, Foxconn used an 8 k camera to capture the changes in the manufacturing process, simulating highly skilled workers' experience and making fine adjustments according to the changes in production. For example, through camera analysis, it is found that there is a loss in the shell mold of mobile phones, which is highly nonlinear and uncertain, and the system automatically adjusts the process parameters such as the feed rate in order to reduce product deviation, rather than to throw away the mold while the mold is achieving a certain amount of production. AI improves accuracy and life in this way.

Researchers at Huazhong Numerical Control collect speed, acceleration, vibration, and fluctuation data of milling cutters during metal cutting through sensors and use the data to find the machining error and its causes. Chromatograms are drawn and observed through tens of thousands of experiments on various body models, such as human face models. It is found that there is a significant vibration and machining error at the corner of the body in the process of tool machining. By suppressing the tool's vibration caused by the fluctuation of the spindle speed, the machining accuracy of the domestic CNC system reaches 0.01 μM.

1.4 Connotation and Goal of IM

1.4.1 Definition of IM

Based on the above research, the definition of IM can be: IM is based on the digital network manufacturing technology, intelligent technology, data-driven and the software-defined technology to build a closed-loop of data to software control data automatic orderly flow to eliminate the uncertainty of complex systems.

We can understand the meaning of IM from the following aspects:

Intelligent mechanism: situation awareness—real-time analysis—human–machine decision—optimization of execution—autonomous adaptation;

System composition: The human–machine fusion system integrates human, physical, and control systems.

Operation objects: data as the digital carrier of information and knowledge;

Enabling: Algorithms, models, rules, and knowledge in software to form the "software-defined";

Essential features: using software to control the automatic and orderly flow of data to solve the uncertainty of complex products;

Objective: To eliminate the uncertainty of complex systems caused by customization;

Basic logic: tacit knowledge → explicit knowledge → software → chip → hardware → physical equipment, forming intelligent products and devices.

Value: Optimize the allocation of manufacturing resources to achieve agile, high quality, high efficiency, low cost, sustainability, user satisfaction;

Mode innovation: IM guides new business forms such as personalized customization, collaborative manufacturing, remote operation and maintenance, and promotes enterprise transformation.

Software definition is the most fundamental characteristic of IM. The core features of IM include a comprehensive digital manufacturing enterprise that enhances interoperability and productivity; Realize real-time control and small-batch flexible production through device interconnection and distributed intelligence; Coordinated supply chain management that responds quickly to market changes and supply chain disorders; Integrated and optimized decision support to improve energy and resource use efficiency.

If the human body is compared to the IM system, the brain is composed of various controllers and industrial software with various algorithms and digital models. The five sense organs and nerve endings are all kinds of sensors such as machine touch, vision, and hearing. Skeleton is the foundation of a network with workshops; Blood is equivalent to data flow, logistics, new product introduction; Limbs are the body of industrial robots and all kinds of intelligent equipment. The integration of these subsystems constitutes the IM system.

[Case: Geely uses simulation to improve welding accuracy]

In Geely Auto's first workshop in China, where conventional, hybrid and pure electric vehicles can be produced simultaneously, 183 data are collected into the simulation system to establish a digital plant. Before the adjustment simulation, the chassis welding accuracy reaches 96% at most. Now data was used to simulate the adjustment of equipment thousands of times for each day, the diameter of 12 positioning holes for welding is reduced by 0.2 mm, and the gap between the holes and pins is only 0.1 mm. The welding point data in the welding production process is entered into the simulation system and compared with the model data in real-time, and the welding accuracy reaches 100%.

1.4.2 The Goal of IM

The overall goal of IM is to realize the intelligence and innovation of the entire manufacturing value chain to promote the deep integration of information and industrialization. According to the US National Institute of Standards and Technology, IM has four goals: agility, quality, productivity, and sustainability. Table 1.1 shows the breakdown and measurement of the critical target capabilities of IM by the National Institute of Standards and Technology. It is necessary to balance four goals and 14 capabilities to become successful in IM.

The IM system is a part of the enterprise management system. Like other elements of the enterprise, the ultimate goal of the system is to serve enterprise production, reduce costs and improve the quality of products and services. The transformation and upgrading of the new-generation manufacturing industry are multi-objective. If the establishment of the IM system deviates from some key goals, it will inevitably lead to the failure of system construction and implementation. For example, if someone wants to set up an automatic assembly line, it can save much labor, but if the daily maintenance cost is higher than the reduced labor cost, it may make the operation cost not reduced but increased, so it is not sustainable. As another example, some people think that the goal of IM is personalization. However, it is prejudice. In many products and services, personalized customization is not the first option. Low cost, on-time delivery, high-quality products, and services are still the mainstream demands. Therefore, mass production, small batch production, personalized customization will co-exist for a long time.

1.5 International Comparison and Experience Reference of IM

See Table 1.2.

1.5.1 International Comparison of IM Strategies

1.5.1.1 U.S.

The advantages of the US include: (1) Microsoft, Google, IBM, and other IT giants and a large number of IT enterprises, and CAX/PLM/ERP and other industrial software, chip, big data, AI. Generally speaking, all software used for industrial purposes is industrial software. Industrial software is the crystallization of industrial knowledge and know-how accumulated over a long period of industrialization. The difficulty of industrial software lies in modeling, and the focus is on simulation. Industrial software has emerged since the 1960s from Boeing, Lockheed, NASA,

Table 1.1 Decomposition and measurement of key target capabilities of IM

Competitive strategy	Key objectives of IM	Target capability decomposition	Performance measurement
Cost leadership strategy	Productivity (P)	Production capacity	A product produced by a machine, line, unit, or factory within a specified period of time
		OEE	Overall Equipment Effectiveness: availability × performance × quality
		Material/energy efficiency	Substance/energy (electricity, steam, fuel oil, gas oil, etc.) used to produce a particular unit or output of a product
		Artificial efficiency	Labor hours per unit of product
Differentiation strategy	Agility (A)	Response speed	Response time, new product introduction rate, engineering change transaction cycle
		On time delivery	The rate at which a complete product is manufactured and delivered on schedule
		Fault recovery	The rate of downtime during operating hours
	Quality (Q)	The quality of the Product	Production capacity, customer rejection/return, and return authorization
		Innovation	Product innovation
		Diversity	Diversity/family of products, each product optional, personalized options
		Customer service	Customer's evaluation of service
	Sustainability (S)	Product	Recyclability, energy efficiency, lifetime, manufacturability
		Process	Primary energy use, greenhouse gas emissions
		Logistics	Transport fuel for use, refrigerated energy for use

Table 1.2 Comparison of strategies, advantages and disadvantages of US, Germany, Japan and China

	Advantages	Disadvantages	Path and strategy
United States: National Manufacturing Innovation Network	(1) There are many IT giants and a large number of IT enterprises, with advantages in industrial software, big data, artificial intelligence, IoT and high-end manufacturing (2) Firmly occupy the top of the global Internet pattern and technology (3) Years of information process has accumulated massive data (4) A strong producer services sector (5) It has an energy advantage	(1) Lack of scale of traditional manufacturing industry (2) High labor cost	(1) Give full play to its systematic advantages in the fields of Internet, big data, artificial intelligence and service innovation to achieve disruptive innovation in the industrial field (2) Emphasis on the upstream of the value chain to draw additional value, oriented to the system rather than components, top-down control of the market (3) Solve the problem by "data" to complete (4) Formed a comprehensive policy synergy

(continued)

Table 1.2 (continued)

	Advantages	Disadvantages	Path and strategy
Germany: Industry 4.0	(1) High-end manufacturing developed, IM has a first-mover advantage (2) With advanced manufacturing equipment industry, its manufacturing equipment has a high quality level and reputation (3) The proportion of small and medium-sized enterprises is high, and the economic structure is conducive to the development of industry 4.0 in the future	(1) The domestic market is small, the solution is difficult to implement on a large scale (2) The IT, Internet and chip industries are backward (3) Lack of data accumulation, product (4) Lack of service integration (5) High labor cost (6) Industrial development is simple, the development speed is relatively slow	(1) In the layout of the industrial value chain, emphasizing the "knowledge solidified in the equipment", solve the problem through the "system" to complete, through the equipment and production system of the continuous upgrade, for Germany's industrial equipment export to explore new markets (2) To change the past only selling equipment and the proportion of service income is small, the focus will be shifted from products to services, through services to enhance profitability and competitiveness, enhance the sustainable profitability of German industrial products

(continued)

Table 1.2 (continued)

	Advantages	Disadvantages	Path and strategy
Japan: Industrial value chain	(1) Long-term accumulation of robot technology has been widely used in industrial production (2) Toyota production system (lean production) and concept has been promoted worldwide (3) Strong automobile industry (4) Attach importance to knowledge intensive industry and heavy chemical industry (5) Attach importance to the accumulation, inheritance and learning of knowledge (6) Pay attention to the cultivation of people	(1) Population aging (2) High labor cost, small domestic market (3) Lack of resources, need a large number of imports (4) Industry hollowing out, relying too much on the government	(1) Promote the industrial competitiveness in the upstream transfer to the value chain; Continue to give priority to 3D printing technology (2) Artificial intelligence and robotics as a key direction of development (3) Promote the application of IT in medical treatment, administration and other fields Support the development of environment-friendly automobiles, electric automobiles, solar power generation and other industries The IM system based on lean production solves problems by "people"

(continued)

Table 1.2 (continued)

	Advantages	Disadvantages	Path and strategy
Made in China: 2025	(1) China has the world's largest manufacturing scale, with engineer bonus (2) The Chinese market is becoming more open, with greater emphasis on key technologies (3) The scale of China's automation technology market has exceeded 100 billion, accounting for more than 30% of the world market share, with a good market atmosphere (4) The largest online population and the largest Internet economy (5) Still have certain cost advantage, have strong government and policy support	(1) The population is large but the demographic dividend disappears, and the manufacturing industry is low-end (2) The cost of IM transformation is difficult to digest (3) Industrial software, chips and other key components are backward (4) Low product quality, low resource utilization efficiency (5) The fusion depth of the two is not enough (6) Insufficient ability to operate globally	(1) Give full play to the advantages: strive for lane change overtaking; Completing the three-step strategy of digital, network, intelligence in parallel; Give play to the advantages of the information technology industry; Set up a number of national cyber physical system network platforms; Enhance the integration of the two industries; To cultivate globally competitive enterprise groups and competitive industries (2) To complement the weak points: the implementation of digital, networking, IM; Improve product design ability; Improve the manufacturing technology innovation system; Strengthen the manufacturing base; Improve product quality; Promote the concept of green manufacturing; Developing modern manufacturing services; Strengthen cooperation with Germany

and other aerospace giants. According to the ministry of industry software and integrated circuit promotion center (CSIP) research conclusions, one Tesla motors has 200 million lines of software code. In contrast, the number of the software used in developing a Boeing 787 aircraft with more than 1 billion lines of code is more than 8000, in which nearly 1000 kinds are commercial software, and the rest of 7000 kinds of software is not for foreign sales (in house) for many years. They have become the central part of enterprise core competitiveness. CADAM, I-DEAs, UG, GE's Industrial Internet, and other famous software are all commercial-industrial software developed by American companies. (2) The US is the birthplace of the Internet and has hence the advantage of it. All Internet root servers are managed by ICANN authorized by the US government, firmly occupying the top of the global Internet pattern and technology. Second, the US has the advantages of "high-end" manufacturing technology, intelligent technologies such as artificial intelligence, cybernetics, IoT most originated in the US. From the early NC machine tools, integrated circuits, Programmable Logic Controller (PLC) to the first smartphone, driverless cars, heavy rockets, large aircraft, and all kinds of advanced sensors, high-end chips, its pursuit of technological innovation and absolute leading spirit are reflected. (3) After years of information, manufacturing enterprises have accumulated massive data, with the advantage of "big data". (4) There are vital producer services and soft power, which benefit from its advantages in industrial software, the accumulation of big data, and the convenience of English-speaking countries. (5) It has an energy advantage. In recent years, the development of new shale gas resources has reduced the production cost of the manufacturing industry, and there is also strong policy support.

Its disadvantages include: the traditional manufacturing industry lacks scale and high labor costs.

To this end, the development path and strategy of IM adopted by the US are as follows: (1) give full play to its systematic advantages in Internet, big data, AI, and service innovation to achieve disruptive innovation in the industrial field; (2) Occupy both ends of the lucrative value chain, firmly occupy the upstream of production, and strive to extend to the downstream. (3) Oriented system rather than the parts, top-down control the market, in particular, trying to create front-end control technology product innovation and demand, production systems and the most basic energy, critical materials with the enabling technology, as well as the use of information network technology product, value-added services, firmly grasp the industrial value chain of the highest levels; (4) In the form of knowledge inheritance, US pays the most attention to the role of data in solving problems and is good at subverting and redefining problems, emphasizing the solution of problems through "data to complete". One example is that Japan chose the Toyota production system, which relies heavily on people and institutions, while American companies generally chose the 6-Sigma system, which relies heavily on data. (4) A series of supporting policies have been formulated around the economic strategy of reindustrialization, forming a complete policy synergy, including industrial policy, tax policy, energy policy, education policy, and science and technology innovation policy, and strengthening the coordination and cooperation between enterprises, universities, and local governments.

1.5.1.2 Germany

Germany's advantages include: (1) high-end manufacturing was developed. IM has the first-mover advantage, Mercedes Benz, BMW, Audi, and other luxury brands, but also has Siemens, Schneider, Bosch Rexroth, Demaghi, and other IM benchmark enterprises; (2) Advanced manufacturing equipment industry. Intelligent equipment has a high-quality level and reputation, mold manufacturing, CNC machine tools, precision instruments, power devices, mechanical transmission, and other fields are in the world's top-level; (3) Small and medium-sized enterprises (SMEs) account for a high proportion. Small and specialized "hidden champions" are substantial, equipment manufacturing enterprises cluster (agglomeration) effect is noticeable, the economic structure is conducive to the "industry 4.0" in the future.

Its disadvantages include: (1) the domestic market is small, IT solutions are challenging to implement on a large scale; (2) the Internet technology, information industry obviously, chip technology and industry are backward; (3) Lack of data accumulation and service value; (4) high labor cost, 20–30% higher than the US. So Germany has had to make up for this by developing more advanced equipment and highly integrated, and automated production lines; (5) Due to the lack of Internet technology advantages, high-tech cost and network security threats are also facing challenges.

Since 2007, it has been falling behind China in sales of its most vital machinery and equipment, and the gap is widening. To this end, the German IM development path and strategy is to: (1) in terms of the industrial value chain of the layout, emphasis on "consolidating knowledge into the device", developing "smart device", constructing "smart factory" to solve the problem through the escalating of "production systems", give full play to the critical equipment and technical advantages, explore new markets for the German industrial equipment; (2) To change the previous status of only selling equipment and a small proportion of service income, shift the focus from the product end to the service end, through the service to enhance profitability and competitiveness, to enhance the continued profitability of German industrial products.

1.5.1.3 Japan

Japan's advantages include: (1) Japan has the world's most significant number of industrial robots installed, its robot industry group has a very competitive advantage, long-term accumulation of robot technology has been widely used in industrial production; (2) Toyota production system (TPS), which is promoted in the global; (3) Strong automobile industry; (4) Attach importance to the knowledge-intensive industry and heavy chemical industry sector; (5) It attaches importance to the accumulation, inheritance, and learning of knowledge, and is good at integrating and innovating different advanced technologies from different countries to form advanced technologies belonging to its own country. (6) Attaching importance to the cultivation of people, Japan has a unique social culture of contradictory restraint, patience, and dedication to the collective, which also affects the manufacturing culture. Its most important feature is to solve problems through continuous optimization of organization and culture, improvement of labor quality, and human training. Both TPM, PDCA cycle, and TPS attach great importance to the inheritance and dependence of human knowledge instead of only consolidating knowledge into machines.

Its disadvantages include: (1) the aging population and the rising cost of manufacturing; (2) High labor cost; (3) small domestic market, Japan's auto industry, and other manufacturing industries to move to China or Southeast Asia with low labor costs, its domestic market further shrink; (4) Lack of resources, the need of imports for a large number of coal, oil, natural gas, metal ore, and other significant resources is more than 95%; (5) Industry hollowing out, relying too much on the government, if the government policy direction is wrong, it will make the industry develop into a dilemma.

To this end, Japan's development path and strategy to adopt the IM are: (1) to promote industrial competitiveness moving upstream along the value chain. Behind Japan's decline in consumer electronics is a shift in the direction of Japanese innovation, as the country begins to become more capable of upstream raw materials and enabling technologies, essential equipment, and components. For example, Panasonic found solutions in batteries and became a supplier of Tesla battery electric vehicles since it has lost its leading position in the home appliance market; (2) Promote the application of IT in medical treatment, administration, and other fields; (3) Take AI and robotics as a key direction of development, but also to strengthen the field of materials, medical, energy and critical parts; (4) Support the development of environment-friendly automobiles, electric automobiles, solar power, and other industries; (5) Emphasis on the solution of problems through "people to complete". Lean production emphasizes "talent training", "full participation," and "obedience to the collective". One of the two pillars of Toyota's production system is "Jidoka" (the other pillar is JIT). For example, if material sorting errors frequently occur on the production line, the solutions of Japanese enterprises are more likely to improve material color and identification, strengthen staff training, and set up a review system. On the other hand, Germany is more likely to design an automatic sorting system based on radio frequency identification (RFID), image recognition, or robotic arms for automatic sorting independent of humans.

1.5.1.4 China

China's advantages include: (1) China has the world's most enormous manufacturing scale, a more efficient industrial system; (2) China is becoming more open to new technologies; (3) The scale of China's automation technology market has exceeded 100 billion, accounting for more than 30% of the world market share; (4) It has the most extensive online population and the most prominent Internet economy in the world; (5) Have cost advantage, have strong government and policy support.

According to the manufacturing competitiveness index model, China's manufacturing competitiveness index has been firmly ranked first in the world, as shown in

Fig. 1.5. While developed countries such as Germany, Japan, and the US are all on a downward path. China's manufacturing industry enjoys advantages in cost, market, and government. However, the German manufacturing industry still has advantages in talent, economy and trade, suppliers, law and medical treatment, etc., as shown in Fig. 1.5.

China's disadvantages include: (1) The population is the largest, but the demographic dividend disappears, the labor quality is not high, the industrial structure is not reasonable, the development of high-end equipment manufacturing industry and producer services lags; (2) On the one hand, the cost of IM in the early stage of equipment investment and technology learning is high, facing the cost pressure of human, financial and material aspects, directly leading to investment enthusiasm is not high; On the other hand, the risk and resistance are high. IM has a subversive change, which impacts the production of enterprises. For example, the failure and downtime maintenance of unreliable industrial robots will cause huge production losses; (3) Weak independent innovation ability, high external dependence on crucial core technologies, and high-end equipment. Currently, 80% of the design software, 50% of the manufacturing software, and 95% of the service software are occupied by foreign brands in China's industrial software market. The "ecology" of domestic industrial software is chaotic, the industrial chain is fragile, and there are few value-added services for domestic industrial software. The absence of high-end industrial software is a Damocles sword that hangs over the made-in-China roof.

Combined with its advantages and disadvantages, China formulated the strategic program "Made in China 2025" and proposed eight countermeasures to achieve the strategic goal of manufacturing power through "three steps". The development path and strategy of IM mainly include: (1) give play to the advantages: strive for lane change to overtake, give play to the advantages of the scale of manufacturing industry, mutual intersection and integration, parallel to complete the digital, networking, intelligent three-step strategy, using AI, Internet to solve the problems existing in the early

Competitive factors	Germany	USA	Japan	China	Brazil	India
Talent Driven Innovation	9.47	8.94	8.14	5.89	4.28	5.82
Economic, trade, financial and tax systems	7.12	6.83	6.19	5.87	4.84	4.01
Cost and acquisition of labor and materials	3.29	3.97	2.59	10.00	6.70	9.41
Supplier network	8.96	8.64	8.03	8.25	4.95	4.82
Legal and regulatory system	9.06	8.46	7.93	3.09	3.80	2.75
Physical facilities	9.82	9.15	9.07	6.47	4.23	1.78
Energy costs and policies	4.81	6.03	4.21	7.16	5.88	5.31
Local market attractiveness	7.26	7.60	5.72	8.16	6.28	5.90
Medical system	9.28	7.07	8.56	2.18	3.33	1.00
Government investment in manufacturing and innovation	7.57	6.34	6.80	8.42	4.93	5.09

Fig. 1.5 International comparison of competitive factors [1]

digital manufacturing; Give full play to the advantages of the scale of the information technology industry, set up several national Cyber-Physical System network platforms, promote the informatization transformation of the traditional manufacturing industry, "adhere to the mutual driving forces between informatization and industrialization", enhance the integration of the two, and cultivate the global competitiveness of enterprises. (2) To complement the weak points in the implementation of digital, networking, IM; Improve product design ability; Improve the manufacturing technology innovation system; Strengthen the manufacturing base; Improve product quality; Promote the concept of green manufacturing; Develop modern manufacturing services. Marked by the China-Germany Program of Action for Cooperation: Shaping Innovation Together, China has shifted from learning from Japan to Germany.

1.5.2 Comparison of Reference Architecture of IM Model

After RAMI4.0 (Reference Architecture Model Industries 4.0) was proposed by Germany and IIRA (Industrial Internet Reference Architecture) by the United States, In terms of intelligent manufacturing Reference Architecture, Japan also released IVRA (Industrial Value Chain Reference Architecture). So far, the three leading countries in intelligent manufacturing have completed benchmarking reference architectures.

The reference architecture guides the development of IM systems, solutions, subsequent designs, vendor selection, and application architecture. It describes the functions and decomposition of IM systems, components and their relationships and interactions, consistent terms and definitions, and serves as a basis for further detailed research and discussion. Just as the reference architecture of buildings specifies the number of rooms, layout requirements, building materials, and performance that the house must contain, users can further study how to use the house to improve efficiency and safety. The main differences between the three reference architectures are shown in Figs. 1.6.

1.5.2.1 U.S.

General Electric (GE) put forward the concept of Industrial Internet in the autumn of 2012, hoping to improve the efficiency of the existing industry significantly and create new industries through the combination of production equipment and IT, Internet, extensive data collection and analysis technology, etc. America's industrial Internet seeks to connect people, data, and machines into an open, global industrial network. It can be understood from three dimensions: network, data, and security. In these three dimensions, the network is the foundation, through interconnection, to achieve seamless integration of industrial data; Data is the core of the model to emphasize the "data services", "analysis of service"; Security is a guarantee. By taking advantage of

Fig. 1.6 Intelligent manufacturing reference architectures and major differences among countries

its "Internet advantages", it can build a security protection system covering the whole industrial system, effectively prevent network attacks and data leakage, and improve interoperability, maintainability, and connectivity. With network, data, and security as the core and guarantee, the development of the industrial Internet can be promoted in a coordinated way from both internal and external aspects, including intelligent transformation and upgrading of production systems and new models/formats relying on the Internet.

As shown in Fig. 1.7, the reference architecture of the US Industrial Internet of Things is the reference foundation for a series of domain architectures, enabling the architecture of each domain to be extended and cross-referenced.

As shown in Fig. 1.7a, based on the hierarchical approach commonly used in complex system modeling, the Industrial Internet reference architecture consists of four hierarchical views: Business; Usage; Function; Implementation. The standard is developed one by one based on four views and discusses the nine characteristics of system security, information security, elasticity, interoperability, connectivity, data management, advanced data analysis, intelligent control, and dynamic combination.

Business view focuses on requirements analysis, considering the value proposition, expected return on investment, maintenance costs, and product liability that must be considered when considering an industrial Internet as a solution to a business problem. In order to identify, evaluate, and respond to these business considerations, the standard introduces the following concepts and relationships: vision, value and experience, key goals, and basic capabilities. It defines how industrial Internet systems can map essential system functions to achieve established goals.

Usage view is to think about how the industrial Internet implements the critical capabilities identified by the business view and map them to basic activity tasks and action units. This view describes some issues in system usage. It is usually expressed as the sequence of people or logical users who realize its essential system functions. It

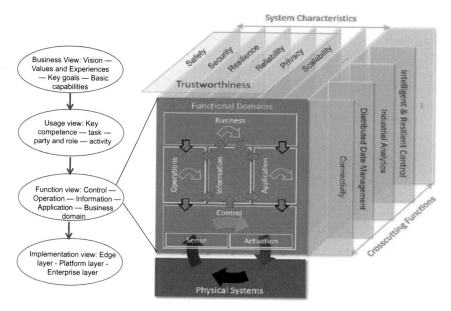

(a) Industrial Internet Reference Architecture Diagram (b) Functional view, crosscutting functions, and system features

Fig. 1.7 Industrial Internet Reference Architecture (IIRA) (V1.8 version) [2]

describes a system through four elements, including roles, actors (parties), activities and tasks, and their interconnections. Usage view guides the design, implementation, deployment, operation, and evolution of the industrial Internet.

Functional view is the front view of the reference architecture. Its main goal is to decompose the functions of the industrial Internet and build the functional architecture of the system based on the activity tasks output by usage view. The identification and functional design are derived from the demand model and use case models of the industrial Internet, which is the key to realizing the industrial Internet. The functional view divides a typical industrial Internet into five functional domains: Control, Operations, Information, Application, Business, and the five function domains are decomposed into more specific small units, focusing on the functional components in the industrial Internet system, including their relationships, structure, interface and interaction, data flow and control flow, and the interaction with the external environment, to support the usage of the whole system. The functions of the control domain include sensor data reading, data recording, and control signal driving, sensor/driver/controller/gateway's communication with other edge systems, entity abstraction, modeling, asset management, actuator, and other functions; The operational domain represents the functional set responsible for function providing, managing, monitoring and optimizing. An information domain represents a set of functions that collect, clean, transform, correct, store, publish, dominate, model, and analyze data from several different functional domains to obtain intelligent information for the whole system. Application domains implement application logic for specific business functions, including logical rules, API (Application Programming Interface), and user interfaces; The business domain enables end-to-end operation and supports business functions including ERP, CRM, PLM, MES, HRM (Human Resource Management), asset management, service life cycle management, account reconciliation and payment, production planning, and scheduling systems.

Implementation view focuses on the technical implementation of the communication scenarios and phases of the life cycle. As shown in the practical case at the end of Chapter One, the implementation view standardizes the general architecture of the industrial Internet: (1) The system is divided into three layers: the edge, platform, and enterprise layers. (2) Edge gateway connectivity and management architecture patterns, including available topologies, edge gateway support functions; (3) The edge cloud architecture pattern, which assumes WAN connection and addressing capabilities for devices and assets; (4) Multilevel data storage architecture mode, supporting storage layer combination: presentation layer, capability layer, archive layer; (5) Distributed analysis architecture patterns. The implementation view also provides technical descriptions of components, including interfaces, protocols, and behaviors, activity mapping from functional components to implementation components, and implementation mapping of crucial system characteristics.

1.5.2.2 Germany

Germany has the world-advanced equipment manufacturing industry, especially in embedded systems and automation engineering. Big data and Internet technology are the weaknesses of the German manufacturing industry. Germany's IM reference architecture focuses on intelligent equipment, which entirely plays Germany's conventional advantages. The IM system architecture in Germany's "Industry 4.0" plan includes three dimensions: activity level, system level, life cycle, and value stream. Figure 1.8 shows that the activity level dimension consists of six levels, namely, asset, integration, communication, information, function, and business. The System-level dimension includes products, field equipment, control equipment, sites, work centers, enterprises, and the connected world. The life cycle and value stream dimension divide the product life cycle into two stages: Prototype and Instance. The focus of the three dimensions is on the underlying equipment level and the production link. Dimension and hierarchy are interconnected and mutually supported to jointly promote intelligent production and realize an "intelligent factory". Germany hopes the framework will make it easier for SMEs to operate.

The strategic points of "Industry 4.0" can be summarized as "123458":

"1"refers to a Cyber-Physical System (CPS) based on a system that links resources, information, goods, and people together. CPS is the common ground between the American industrial Internet and The German "Industry 4.0".

"2" means to focus on two major themes, namely, "smart factory" and "smart production".

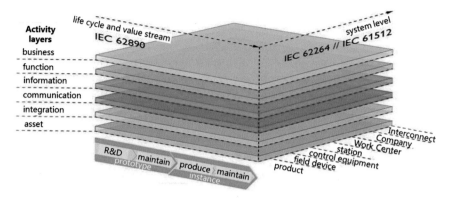

Fig. 1.8 "Industry 4.0" Reference Architecture Model (RAMI4.0). *Data source* German Committee for Standardization of Electrical, Electronics and Information Technology (DKE), 2015

"3" refers to the realization of three significant transformations, namely, the transformation of production from centralization to decentralization, the transformation of products from convergence to individuality, and the transformation of service from "customer orientation" to "customer participation".

"4" refers to the achievement of four kinds of goals: the development of new methods of intelligent production, the optimization of new automated technologies, the meeting of new demands of labor force changes, and the formation of new industrial production models.

"5" means to promote five major tasks, namely, the completion of manufacturing process integration and network production system, strengthening the production of innovation and application of information and communication technology (ICT), building a standardized and normalized model, building a new enterprise organization model based on human–computer interaction, strengthening security and proprietary technology, R&D and promotion.

"8" refers to the eight actions. Industrial 4.0 working group put forward actions in eight key areas to promote industrial 4.0: standardization and reference architecture, management of complex systems, the establishment of comprehensive broadband infrastructure, safety and security, work organization and design, training and sustainable professional development, rules and regulations, the efficiency of resources usage.

Germany's "industrial 4.0" shows a picture of a new blueprint for industrial vision:

- The process of creating new value gradually changed.
- Industrial chain division of labor will restructure.
- Traditional industry boundaries will disappear.
- Many kinds of new fields of activity and forms of cooperation will be produced.

In the German industry 4.0 strategy, vertical integration, horizontal integration, end-to-end integration are the key to achieving intra-enterprise and inter-enterprise value stream integration.

1.5.2.3   Japan

The Industrial Value Chain Initiative (IVI), whose core content is "connected factories" or "connected enterprises", is becoming the core layout of IM in Japan.
The current IVI includes more than 200 global companies, including Siemens. An
autonomous innovation system is no longer adapting to the development of the new
situation, especially IM is a complex system. IVI aims to support cooperation between
enterprises and exert cooperation. IVI launched Industrial Value Chain Reference
Architecture (IVRA), the basic Architecture of intelligent factories.

IVRA is similar to RAMI 4.0 of the industry 4.0 platform, also a 3D model. Each
chunk of the 3D model is called a Smart Manufacturing Unit (SMU). SMU comprises
three axes, longitudinal as the "resource axis," divided into the personnel, process,
product, and plant layers. It is worth noting that personnel is valuable assets of the
enterprise, and they play the role of decision-makers whether they are managers
or not. The horizontal of SMU serves as the "execution axis", which is a standard
Deming ring, namely PDCA cycle, divided into four stages: Plan, Do, Check, and
Action. The "management axis" is the operation and maintenance of core output in
the production process, including Quality, Cost, Delivery, and Environment (QCDE
activity), as shown in Fig. 1.9.

In terms of the three angles of resource, management, and execution dimensions,
SMU has the ability of module combination and software connection to link enterprise resource management and R&D management software to form the system's
integration. It can start from the smallest IM unit, expand to digital production lines,

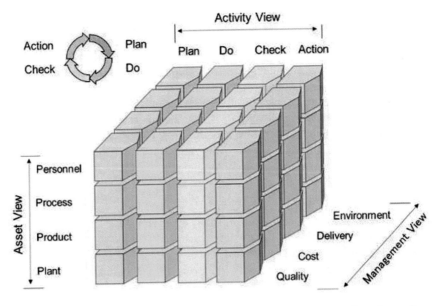

Fig. 1.9 SMU in the Industrial Value Chain Reference Architecture (IVRA). *Data Source* Japan
Industrial Value Chain Promotion Association (IVI), 2016

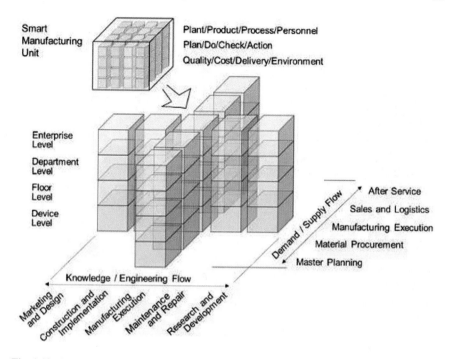

Fig. 1.10   GFB in IVRA. *Data Source* Japan Industrial Value Chain Promotion Association (IVI), 2016

locally interconnected systems, and finally build a giant system. SMU is the primary component and autonomous unit that describes micro activities. If an IM system is a skyscraper, an SMU is a prefabricated board.

Figure 1.10 shows that General Function Blocks (GFB) are constructed from multiple SMU combinations from three views of knowledge/engineering flow, demand/supply flow, and enterprise hierarchy. The Industrial Value Chain Initiative (IVI), whose core content is "connected factories" or "connected enterprises", is becoming the core layout of IM in Japan. The current IVI includes more than 200 global companies, including Siemens. An autonomous innovation system is no longer adapting to the development of the new situation, especially IM is a complex system. IVI aims to support cooperation between enterprises and exert cooperation. IVI launched Industrial Value Chain Reference Architecture (IVRA), the basic Architecture of intelligent factories (Fig. 1.11).

The characteristics of the reference framework are:

1. The concept of interconnected IM units is proposed as an element to describe manufacturing activities and is defined explicitly from the view of Asset, Activity, and Management.

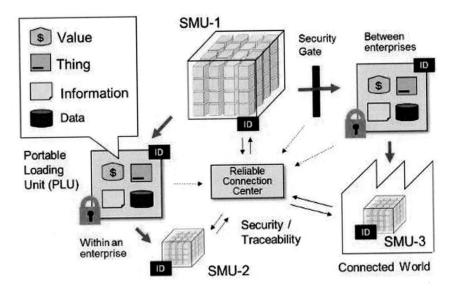

Fig. 1.11 Mobile value of PLU. *Data Source* Japan Industrial Value Chain Promotion Association (IVI), 2016

2. It integrates the Japanese manufacturing industry's unique value orientation and management methods, including the PDCA cycle, lean manufacturing, and continuous improvement, and reflects the management thinking of interconnected manufacturing, loose coupling, and personnel first.

3. Demonstrate the manufacturing value chain through standard functional modules. Through the combination of intelligent manufacturing units (SMU), the construction of the whole intelligent factory can be freely upgraded through modularization and partitioning to improve the construction efficiency of the intelligent factory.

4. Highlight the significance of expert knowledge base, emphasize that personnel is a critical factor in the manufacturing process, and embody the concept of "people-oriented". The knowledge/engineering process is discussed as a separate dimension in the GFB modeling process, including the expertise and experience accumulated in all processes. It not only realizes the real-time and functional correlation between physical devices and information data based on the "IoT", but also regards people as an essential element in the mapping process of Cyber virtual space and physical entity space, fully considering the status and role of people in manufacturing activities, enabling "employees" to participate in manufacturing activities organically.

5. Provide reliable value transfer media. The association between innovative manufacturing units (SMU) is defined as "Portable Loading Unit" (PLU). Specifically, it can be divided into four parts: Value, Thing, Information, and Data. Under the premise of ensuring safety and traceability, PLU can be used to realize asset transfer between different SMUs and simulate the transformation process of

high-value assets such as materials and data in manufacturing activities to truly reflect the flow and transformation of value within and between enterprises and reflect the value chain thought of lean manufacturing.

6. Propose framework standards with extensible definitions. Considering the complexity of interconnect manufacturing system interfaces and recognizing that the IM system is complex, IVRA proposed an extensible definition of standard structure to facilitate the establishment of interconnecting among various types of enterprises in an open international environment.

1.5.2.4 China

In the Guideline for The Construction of National IM Standard System jointly issued by the Ministry of Industry and Information Technology of China and Standardization Administration of China, the IM system architecture is constructed through three dimensions, namely life cycle, system-level, and intelligent function, as shown in Fig. 1.12.

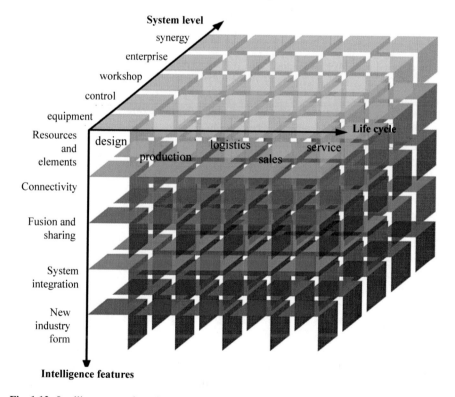

Fig. 1.12 Intelligent manufacturing system architecture [3]. *Source* Ministry of Industry and Information Technology of China, Standardization Administration, 2015

The system-level dimension has five layers from bottom to top, namely, equipment, control, workshop, enterprise, and synergy layer. The dimension reflects the intelligence of equipment and the Internet protocol (IP), and the trend of flattening networks. We discuss this dimension in detail in Chap. 2.

The life cycle dimension is a chain set of interrelated value-creating activities such as design, production, logistics, sales, and service. The model ignores the difference between sample development and product production. The life cycle of products and services varies from industry to industry. We discuss this dimension in detail in Chap. 3.

Intelligent functions dimension including five layers of resources, system integration, connectivity, information fusion, and new industry form, highlighted system integration, data integration, information integration at all layers, its key is to solve the bottlenecks in data integration and connectivity, emphasize the emerging formats such as collaborative network manufacturing, mass customization, remote operational service. This dimension is discussed in detail in Chap. 4 of this book.

The main advantage of China's IM lies in its traditional solid manufacturing base. Therefore, the IM system architecture standards focus on matching the "Made in China 2025" and "Internet plus" action plans, which promote the integration of the mobile Internet, cloud computing, big data, IoT, AI, and traditional manufacturing industries, to accelerate the transformation and upgrading of the manufacturing industry.

Germany's "Industry 4.0", the industrial Internet strategy of the US, and the "Made in China 2025", aims to accelerate the in-depth integration of new generation of IT and manufacturing, and promote IM, coincide and share the same goal. They have in common the combination of IT and advanced manufacturing with CPS as the core, or the combination of Internet + advanced manufacturing to drive the development of the whole new round of manufacturing industry.

1.6 IM Maturity Model

In Fig. 1.13, to help enterprises identify gaps, establish goals, and implement improvements in IM's transformation and upgrading stage, the China Academy of Electronic Technology Standardization issued the "White book on IM Capability Maturity Model (1.0)" in September 2016. In the book, the architectural model consists of dimensions, classes, domains, levels, and maturity requirements. Dimension, class, and domain are the expansion of two dimensions of "intelligence + manufacturing" and the decomposition of core competence elements of IM. The two dimensions of "intelligent + manufacturing" are the starting point of discussing the Capability maturity model, which can also be interpreted as the application of OT (operational technology) + IT (information technology) in the manufacturing industry. The maturity requirement is the description of the characteristics of the class and domain at different levels.

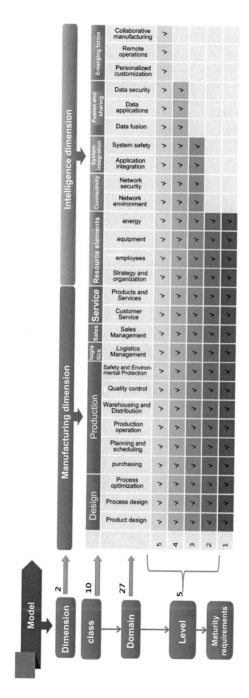

Fig. 1.13 Relationship diagram between model architecture and capability maturity Matrix [4]

Based on two dimensions of "intelligence + manufacturing", that is, manufacturing dimensions including design, production, logistics, sales, and service; It also includes intelligence dimensions of resource elements, connectivity, system integration, information fusion, and emerging industry forms, including ten classes of core capabilities and 27 detailed domains. Five levels (planning, specification, integration, optimization, and leading level) are applied to the relevant domains. Please refer to the appendix for a detailed definition.

This book refers to the 3D model of IM capability maturity, first analyzing one by one from three dimensions, putting forward the corresponding implementation path combining each dimension. The implementation path tries to help enterprises in the following five aspects:

1. How to make a unified plan and implement it step by step?
2. What level of IM is the manufacturing enterprise currently at?
3. How should the enterprise establish the scope of investment and do appropriate planning?
4. What is the key content and focus of each step?
5. What is the logical relationship between the steps?

1.7 Implementation Guidelines for IM

1.7.1 Standards Lead and Specifications First

The standardization of IM should go first. The Guideline for the Construction of National IM Standard System (2015 edition) was issued to guide the work of IM standardization.

In our country, the problems of lack of connectivity are still not solved. The standardization demand for cross-trade and cross-domain is becoming more and more urgent. It is because:

First, a consensus can be established through the formulation of standards so as to solve the fundamental bottlenecks, such as data integration, connectivity, cross-industry, and cross-field integration.

Second, due to the uneven development of the industry and the large gap in enterprise-level and other national conditions, the majority of small and medium-sized enterprises lack methodological guidance. They need to keep pace with The Times, flexible, compatible, and open standard system as a working criterion.

Third, as new models and new forms of IM will continue to emerge, it is necessary to build a normative framework of IM system, define the connotation and extension, extract the common abstract characteristics, and clarify their operating rules, so as to provide a theoretical basis for new forms of manufacturing.

[Case: Midea's pain point for IM]
The real-time data of machine tools in the production site is the most basic information in the manufacturing process, such as torque, position, load, force, vibration, and

other built-in sensor information. In order to obtain such information, it is usually necessary to buy and adopt software packages compatible with equipment suppliers or designated, and pay extra data fees in order to gain the right to go deep into the bottom layer of hardware and software of CNC machine tool system and collect stable and reliable data. Due to the lack of an open standard, the underlying protocol, for example, we can stipulate what OPC UA (Object Linking and Embedding for Process Control Unified Architecture) field represents tricolor lamp, but how many lights are on is decided by equipment manufacturers, which is often a black box, to this end, when Midea want to go through some of the equipment data, it need to purchase data from foreign equipment manufacturers, and pay another fee for protocol opening. For each NC equipment, they need to pay about an average of ten thousand yuan for a year.

The head of Midea's IT department believes that if the situation is not improved, the manufacturing industry in China will be controlled by foreigners like the situation in the robot industry. The costs for IM would be unbearable, so we call for the government to establish enterprise alliances, then develop related protocol standards and keep source sharing, making the domestic manufacturing enterprises obtain the required data.

When Huawei purchases equipment, it requires equipment suppliers to provide not only products but also process data packets. Furthermore, many small companies still do not understand why Huawei has such a requirement.

1.7.2 Lean Production and Foundation Strengthening

Before realizing intelligence, the foundation of lean, standardization, and digitization should be laid well. Do not implement automation on the basis of backward process, need to supplement advanced process and automation; Do not implement digitization on the basis of backward management, but need to supplement the course of information on the basis of modern management; Do not implement intelligence without a lean and digital basis, just as do not information without process optimization.

In the 1990s, both IM and AI experienced a research upsurge, but they fell into a long-term silence. An important reason is that the foundation of digitization and data volume was not accumulated enough, and the level of good management was not high enough so that IM and AI could not be really developed.

The supporting foundation includes the digitization and networking foundation of three dimensions: the underlying equipment, the design and development of the life cycle source, and the resource elements. Small and medium-sized enterprises can choose one of the advantages based on their ability, from low to high, step by step, to achieve IM. Even foreign advanced factories, most of them are just realized information, automation, so IM project, to adjust measures to local conditions according to their own needs and foundation to implement. Some enterprises still have a "black hole" in production feeding link and equipment unplanned downtime often occur, but introduce intelligent AGV, automatic feeding system; Some enterprises introduce intelligent warehousing system when the employees in the warehousing link still lack training; Some enterprises have not yet realized the custom design, but implement a production for personalized product system. The implementation of these systems carries a high risk.

1.7.3   Unified Planning from Top to Bottom

IM is not only the introduction and application of new technology or new software but also a new business strategy that affects the whole company at all levels. Therefore, it must be consistent with the strategic goals of the company. Intelligence should not be the only focus, should according to the company's vision and internal and external environment, look for a clear corporate strategy and sustainable core competitiveness, find the matching of core competence, and then optimize the organization, process, technology, then accumulate data in the optimization cycle. Overall, design from the top down, and implementation from the bottom up, and continuously dynamic cycle to meet strategic consistency.

Due to the lack of overall strategic planning, many IM projects are not clear about the specific needs of digitization, networking, and intelligence in the future, and lack of cognition of the strategic needs of enterprises and the current level, so it is impossible to objectively judge the gap between them and determine the new capabilities that need to be strengthened. In many cases, many Chinese enterprises did not solve "why to build" the fundamental question on a strategic level.

Therefore, enterprises should first design from the top-down, implement layer by layer, consider problems from the overall perspective of strategy, product design, changes in operation mode, select appropriate technology according to their actual situation and goals, and start from the bottom of the local implementation. For example, Haier's development strategy of "integrating people and orders" with Internet factories as the core conforms to the group's development direction of mass customization. It fits Haier's rich experience in moderation and digitalization, thus successfully creating the ecological system of Internet factories.

1.7.4   Step-By-Step Implementation and Concerning the Order of Difficulty and Ease

It should be carried out step by step following the principle of "urgent use first, easy before difficult". There are many stages in implementing IM, the sequence of which depends on the company's foundation. The ultimate goal is to achieve full integration of multiple dimensions. First of all, we can start from the most helpful link and the most significant pain point so that the system can be applied as soon as possible and begin to have economic returns, which is the most significant incentive for the company and the team. Second, the first module and the target should be relatively easy, leading to the overall success of the module, a team with confidence, and providing the buffer time and then gradually into the hard core links such as production plan, MES.

In addition, IM systems need to be flexible, the business strategies and models may need to be adjusted every five years or less, and organizational structures and technical systems implemented need to be updated and adjusted promptly. It is part of continuous improvement and needs to be continuously improved to adapt to market changes and demands. Therefore, it is enough to meet the needs, but not too complicated.

[Case: ERP implementation of Dell Company]
In 1994, Dell spent $100 million implementing R/3 of SAP to transform production and manufacturing. Two years later, Dell had to abandon SAP and declare defeat. In 1997, after paying $100 million as tuition and two years of work, Dell first selected i2 to transform its procurement and Oracle systems to transform order management. A year later, Dell chose Glovia to reinvent its manufacturing management and finally completed the ERP strategy of Dell Company. "Little by little, we uncover the answers and get faster returns than one-off ERP," said Terry Kelly, CIO of Dell, introducing the company's ERP implementation strategy.

The key to Dell's success is the innovation of its business model, namely, the direct ordering model. The direct order approach of Dell company has the following four characteristics: (1) Make to order, no inventory; (2) One to one customer; (3) Efficient production management; (4) Standardization of product development. Although the direct selling order model brings excellent value to users, it puts tremendous pressure on the cost of human resources and production and manufacturing efficiency. Dell's unique information system supports this model. In 2017, Dell ranked 41st on the Fortune 500.

Q&A: Why is the ERP module of Dell implemented in the order of purchase, order management, and then production and manufacturing management?

1.7.5 Supporting from Head and Training of Staff

First of all, the IM project is a "top management project" that needs top leadership's support, which is embodied in fully authorizing the management representative, internal auditor, project team, and issuing documents. The head should regularly listen to reports during the project process, participate in management review, and determine to remove all kinds of obstacles.

In addition, the qualities required for talents are also undergoing significant changes. For example, the demand for assembly line workers will be reduced, and more data analysts, programmers, and robot maintenance engineers will be needed instead. In addition, the implementation will have a profound impact on organizations and people, thus changing the way they work, so they need enough time and training to adapt. The more automated production and decision-making are, the more evaluation and deliberation are required, and the more training is required. The goal of IM should not be to replace machines but to inspire people to contribute as much intelligence as machines or software can produce and become more critical, not less, in a human–machine integrated system.

Enterprises should design novel knowledge learning and acquisition mechanism for future "knowledge workers". Use more interactive electronic learning tools to help students, apprentices, and new workers acquire advanced knowledge of IM technology, giving them more opportunities to continuously develop their skills and abilities and convey novel skills to a new generation of workers, support aged, weak, or disabled multicultural workers.

In a word, IM is not the purpose. Its purpose is to improve product competitiveness and quality; IM cannot be accomplished overnight and requires long-term efforts and reform of enterprises. IM is not an automatic transformation but a fundamental change in operation mode. It should be carried out according to the principles and steps of standard first, foundation strengthening, unified planning from top to bottom, ease and urgency first, supporting from the head, and strengthening training.

1.8   Practical Cases: GE and the Industrial Internet

1.   Project background and status quo

  1.1.   The proposal of industrial Internet

We are the oldest remaining company in the Dow Jones Industrial Average. It is not because we are a perfect company, but because we adapt. Over the years, we have remained productive and competitive. We have globalized the company while investing massively in technology, products, and services. We knew we must change again.—Jeff Immelt, CEO of General Electric

In early 2014, General Electric (GE) CEO Jeff Immelt sat in his office with Chief Marketing Officer Beth Comstock and GE Software's new Vice President Bill Ruh. They reviewed a report on customer contracts for GE's new Industrial Internet Initiative. More than two years after the initiative was announced, betting more than $1 billion and GE's $800 million in sales is directly attributed to the effort. Nevertheless, these numbers represented a tiny portion of GE's annual revenues—close to $146 billion for 2013. Immelt and his team worry: Were they doing enough to give the initiative traction within GE? Could GE, as an industrial machines manufacturer, sell outcome-based services based on analytics and software?

GE's Industrial Internet initiative proposes an open, global network to connect machines, data, and people, and provides data synthesis and analysis for real-time and predictive solutions to optimize the complex operations of GE's varied customer base, including predictive maintenance and repair needs and informing performance and operational decisions. GE's suite of industrial Internet product offerings was designed not only to create and sell "smart" software-enabled machines but also to provide outcomes-based services contingent on improving operational performance via data collected and analyzed in cooperation with customers. Wall Street and technology analysts projected that the Industrial Internet (also referred to as IoT) would generate tremendous value in both increased revenues and decreased costs. Analysts estimate that the industrial Internet would create $1.4 trillion in economic value between 2013 and 2022. By that time, they predicted that Industrial Internet-related technology spending would exceed $514 billion. Some GE customers already see benefits from their connected machines. A 1% increase in efficiency from the industrial Internet would be "unprecedented": improving airline engine efficiency by 1% equated to $2 billion in annual savings; A 1% increase in the efficiency of coal-fired plants would mean $60 billion a year in fuel savings; A 1% increase in oil exploration capital efficiency would mean savings in the oil and gas industry more than $90 billion a year. Through the Smart Shopping suite, GE reduces the stay time of trains for Ferromex, the largest railway operator in Mexico, realizes real-time monitoring and analysis of the health and performance of 100 trains within 7×24 h, and realizes operation and maintenance prediction before the trains enter the maintenance workshop, to reduce downtime and maintenance costs. The engine blade damage analysis established by GE based on big data can provide an accurate prediction of up to 80% for engine maintenance arrangements. GE launched the aviation big data platform, focusing on three critical areas of flight risk analysis, fuel management, and engine analysis. GE and China Eastern airlines share vast amounts of data, fully release GE in big data analysis techniques and best practices in the field of the engine and the value of innovative technologies, help improve airlines safety management level, reduce fuel consumption and emissions, and to respond effectively to the unplanned maintenance and wings in time. In the industrial world, any minor change brings a significant advantage. GE projected that efficiency gains as small as 1% could have huge benefits over time when scaled up across the economic system, so GE calls it "the power of 1%."

1.2. Earthshaking change

Since the announcement, GE has seen intense changes in the past 12 months, including the build-out of a new software headquarters; the launch of a standard technology platform across GE's diverse industrial businesses; a thorough assessment across organizations of GE's software development expertise, and readiness evaluation of its sales talent capable of supporting this new direction; and new and expanded partnerships with companies such as Intel, Cisco, and Accenture.

GE had signed several promising agreements, including a $300 million contract with a utility company, a reliability-based service for oil and gas customers for $20 million, a wind farm deal for $35 million, a $100 million deal with the U.S. hospital chain, and a possible $1 billion rail deal. A GE/Accenture joint venture, Taleris, which provided intelligent operations for aircraft and cargo carriers, had also just announced its first signed customer, Etihad. These agreements provide a range of benefits, including a monitored flow of oil rigs, optimizing wind turbines to adapt to changes in weather, optimized patient intake at hospitals, and predicted replacement of air conditioners in a fleet of aircraft to avoid downtime. Each deal was highly idiosyncratic, relied on deep familiarity and expertise with the specific customer and its sector, and required GE to be innovative and customize how it partnered with the customers and sold software-enabled, outcomes-contingent offerings. In contrast to GE's more traditional contract service agreements, many of these deals also require customers to allow significant access to internal operational data and some value/revenue/profit-sharing arrangement.

Immelt felt the initiative was an opportunity that GE could not ignore. "We have new non-traditional 'competitors' starting to approach our long-term customers, IBM mostly, but SAP and big data start-ups are telling our customers they can provide these analytics and services, on GE assets," said Steve Liguori, executive director of GE's Global Innovation Center. Comstock added, "Our customers are under intense pressure, given the uncertainty of the current economic environment. We are not selling as much hardware ourselves." Immelt and his team had to ask if GE's customers were ready for the Industrial Internet. In October, GE took an informal poll of customers about their readiness and adoption of the Industrial Internet and learned that 63% of customers polled said their machines were connected to the network, but they were not yet using these data, and 13% claimed they used the data to achieve a competitive advantage.

2.   Business model selection decision and kinetic energy conversion

Internally, debates remained heated over which business model the initiative should pursue. Some argue that GE should develop software capabilities and give them away for free as part of equipment sales and service contracts. "In the past, our mentality was to build and ship boxes," says John Magee, Chief marketing officer of GE Software, "Any software we have was often given away as part of a hardware sale." A second camp saw an opportunity in the software capabilities themselves, believed that these software should be licensed as a separate product. Finally, others argued that GE should embrace the initiative and pursue software and analytics investments that enabled new, outcomes-based service offerings that would mean rich, deep integration with GE customers and their data.

Pursuing the third suggestion meant a host of changes. "This initiative creates a brand-new business model for us," Magee said, "Software as a service represents a whole new beast for GE." "GE had to identify and develop new opportunities, source and hire developers and sales expertise and talent, build the offerings, and price and coordinate sales of each offering through each division's sales mechanisms. At GE, 99% of our sales force is in 'big iron,'" Liguori said. "They are used to selling capital goods versus gaining share or revenue-share arrangements with our customers. They are used to talking to the operations managers who run our equipment's plants. Now we need to expand our message to the entire C-suite, showing how we can help run all their assets and, ultimately, their enterprise better." The Industrial Internet called for a different approach. However, what levers could the team pull to accelerate the initiative? Immelt was well known for his metric-driven management; Which metrics have the most impact on GE's ability to speed up the initiative? Immelt turned to Comstock and Ruh and asked, "Are we moving fast enough? Can we be faster in this or not?".

As for GE's industrial Internet strategy, Marge said, "It involves the emergence of new business models based on both the business and consumer sides. On the business side, the Industrial Internet is a service that encompasses a range of things, such as remote monitoring and diagnosis, information services, platform-as-a-service (PaaS), or data management. There are results-based solutions, benefit-sharing, risk-sharing, and even flexible service contracts on the consumer side. However, we are not Microsoft. The consumption model is different."

In 2012, GE opened a software center in San Ramon, Calif., 24 miles from San Francisco. There was no announcement about the software center until 2013, when a handful of commercial media picked it up. The New York Times visited in 2015, but the people at GE did not seem to want to say much about it, so the Times devoted much space to describing GE's software center, but not much about what it was for. In 2016, the Harvard Business Review also talked about GE's Software Center. However, the media has not explained what activities and projects GE Software Center are responsible for to this day.

In 2013, GE's scale and scope meant that billions of devices and machines were spinning in operation around the globe. Total assets of global operations continuingly were $337.6 billion in 2012. GE produced aircraft engines, locomotives and other transportation equipment, kitchen, and laundry equipment, lighting, power distribution and control equipment, generators, turbines, medical imaging appliances, lighting, electric distribution and control equipment, generators, turbines, medical imaging equipment, mining equipment, oil and gas equipment, along with a host of commercial finance, insurance, real estate, energy-leasing products, touching every corner of the globe.

3. Development of industrial Internet platform

Will the Industrial Internet be a new opportunity? "With the early Internet, we never imagined the implication of a billion people being connected," says Bill Ruh. "So when 50 billion machines become connected…".

Most of the software that GE runs worldwide is embedded and customer-facing, with developers behind the scenes. Research shows that only 17 of the 136 software GE offers are profitable, says Mr. Ruh. The problem is: it takes us a few years to develop the software and a few years to get it to market. And customer needs are changing too fast for us to keep up. Another problem is that part of the development cost is spent on the core platform rather than the application itself. For example, in the oil and gas industry, any equipment on an oil rig is monitored locally, and the data is on the rig. A customer asked to develop remote monitoring of underwater switches via the cloud and measuring sensor switching status. The engineers wrote a report promising to deliver the software product within 18 months. Three years later, and not yet to be delivered, the latest version of the product has expanded to include 5000 cool features. Three months after turning to GE Software HQ, the customer got a low-cost solution.

Typically, software engineers get their experience in one of two ways: mechanical engineers or computer scientists, and GE software experts do not speak the same language. As a result, GE experts are skeptical about moving services and capabilities to the cloud. "Let's move more and faster," Surak says. "GE's software is more culturally different than apps and services on the cloud. It is too conservative." "We have much difficulty in explaining to people the importance of moving some of our capabilities to the cloud, so Immelt's input and support are critical." As for the question of whether to force employees to change, Ruh said, "We do not force everyone here, we will do it, and you can decide whether to get on the bus or not." Maggie said we do not have sticks, only carrots.

Integrating GE's products to different segments and customers in a unified, intelligent way is another challenge. Some business units, such as health care, have thousands of different products, machines, and devices, each with its complex software requirements and legacy systems. "We realized very early on that our user experience, or user interface, would be different across different products, even from the same headquarters," says Ruh. Sometimes people do not even know if it is a GE product. Ruh's team had to decide how to develop a single platform for all GE business units, creating a design team to accomplish this task. "I knew we had to come up with a single, universal platform and then make sure everyone could build systems on top of it," Ruh said.

A common platform drives GE software to develop a common language and standard practices across businesses, even though the customers and industries that GE serves are broad and diverse. This platform will have to address multiple requirements, such as providing mobility, ensuring security and compatibility, providing a seamless customer experience, connecting GE's many machines, and managing distributed computing and analytics, all in the cloud. However, Marge notes, "If GE were isolated, developing such a universal platform would have been a pretty radical decision. We cannot do it alone." RUH used all of its business technology architecture to develop the platform, although the work was not directly funded by those businesses.

However, the significant decision was to allow each business unit to continue to maintain its existing platforms and services while developing an underlying platform to support the building of GE's evolving software services ecosystem, creating a unified user interface for the industry's diverse environmental and data analytics needs. "It was critical to have and build the platform," Marge notes. "It was also important that we decided it should be a middleware platform that does not make software decisions for individual business units but helps teams deliver products faster. It is to our advantage to create an ecosystem to share data based on each person's assets."

"GE grew up in a linear, process-based world of selling machines that separated customers in parallel," Comstock added. "How do we get our customers to adopt a results-based approach instead of just selling their goods? How do we educate them? What capabilities do we need to build in the sales force?"

Business units' sales and delivery capabilities have to grow and transform. Selling expensive goods, such as GE's spinning machines, has a sales cycle of varying lengths, ranging from three to 18 months, depending on the equipment.

"It is going to take much organizational change, putting the right people in the right places," says Maggie, "In the development phase and also in getting the

goods to market as an extension of the value of the goods. Services can also exist independently. Every decision has an impact on investment and sales."

4.   Industrial Internet solutions and ecosystem construction

The IoT encompassed all connected devices from consumers to industry. Cisco estimates that there were about 9 billion connected devices in 2010, which would grow to 50 billion by 2020. Quantifying the size of the Industrial Internet was challenging. Most industrial assets already have microprocessors, sensors, and other software components. By 2011, GE's assets also had powerful embedded software, along with sensors and microprocessors, running power plants, jet engines, hospitals and medical systems, utility companies, oil rigs, and rail and other industrial infrastructure worldwide.

GE's various business units collectively employed over 8000 software professionals that helped generate several billion dollars in software revenue alone. However, no overarching strategy guided their technical choices and commercial offerings. Thus, each business unit and even product leader optimized software choices to local conditions, resulting in significant technical and commercial success heterogeneity. The Industrial Internet initiative required a new type of software technology and required a more coordinated approach. The advent of cloud computing added more opportunities and threats. As the scale and scope of the Industrial Internet opportunity across GE's businesses became clear, two things became evident to headquarters staff: GE needed a global software center to develop and support emerging software applications uniformly across the businesses and needed new and innovative approaches to managing customer relationships, including how to sell and service the new offerings.

The architecture design of the industrial Internet system was completed in 2015. Figure 1.14 shows that Predix System is a primary system platform launched by GE for the whole industrial field. It is an open platform, which can be applied in industrial manufacturing, energy, medical and other fields. Through the industrial Internet, the man–machine connection is finally realized. Combined with software and extensive data analysis, continuous accumulation, flow, and application of data cycle are promoted, as shown in Fig. 1.15.

GE's Predix platform started as a PaaS (Platform-as-a-Service) Platform, but as GE has refined it, it includes a three-tier structure of edge (device side) + cloud

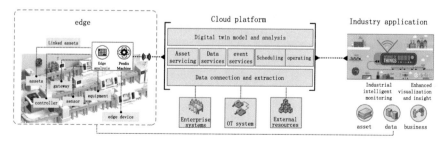

Fig. 1.14   The architecture of GE's Industrial Internet platform PREDIX System solution

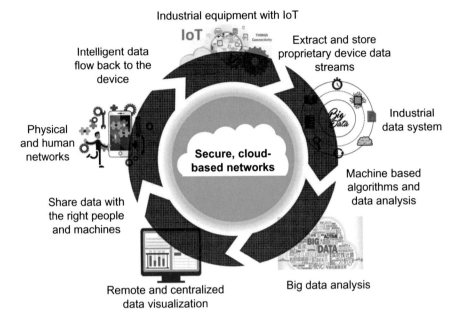

Fig. 1.15 Big data cycle of GE

(platform side) + application (enterprise side). At the edge layer, it is the gateway of the edge to collect various data. After being sent to the platform layer, the platform layer does the necessary processing and analysis of the data; after the analysis, it is sent to the enterprise layer and sent to the enterprise application system. The enterprise will make further analysis, judgment, and decisions. Then, it sends the data back to the platform and edge layers until it reaches all departments and units connected inside and outside the enterprise. The details of these three levels are as follows.

1. The edge layer

The current industrial equipment connection and protocol has the characteristics of complexity and diversity, and many of the major manufacturers have competitive relationships with GE, Siemens, ABB, etc., the dominant enclosed agreement. Therefore, Predix does not directly provide data-acquisition gateway hardware but provides a gateway framework—Predix Machine, to realize data collection and links.

Predix provides the development framework of Predix Machine, supports access to open field protocols, and enhances the function of edge computing. By developing the corresponding device access and edge computing functions by partners, Predix covers almost all problems that need to be solved by edge devices: (1) industrial protocol analysis; (2) Flexible data acquisition; (3) Cooperation with the platform; (4) Local storage and forwarding; (5) Supporting the application on the platform; (6) Rich security strategies; (7) Local device communication, and a large number of partners have developed many edge gateway products based on this framework.

2.   The cloud platform layer

Predix Cloud on the platform is the core of the entire Predix solution. It is centered around industrial data and provides rich industrial data collection, analysis, modeling, and industrial application development capabilities.

Because GE itself is the production of large, complex industrial products, such as aircraft engines, gas turbines, wind generators, locomotives, a wide range of high-end equipment, so Predix Cloud building is also starting from the characteristics of the business of GE itself, which is closely related to the discrete manufacturing industry in large-scale high-end equipment design, production, and operations, provide to industrial equipment data analysis as the mainline of ability, convenient to construct the application of high-end equipment industry. However, due to the excellent openness of the platform, many other industries, including customers of many process manufacturing and services, are leveraging Predix Cloud to develop related applications.

Predix Cloud integrates industrial big data processing and analysis capabilities, Digital Twin rapid modeling, industrial application rapid development, and a range of shelf microservices that enable rapid integration. There are mainly the following parts:

(1) Infrastructure. Predix provides three deployment architectures: public Cloud (AWS, Microsoft Azure), private Cloud, and hybrid Cloud; (2) Security. Predix Cloud provides a lot of security mechanisms, including identity management, data encryption, application protection, logging and auditing, etc. (3) Databus, which includes the functions of data injection, processing and heterogeneous data storage, supporting the import and processing of stream data and batch data; (4) High-productivity development environment, providing visual application development environment including Predix Studio, and supporting Citizen Developer to quickly build industrial applications by dragging and dropping; (5) High-control development environment, code-level development environment (based on Cloud Foundry, the industry's first open-source PaaS cloud platform), industrial application development environment with the highest degree of control, and a series of microservices that can be quickly integrated; (6) Digital twin development environment, providing fast modeling tools to realize model development including device model, analysis model and knowledge base combination.

The most potent part of Predix is the extensive industrial data analysis function based on digital twins, which reflects the various original states of physical devices through data collection and storage in virtual Cyberspace. It enables the control and prediction of devices by constructing holographic models of devices. Internal management information systems fall within the product range of enterprise application software vendors, not GE.

Predix provides a catalog of models that GE and its partners have developed to publish as APIs and provides test data that allows users to stand on the shoulders of giants, train models with existing models, and quickly implement instantiations. At the same time, user-developed models can also be published in the model directory for Shared use by more customers. The model here includes conventional anomaly detection and text analysis, signal processing, quality control, and operational optimization. According to the generally recognized industrial significant data analysis types, it can be classified into four categories: Descriptive, Diagnostic, Predictive, and strategic.

In addition to these analytical models, GE offers more than 300 asset and process models relevant to GE's diverse portfolio of products, including attributes and 3D models that allow customers or partners to build digital twins quickly. Tens of thousands of digital twins have been built on the platform, both by GE and its partners.

3. The application layer

What is needed for industrial customers is the ability to solve problems, not the tools to solve them. GE's primary goal of launching Predix is developing industrial applications and analyzing industrial problems more efficiently and quickly.

Predix applied for is not the traditional MES, ERP, PLM, and other traditional IT applications, but for all kinds of industrial equipment to provide complete equipment health and fault prediction, optimization efficiency, energy consumption management, scheduling optimization application scenarios, with the method of data-driven and mechanism combination, aimed at solving the traditional industry for decades have failed to solve the problem of quality, efficiency, and power consumption will help transform the digital of industrial enterprises.

Predix works very hard on ecological construction, and good platforms have a black hole effect. More than 33,000 developers, 300 partners, and more than 250 applications developed on the Predix platform. The industry also increasingly recognizes Predix's influence in the industrial Internet ecosystem. Its partners include (1) horizontal partners: including giant IT companies like Microsoft and Apple and many start-ups. Microsoft Azure provides Predix with an IaaS (Infrastructure-as-a-Service) platform, machine learning, and Power BI business analytics tools, while Apple enriches Predix's ability to develop industrial-scale mobile applications. (2) Vertical partners: including many consulting agencies, integrators, and independent software developers. Based on the GE APM + Predix platform, these partners can provide their industrial customers with customized industrial application development and data analysis solutions, including Infosys, Wipro, Accenture, Capgemini, TATA, Tech Mahindra, and other global companies. The Predix platform has developed many applications for industrial equipment performance improvement, predictive maintenance, and supply chain management.

Thinking Exercise

1. Why did a traditional hardware company like GE develop Predix Cloud, an industrial Internet software platform, rather than being led by a IT company?
2. Why is ecological construction important for industrial Internet platforms?
3. Should the platform be open?
4. What are the advantages and disadvantages of edge computing and cloud computing?
5. What is the most difficult thing about the company's transformation?

Homework of This Chapter

1.   Why is it essential to build new ecosystems for IM (Intelligent Manufacturing)
2.   Please discuss the relationship between the Five Development Concepts and IM.
3.   What are the crowdsourcing platforms, sharing economy, and new ecology platforms and examples? Please list some examples.
4.   Regarding the sharing economy innovation, what are your comments and suggestions on the current model of shared bikes and the problems caused by excessive imitation, such as large amounts of urban garbage, and disorderly parking?
5.   Why is manufacturing the main focus of "Internet plus" rather than other industries?
6.   What is the relationship between IM and the "Internet Plus" and "Made in China 2025" vigorously promoted by the Chinese government and Germany's "Industry 4.0"?
7.   What is the relationship between an IM and smart city and smart home?
8.   What is the relationship between Advanced Manufacturing and Intelligent Manufacturing?
9.   How to coordinate the development among the cities in a big bay area?
10. What is the essence of intelligent manufacturing? Where does uncertainty come from?
11. Why can Geely company improve welding accuracy?
12. How to look upon this matter: after some enterprises use industrial robots, the manufacturing cost increases instead?
13. What kind of resources are American companies good at using to solve problems? What are the significant advantages? Please give some examples.
14. What kind of resources are German companies good at using to solve problems? What are their significant advantages?
15. What kind of resources are Japanese companies good at using to solve problems? What are the significant advantages?
16. What strategies should Chinese companies or government adopt to play their advantages and overcome their disadvantages fully?
17. The Industrial Internet reference architecture (the US IIS) has rich connotations in technology and management. Please (1) analyze the focus of the four views in Industrial Internet reference architecture. (2) discuss the steps of intelligent manufacturing system planning.
18. Try to analyze the correspondence between the United States IIS and German "Industry 4.0" reference architecture.
19. According to the differences between the "Industrial Internet" strategy of the United States and the "Industry 4.0" strategy of Germany and the reference architecture, try to list the representative schemes respectively and point out the characteristics.
20. Please indicate the position of industrial Internet, industrial robots, industrial clouds, and big data in the reference architecture model of IM.

21. After Midea purchased the CNC machine tool, did it have all the relevant resource elements? What are the problems with the lack of ownership of data? How do we break this yoke? What is the function of OPC?
22. Do you think it is correct to say that "industrial Internet platforms are blooming everywhere" and "Industry 4.0 platforms are very rare"? Why is that?
23. Why is the ERP module of Dell implemented in the order of purchase, order management, and then production and manufacturing management?
24. Introduce the connotation of an advanced manufacturing enterprise.
25. What is the subversive significance of intelligent manufacturing to the development of the industry?
26. What is the relationship between the "One Belt and One Road" national policy and the development of the intelligent manufacturing industry in China?
27. How should we promote the development of advanced manufacturing industry?
28. What impact do you think intelligent manufacturing will have on our life?
29. What about the two concepts of intelligent dimension and manufacturing dimension in maturity?

References

1. 2016 the global manufacturing industry competitiveness index [EB/OL]. Deloitte, American Competitiveness Council; 2016. https://www2.deloitte.com.
2. Industrial Internet Consortium (IIC), Industrial Internet Reference Architecture [M]. 2018.
3. National Intelligent Manufacturing Standard System Construction Guide (2015 Edition) [EB/OL]. Standardization Administration of China; 2015.
4. Intelligent Manufacturing Capability Maturity Model White Paper (1.0) [EB/OL]. China Institute of Electronic Technology Standardization; 2016.

Chapter 2
System Level of Intelligent Manufacturing

2.1 Inspiration Case: Dongguan Tianxiang Clothing Company's Integration of Informatization and Industrialization

Integration of informatization and industrialization refers to enterprises moving around the strategic target, strengthening the foundation of industrialization, promoting interactive innovation, and continuously optimizing the data, technology, business process, organization structure. It also requires a full tap of the allocation of resources potential, and constantly creating new information environment ability, form sustainable competitive advantage and realize the innovation and development, intelligent and green development process, as shown in Fig. 2.1 Management architecture of binarization fusion. The management system of information and industrialization is based on the methods and rules extracted from implementing and evaluating more than 10,000 enterprises.

Dongguan TianXiang clothing co., LTD. (shortened to TX), was founded in 1996. The founder hopes that she can give her children safe and environment-friendly clothing. The company has its marketing network spread to more than 30 provinces all over the country, set up eight offices, with more than one thousand terminal stores. The company ranks Top 10 according to the comprehensive strength of the children's clothing industry in China and Top 3 in South China. It has a successful multi-brand operation platform, with "Pencil club" and other famous brands. It is the governing unit of the Guangdong clothing industry association. The company has transformed from single-channel marketing to multi-channel marketing, including network marketing, chain stores sales, supermarkets, and other channels.

In China, with the implementation of the "two-child policy", the size of the children's clothing market China has maintained an average annual growth rate of 12.6%. At present, foreign-funded or joint-venture children's clothing brands are temporarily in the lead in sales, such as Mickey Fun, Lenfoo, PacLantic, etc., but the industry concentration is low, and the gap between brands is relatively unnoticeable. The total

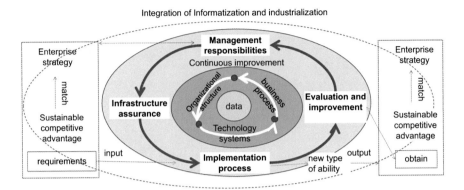

Fig. 2.1 Management architecture of binarization fusion [1]

market share of the top eight independent brands is only 5.33%. There are many channels to enter the children's clothing industry. The initial capital investment flexibility is incredible. Due to the rapid change of children's clothing popularity, there are few patent applications in the clothing industry. Children's clothing is lucrative, with sports, adults, and children's brands also entering the market. Internationally, with the substantial appreciation of RMB, the competitiveness of enterprises in Southeast Asia has been greatly enhanced.

The cluster trend of China's children's clothing industry is apparent in the industrial chain. Foshan in Guangdong and Zhejiang provinces are the two large children's clothing industrial clusters that develop rapidly. There are many alternative raw materials or spare parts for upstream garment enterprises, and the conversion cost of the supplied products is not high. Online shopping has reduced the cost of product search, and the proportion of online clothing has gradually increased.

The company's problems include: (1) insufficient control over terminal stores, irregular use of POS machines, payment problems in sales, and slow payment collection; (2) the WeChat platform of the company has no two-way interaction and only serves one-way advertising, which cannot enhance user stickiness; (3) Automation and intelligence of product layout and discount optimization have not been realized; (4) The online and offline warehouses are not integrated, and the inventory and warehouse management staff are redundant; (5) The balance rate of the production line has not reached 80% stably, and there are difficulties in data sharing and insufficient data utilization among software systems; (6) With a long product development cycle of more than 200 days, it is difficult for new products to keep up with the fashion trend.

After becoming a pilot enterprise for standard implementation in 2015, the company set up a project organization, formulated a project plan, and held a project launching meeting. The chairman announced the division of tasks and authorized the project team, management representative, and several internal auditors. Finally, the project team provided the first training for all the staff at the meeting. After that, the project team carried out a 5-day on-site investigation, evaluation, and diagnosis. The team made an improvement plan based on diagnosis analysis and extract improvement project; based on the consideration of daily operation and practical constraints, the team assisted enterprises in writing and modifying the management system documents at four levels and released them. Through the document description, the management system process was optimized so that the input, output, activity content, control, and evaluation methods of each process could be solidified by the formation of documents. After the trial operation of the system for some time, based on the data generated from the operation, conduct an internal audit of the management status quo according to the management system, and rectify and improve the deficiencies. The top management of the enterprise will conduct a management review on the management system and judge whether the management system can support the current enterprise strategy. After several cycles of audit, it entered the system identification stage and system maintenance and improvement stage and carried out regular and continuous monitoring of the proposed quantitative indicators, and finally completed the whole process. Its main feature is "circular improvement, the document first", and its focus is the matching among "enterprise development strategy—sustainable competitive advantage—new capability—system (information technology, business process, organizational structure, data)", it should keep top-down strategic consistency and dynamic adjustment.

In the process of investigation and diagnostic analysis, the project team used Porter's Five Forces Analysis Model, SWOT analysis, and other methods to identify the direction in which the company is most necessary to strive for market advantages, and formulated an 8-word enterprise strategy: brand promotion, full chain integration, rapid R&D, and wisdom enhancement. In the planning process, the team identified and confirmed the sustainable competitive advantage that the company needs to obtain and match with the strategy, which includes five aspects: rapid R&D; Intelligent operation; Terminal control; Channel coordination; Lean manufacturing. Among them, rapid R&D is to apply for the vital certification of the new ability.

1. Rapid R&D

The key to rapid product development is to shorten its development cycle, which includes knowledge base construction, concurrent engineering, design standardization, and product planning.

As shown in Fig. 2.2, the rapid R&D and intelligent operation process are based on a unified database and several knowledge bases, and each link is carried out in parallel. All departments cooperate fully and promote efficiency. The sample making, new product introduction, and display scheme design are carried out to shorten the product development cycle. The idea of concurrent engineering is applied to two fundamental design and manufacturing processes. Specific measures include:

(1) Make product planning before product design. Planning includes consideration of price dimension, age dimension, style dimension, or dress dimension. The clothes in one's wardrobe for spring, summer, autumn, and winter are related to one's lifestyle. Therefore, we should make a detailed study of the lifestyle

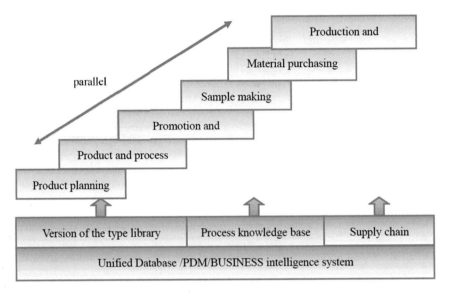

Fig. 2.2 Framework of rapid R&D process

of target customers, and then make a good product plan and standard, and then follow the follow-up procedures. Design is concerned with the enterprise's survival tomorrow, while planning is about the enterprise's survival the day after tomorrow.

(2) Parallel design and manufacturing. Through the statistical analysis of sales data, predict the future sales of products, prepare for the production of products with optimistic sales expectations in advance, to support the parallel development of products, production and purchasing business, shorten the development cycle of new products, and speed up the launch of new products.

(3) Design standardization and knowledge management. The company has developed a database of nearly 30,000 garment models and a knowledge base of more than 8000 fabric solutions, which can be digitized for easy access by designers. It can be flexibly combined according to the user's body shape data, and the automatic cutting equipment can also flexibly adjust parameters. For example, the size of the T-shirt, position of Logo printing embroidery, the sides where buttons are woven, and the front or the back of the neckline—the company did not have any regulations before. Employees could do whatever they think is convenient. Finally, the company formulated detailed process standards and solidified them in the version database and fabric knowledge base.

(4) Software application. Introduced PDM (product data management) system and CorelDraw CAD software to manage design knowledge base, document, version, product configuration, and supervise and speed up the development task, achieve the purpose of efficient data security and sharing.

(5) The application of new logistics technologies. Clothing sorting is an essential link in the logistics and distribution system. By using RFID automatic identification technology, finished products can be sent to the final inventory location or even directly to the truck, instead of being put into a temporary warehouse and then sorted to the target location as before, which reduces the sorting operation time by about 50% and also shortens the product development cycle.

(6) Organizational reform. The technology development Department moved from the commodity development center to the supply chain center so that the sample production and trial production of new products can better integrate with the large-scale formal production and speed up the process of new product development.

A brand is critical to clothing. Shorten the product development cycle, realize the rapid launch of new products, shorten the market demand response time, and improve product quality, which will ultimately improve the brand image.

2. Intelligent operation

As shown in Fig. 2.3, the business intelligence system is used to analyze the big data of physical sales outlets and online sales all over the country, to assist the operation decision of purchasing, promotion, goods adjustment, goods layout, and other aspects, generally improve the operation decision level of each sales outlet, promote sales and reduce inventory. In terms of system matching and specification, Yum BI (Business intelligence) solution is used to automatically integrate, process, and summarize massive business data of enterprises. Using the management cockpit for business extraction and threshold monitoring; The intelligent replenishment module can comprehensively score the goods based on indicators such as the sold-out rate,

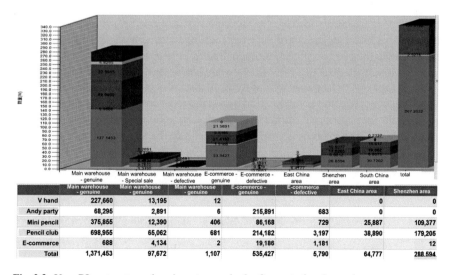

	Main warehouse - genuine	Main warehouse - genuine	Main warehouse - genuine	E-commerce - genuine	E-commerce - defective	East China area	Shenzhen area
V hand	227,660	13,195	12			0	0
Andy party	68,295	2,891	6	215,891	683	0	0
Mini pencil	375,855	12,390	406	86,168	729	25,887	109,377
Pencil club	698,955	65,062	681	214,182	3,197	38,890	179,205
E-commerce	688	4,134	2	19,186	1,181		12
Total	1,371,453	97,672	1,107	535,427	5,790	64,777	288,594

Fig. 2.3 Uses BI system to analyze inventory and sales forecast of each warehouse

average daily sales volume, and available days of inventory, identify best-selling goods, and conduct sales tracking analysis to realize intelligent allocation.

3. Terminal control

Including: enhance the control ability of more than 1000 chain terminals, realizing information exchange with terminal stores, grasping the real-time situation of store sales, quickly obtaining market information, and improving sales efficiency; Enhance the control over the final consumers, interacting closely with them, push relevant information to them, and obtain dynamic information of consumers, to enhance the brand loyalty and user stickiness of the final consumers.

Specific measures for terminal control include:

(1) Through the application of DaoXin ERP store terminal system, POS machine, and other information systems, standardize the cashier desk operation, order return operation and inventory operation of terminal stores, and realize information exchange with terminal stores, master the real-time situation of store sales, and realize real-time payment collection.

(2) promote product knowledge and provide two-way interactive services to end consumers through emerging tools such as the WeChat platform to improve brand loyalty.

(3) Data mining analysis, building models, looking for potential users and carrying out targeted database marketing.

(4) Expand the immediate office environment for business personnel through the mobile platform, and enhance the market response-ability and serviceability.

4. Channel coordination

Including: online and offline inventory coordination and business coordination, coordination of supply chain, coordination of sales channels in different regions. Improve resource efficiency and reduce costs through collaboration.

Based on the means of logistics warehousing system (WMS), inventory redundancy can be reduced through sharing between online and offline inventory to realize inventory coordination and business coordination between physical stores and e-commerce, thus reducing the number of employees, improving efficiency, reducing inventory, and finally reducing cost. Due to the unified coordination of online and offline sales, the logistics Department and e-commerce logistics Department were merged into the "Logistics Department".

5. Lean manufacturing

(1) Establish a unique bar code for all used clothing materials, decompose the product production plan into the material plane through the material list of each type of product, reasonably calculate the shear material allowance, and control the material amount of each batch of products and the deviation from the standard cost.

(2) Use RFID radio frequency identification technology in data acquisition, find the bottleneck constraints of the production line, correct timely,

increase the rate of the balance of the production line. It can also automatically carry out output statistics and scrap statistics to investigate the output for relevant incomplete. Promote the upgrade of lean financial control by focusing on "precision", especially strictly managing the financial process of chain stores; Reduce cross payments (refers to the goods of Type B with the tag of Type A, or the sales of type A are recorded under the name of Type B) to improve management efficiency and economic benefits.

(3) In the aspect of system matching and specification, the introduction and interface optimization of ERP system (U8) were carried out, and the data standards and structure of ERP system were strictly observed in the planning, architecture design, system development, and other stages so that the data of the platform could be seamlessly connected to other systems of the company to achieve integration.

The quantified target was established at the end of the project, as shown in Table 2.1.

The company's standard implementation work has reached the request-target including the top management support, the system scope is clear, the system construction is sustainable, the strategy is appropriate, the information planning is suitable, the

Table 2.1 Stage targets of intelligent manufacturing (2017–2018)

New capability name	Quantitative indicators	Monitor and measure targets for 2017	Monitoring and measurement targets for 2018
Rapid R&D and design ability of children's clothing products (this key point)	The average development cycle for new products	30 days shorten	Two months shorter
	Success rate of new product development	To more than 70%	To more than 75%
	The number of knowledge items in the knowledge base	More than 9000	More than 10,000
	Increase in profit margin	More than 5%	More than 9%
Intelligent operation capability	Store inventory	By more than 2%	By more than 5%
	Return rate	By more than 2%	By more than 5%
Sales terminal control ability	WeChat number of registered or following users	More than 50,000	More than 100,000
Logistics channel coordination capability	inventory	By more than 2%	By more than 5%
	Inventory manager	By more than 5%	By more than 10%
Fine production control and financial control ability	Production line balance ratio	75%	85%
	String section rate	Reduction of about 2%	Reduction of about 5%

system is integrated, the implementation scheme is systematic, the business process is appropriate, the organization structure is suitable, the information technology is reasonable, the data utilization is effective, the implementation process can be monitored and so on. TX, through the integration of two management system review and recognition, Finally, it passed the national standard certification of two management system integration and is on the way forward to the vision of "becoming the world-famous, China's most competitive, innovative, high-efficiency children's clothing group company".

Thinking Exercises

1. What are the working steps for implementing the standard certification of the integrated management system? What are the key points?
2. What matching and improvement work has TX made in terms of the company's strategy, competitive advantage, new capability, technology, business process, organizational structure, and data during the certification process? How does strategic coherence come about?
3. Try to use five forces analysis and SWOT analysis methods to carry out strategic analysis for TX Company.
4. What is the difference between internal audit and management review?
5. How to understand the phenomenon of "form adherence standard" and "two skins" of the management system? What can be done to prevent this tendency?

2.2   The Introduction

Since the concepts of "Industry 4.0" in Germany and "Made in China 2025" were proposed, various practices have emerged around the upgrading and transformation of the manufacturing industry. According to the IM system architecture standard, the system has five layers from bottom to top, namely equipment layer, control layer, workshop layer, enterprise layer, and collaboration layer. It is a common and easy way to realize IM from the bottom up.

The system-level of IM reflects the intellectualization of production equipment and the Internet protocol (IP), as well as the flattening trend of the network. The five levels are the traditional path and view of manufacturing information for enterprises. Each level is explained as follows:

1. The equipment level includes sensors, instruments and meters, bar code, Radio Frequency Identification Detection (RFID), CNC machine tools, robots, and other perceptual and implementation soft and hard equipment, which are the material and technical basis for the production activities of enterprises;
2. Control level includes PLC, SCADA, DCS, Fieldbus control system (FCS), industrial wireless control system (WIA), etc.
3. workshop level Realize factory-oriented production management, including manufacturing execution system (MES), etc.

4. enterprise-level implement enterprise-oriented operation management, including enterprise resource planning (ERP), supply chain management (SCM), customer relationship management (CRM), PLM or Product data management (PDM), etc.;
5. Collaborative level: collaborative R&D and collaborative manufacturing services are realized by sharing information on the Internet between enterprises in the industrial chain.

With the unification of product design standards, the comprehensive integration of the new automatic control architecture, and the comprehensive upgrade of MES to the operation management system, various links such as ERP/SCM/CRM/PLM, which used to be relatively isolated in the manufacturing industry, have been connected and the "island" of information has been eliminated. Understanding the interlocking five stages of the IM implementation path can better meet the needs of enterprise customized production and improve operational efficiency, define the main responsibilities of each department and cross-department and cross-enterprise cooperation, and effectively promote the enterprise's transformation.

2.3 Equipment Layer

Equipment, also known as devices, is the first fulcrum for IM.

Intelligent equipment is an important means and tool for intelligent factory operation. Manufacturing equipment has experienced the stage from mechanical to numerical control equipment and is now gradually developing to intelligent equipment. Intelligent equipment mainly includes intelligent production equipment, intelligent detection equipment, and intelligent logistics equipment. The equipment layer also has decision problems. When the product is handed over to the "intelligent" machine tool, the digital product definition and human knowledge and experience is input to the machine tool, the intelligent software of the machine tool will automatically process in accordance with the instructions, and even the software can optimize the processing path to achieve the purpose of saving time and effort. At the equipment level, the lowest level of execution, decisions are usually made with clear goals and identified resources.

The critical step to realizing intelligent equipment is to transform a "deaf and dumb" physical asset (such as non-digital traditional machines, equipment, etc.) without communication ability into a digital "new asset" that can be defined, expressed, and exchanged in the digital world, also known as "IM equipment". The words "deaf and dumb" means that the machine can neither speak nor listen. Manufacturing equipment is generally relatively complex, and the batch may be small, making the production cost usually high. The equipment development cycle is long, which leads to equipment development risk. In addition, the difficulties in equipment manufacturing are essentially in soft equipment. That is, the soft equipment represented by industrial software, including software tools such as CAD/CAE, which are high-end industrial products themselves and are often very complex. Without soft equipment, there can be no "digital, networking, intelligent". Therefore, intelligent equipment is the first and most challenging step in the implementation of IM.

Intelligent equipment can be divided into three stages:

1. Digital devices: The hallmark of this phase is that individual machines and devices can continuously collect data through sensors or RFID tags. Field data is the source of the IM system. China's manufacturing enterprises are generally still in the initial stage of this inevitable experience. In this stage, the pursuit of "black lamp factory", robot substitutions, high-end CNC equipment, but not with the corresponding industrial software system, is aggressive. As a result, the data generated by the production equipment is still not fully utilized, and the equipment still fails to express its health state or conduct fault warning and is still in the "semi-dumb" state. Due to the equipment failure, an unplanned shutdown will be caused, affecting the production. Therefore, in this stage, we should focus on the use of sensor technology and communication technology to achieve equipment data acquisition, including production process data and equipment health data, in order to enter the second stage smoothly.

2. Networked equipment: In this stage, information network technology and data analysis technology should be mainly used to integrate CNC, industrial robots, processing centers, and equipment with a low degree of automation to make them more flexible and improve production efficiency.

For example, an intelligent sensor launched by ABB can be directly installed on the motor's enclosure and directly connected to a smartphone via Bluetooth. The mobile phone ACTS is a gateway to upload motor data to the ABB Ability cloud platform for analysis. Siemens has realized the "Internet of things" of SIMOTICS series motors. Through WiFi communication, the motors are connected to the MindSphere cloud platform. Based on the shell's vibration data, the motors' running state is monitored to achieve maximum transparency and the highest productivity. Rexroth company of a torque wrench, it will automatically record automobile assembly line workers operating on the number of artifacts, torque capacity, tighten the process and various parameters, when charging wrench to upload the data to the clouds, do each parts, each station operation can be recorded, traceability, will originally offline "dumb" operation to realize the digitization, as shown in Fig. 2.4a; Installed in the industrial sewing machine "industrial sewing machine data acquisition and management analysis platform" and "RFID industrial sewing machine control system (driver + controller)", "Internet + sewing machine" can be formed by the real-time monitoring of car density of suture and dosage, clothing production, such as data and upload data center, through the analysis of the data center, can provide the factory with garment sewing operation Suggestions to improve the, production scheduling, etc., can also be visualized remote monitoring by mobile phones, as shown in Fig. 2.4b.

Two-way communication between devices for data exchange and instruction interaction is usually realized through "ask/be asked" and "notify/be notified" functions, as shown in Fig. 2.5. In the ask/be asked function, the equipment can answer whether the work status can be used to produce the current or subsequent order, the time

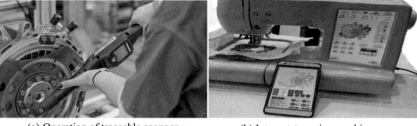

(a) Operation of traceable spanner   (b) Internet + sewing machine

Fig. 2.4   Realize digital and networked upgrade of offline "dumb operation"

(a) Direct communication between products and equipment

(b) Communication schemes between products, trays and equipment

Fig. 2.5   "Ask/be asked" and "notify/be notified" functions of the device

required to adjust tooling, or to issue an order, material information, or maintenance details to the production system. In reporting/notification function, the equipment can be active on the device will as planned or emergency stop, which will help planner or planning system for production scheduling, this equipment to the machining process of self-tracking with deviation, if fluctuations or deviate more than the limit, can signal, such as change the tooling. In order to meet the requirements of this stage, enterprises should require equipment suppliers and manufacturers to open the data interface when purchasing equipment, which is the premise of realizing automatic data collection and workshop networking. At present, major automation manufacturers have their industrial bus and communication protocol. OPC UA architecture standard application is not popular.

3.   Intelligent equipment. The mark of this stage is to realize the control of reconfigurable machine tools, reconfigurable control system, and adaptive control through electromechanical equipment and a built-in controller with a learning function so that the system's behavior can adapt to the changing environment more. It means the development of embedded industrial software to support the cognitive functions of devices and robots to adapt to the unstructured workplace environment; The equipment has more advanced sensing and perception and maintains stability under unpredictable changes, thus working well in an uncertain environment; The equipment has the ability of self-monitoring and adaptive recovery.

2.4 Control Layer

The second fulcrum is the production process control. The developed countries have taken the lead in intelligent equipment. Japan and Germany, in particular, have monopolized mainly the world's market for significant manufacturing equipment. The subsequent development of IM is from the "point" of equipment to the "line" of process control, which is a technological opportunity for China's equipment manufacturing industry to "change lanes and overtake". In the control layer, two main problems that need to be paid attention to are man–machine fusion and control systems.

2.4.1 Man–Machine Fusion

The ultimate goal of an intelligent factory is not to build an unmanned factory but to meet the needs of market customization under the premise of pursuing reasonable costs. Therefore, man–machine fusion will become the primary trend of intelligent factory development in the future. The most significant characteristic of man–machine fusion is that it can use human flexibility, intuitive judgment, and decision-making ability to complete complex, changeable and uncertain work tasks. At the same time, robots are good at repeated labor and fast calculation. The approaches of man–machine fusion mainly include man–machine collaboration and augmented reality.

Man–machine labor division is a dynamic evolution process. Traditional robots need to take safety isolation measures to guarantee their safety, while Human–Robot Cooperation (HRC) can establish a mutually reinforcing relationship to achieve more flexible, efficient, and quality production, as shown in Fig. 2.6.

In the modern automotive industry, this concept of HRC is already realized where human operators and robots work together to manufacture a car product. The concept requires studying several aspects as collision, safety, and respect the robot workspace

(a) collaborative robots that can be taught and guided   (b)Traditional robot safety isolation

Fig. 2.6 Process of human–robot co-evolution

aiming to enable versatile automation steps and increase productivity. It is an additional element that combines human capabilities with the efficiency and precision of machines.

According to the degree of cooperation from low to high, there are five levels of man–machine cooperation:

1. The security level of monitoring stops, which is the most basic collaborative mode, that is, when people enter the collaboration area, the robot stops moving, and when people leave the collaboration area, the robot automatically returns to regular operation;
2. Instruction guide. The operator will teach the robot hand in hand through a manual guide device. Collaborative robots do not only perform pre-programmed tasks but also can be "trained" interactively. Instead of spending much time on programming, they just repeat their actions, as shown in Fig. 2.6.
3. Monitoring of speed and distance. When the robot judges that the distance between man and machine is less than the safe distance through the sensor, or when the number of people in the cooperation area exceeds the limit, the robot will stop immediately; If the moving speed of the robot is reduced, the safety protection distance can be correspondingly reduced. The robot will choose a motion path that will not violate the minimum safe distance rule.
4. Limitation of power and force is a more active, advanced, and safe collaborative function. It limits the kinetic energy and force output to avoid the occurrence of injury events from the root. In this mode, slight planned or unintentional contact between the robot, workpiece, and the human body is allowed.
5. Self-organization: Robots break the barrier of passive receiving instructions. Robots can work continuously and adjust goals and plans in real-time to implement new daily processes and adapt to and learn from others.

The Automated Production Assistant System (APAS) introduced by Bosch is a safety-certified collaborative robot. It is based on the universal robot, covered with a "skin" of built-in sensors, so it has sensitive force feedback characteristics and can sense the location of people around. When people get close to it, it will automatically slow down, and when it reaches the set force, it will stop immediately, which is equivalent to the existence of a kind of invisible protective net, which can ensure the safety of people. After the person leaves the area, the robot will automatically return to average speed. In this way, after risk assessment, there is no need to install the isolation fence, as shown in Fig. 2.6a, so that humans and robots can work together safely. One model of APAS can be moved and added to the production line as needed. Second, collaborative robots perform pre-programmed tasks; workers can also "train" the robots interactively. Instead of spending much time on programming, they just repeat their actions, as shown in Fig. 2.6b. Finally, the evolution between humans and robots can be coordinated. For example, the robot can evaluate various holding positions, find out the most appropriate holding position, and directly deliver the products to the hands of workers. The robot adjusts heavy doors to open or close, allowing human employees to spray glue into the correct position. In addition, the robot should work silently, without disturbing noises. A safe and comfortable working environment should be created to get along with the robot. Integration of vision, touch, hearing, and other sensors are the technical basis for collaborative robots to cooperate with humans.

Augmented Reality (AR) combines digital camera image technology, 3D model, and other IT technologies to project the virtual world to the real world. Its core objective in manufacturing is to support people in the production plant to deal with increasing technological complexity and uncertainty. New tools, including augmented reality devices based on wireless networks and wearable technology, will lead to the increasing use of multimodal human–computer interfaces on production lines, allowing workers to focus on decision-making and control production.

In the future, AR technology will be widely used in equipment maintenance and personnel training in factories, as shown in Fig. 2.7. When workers wear AR glasses, they can "see" where they need to operate. For example, where the bolts need to be tightened, the worker can see the correct twist of the wrench, the correct installation position, making it easy for the worker to face various changes in the production process brought by personalized products. Maintenance personnel can make the virtual model superposition with the actual model by scanning the code in natural objects. Meanwhile, information such as equipment model and working parameters can be displayed in the virtual model, and maintenance operations can be carried out according to the prompts in AR. AR technology can also help equipment maintenance personnel to compare physical operation parameters with digital models, locate problems as soon as possible, and give possible fault cause analysis.

Google launched Google Glass, an AR eyewear, in 2012, but sales were lackluster because it could not find the exemplary application scenario. Later, Google released an enterprise AR glasses, mainly targeted at manufacturers. Microsoft HoloLens 2

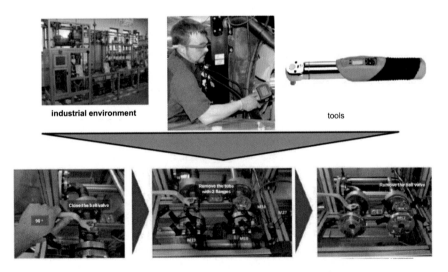

Fig. 2.7 Provides an intelligent augmented reality system for field workers

is an utterly holographic computer, providing a more comfortable immersive experience by combining options for more collaboration in mixed reality. AR technology will be increasingly popular in production and manufacturing, installation, training, maintenance, and other links.

2.4.2 Control System

The control system is the central nervous system of the production line. It can assist in micro human factors engineering action analysis, mesoscopic human–machine matching and interaction analysis, and macro production line balance analysis, as shown in Fig. 2.8.

The technological transformation of many enterprises focuses on the establishment of automated and intelligent production lines, assembly lines, and inspection lines, such as the U-shaped pulsating final assembly line established by the aircraft assembly plant of Boeing Company. Automatic production lines can be divided into rigid and flexible automatic production lines. In order to improve production efficiency, industrial robots and suspension systems are more and more widely used in automatic production lines.

The control system selection needs to consider the application scale, scalability and enterprise development plan, integration requirements, functions, availability, investment, and many other factors, currently mainly the choice between PLC (Programmable Logic Controller) and DCS (Distributed Control Systems).

PLC is mainly used in discrete manufacturing, such as parts assembly production. DCS is mainly suitable for process manufacturing in the process industry and relatively large factories with thousands of control points, with a significant coefficient of control difficulty, high complexity, and strict safety requirements, where an

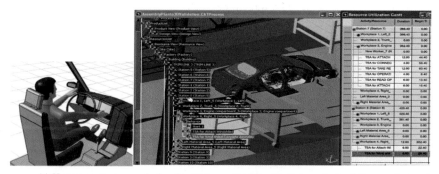

(a) Human-computer analysis of vehicle availability (b) Balanced simulation analysis of vehicle production line

Fig. 2.8 Balance simulation of ergonomics and production line

expensive DCS control system has to be adopted. Hybrid applications can use both PLC and DCS.

[Case: Intelligent wireless production Monitoring system of Usiminas Company]

Usiminas is one of the world's leading producers of steel. The company's heavy-duty steel plant needed a plan to protect the plant's valuable equipment assets from unexpected shutdowns. In general, when the roll is damaged, the steel plate manufacturing process must be stopped for at least 6 h to replace the backup roll. In the past, the company has spent \$40,000–\$175,000 to repair a bearing. An accident like this can cost the company at least 600 tons of production.

The company's production monitoring system decided to use DCS with an integrated database. A multimedia wall has been built in the production and operation center, which is installed with more than 40 plasma displays for visual graphical display. The company decided to adopt a wireless solution. Eight wireless temperature sensors were installed on the rollers used to produce steel plates to measure the oil temperature of the roll. An intelligent wireless gateway was used to collect this critical information and transmit it to its DCS. Operators use the wireless data collection and network roll back to the oil temperature information-bearing running condition, the wireless solution of reliability in a harsh factory environment to achieve more than 99.99% of the industrial requirements, more precise information and information redundancy allows companies to safeguard roll bearing better, avoid unplanned downtime, to maintain the steel manufacturing process smoothly. Self-organizing wireless networks provide stable data even when the temperature sensor is extremely hot, with water and grease present. The wireless solution is easy to install and debug. It takes 4 h to debug the wireless equipment, while it used to take 2–3 days to debug the wired equipment. Based on the SAP MII scheme, it has been deployed in 17 factories, integrating four real-time database products, dozens of local systems and databases, and data scattered in various production and operation systems. The company plans to use wireless temperature, pressure, and vibration sensors at monitoring points throughout the plant, as well as pH analysis sensors, to monitor sewage discharges.

2.5   Workshop Layer

In the whole IM system, the intelligent processing workshop undertakes the production task of products and is the third fulcrum of IM. A workshop usually has several production lines, which are linked together by the similarity of parts or products, forming a workshop. The workshop control needs to be based on the device interconnection, the use of manufacturing execution system (MES), based on advanced planning and scheduling (APS), improve the efficiency of OEE and employees, realize the traceability of the production process, reduce the WIP (Work In Process) inventory; It is necessary to make full use of intelligent logistics equipment to realize the timely delivery of materials in the production process. DPS (Digital Picking System) can be used to realize the automation of material Picking. Human–machine interface (HMI), as well as industrial tablets and other mobile terminals, using Digital Twin (Digital mapping) technology not only provides workshop of VR (virtual reality) environment but also can display equipment's actual state, realize the paperless production process, as shown in Fig. 2.9.

From the perspective of a digital twin of "physical element + virtual Cyber", the intelligent workshop mainly includes intelligent processing center and production line, intelligent production control center, intelligent production control system, intelligent storage/transportation, and logistics system, which is the integration of physical workshop and digital virtual workshop. Before manufacturing, modeling and simulation verification are carried out for the whole product life cycle in the digital workshop based on the digital model of each resource and process. In the manufacturing process, the simulation results of the virtual workshop are used to guide the manufacturing process of the actual physical workshop, and real-time monitoring and adjustment are made to realize the intelligent characteristics of the process, such as self-adaptation, self-control, and self-optimization.

(1)　Intelligent processing center and production line

It mainly includes intelligent processing equipment, intelligent tool management system. There are several different types of processing equipment, such as welding, painting, assembly intelligent robot, 3D printer, CNC machine tool, etc. Different processing equipment is responsible for different types of processing products. Each intelligent processing equipment has an RFID tag, RFID reader, sensor equipment. The intelligent tool management system provides overall flow management for tools, fixtures, and measuring tools in the production process, and helps the manufacturing system to operate effectively through real-time tracking of the process of tool purchase, in-and-out storage, grinding, scraping, and calibration.

(2)　Intelligent production control system

It mainly includes APS, MES, digital quality detection systems. APS is responsible for guiding enterprises' production, procurement, and inventory. According

(a) Production modeling and simulation of Geely Automobile Company (b) real-time statistics and digital display of production status

Fig. 2.9 The virtual space using industrial flat plates

to the workshop production plan, equipment, materials, tools and fixtures, inventory, and other resources, APS optimizes the allocation of production tasks to each production line or manufacturing unit according to the process steps to achieve stable, balanced production and optimize equipment utilization. MES collects, analyzes, and processes all kinds of process data, execution status, and other process data in the production process through PLC/DCS in real-time, and allocates tools, materials, tools, processing, and other resources according to different manufacturing requirements, to ensure the smooth progress of production according to the plan. At the same time, the production process refers to digital technology, digital quality detection to ensure the quality of products. MES and APS are detailed in Sects. 2.3.1 and 2.3.2, respectively.

(3) Intelligent warehousing/transportation and logistics system

It mainly includes WMS (Warehouse Management System), intelligent material handling equipment (AGV intelligent car), and an automated three-dimensional warehouse. The system optimizes the allocation according to the production tasks assigned, distributes the material distribution requirements to the appropriate AGV, does intelligent scheduling planning, and optimizes the handling container and handling path; At the same time, the AGV navigation/guidance system is responsible for the control of position, movement direction, speed, obstacle identification, etc., and the AGV load shifting system is called to complete loading and unloading operations. The functions of the three-dimensional warehouse are mainly warehouse management and warehouse schedule to realize the optimal management of warehousing and storage of materials.

(4) Intelligent production control center

It is the core of the whole intelligent workshop, mainly including the central control system, field Andon system, field monitoring device. The Andon system is used to realize the work station management and equipment operation management on the production site and to transmit relevant work information to the central control system, which can be regarded as an extended management function of MES. The field monitoring device is used to monitor the working state of the field production process and is an integral part of the production equipment. The central control system receives feedback information through a wireless or wired sensing network, adjusts and modifies relevant models in real-time according to the feedback information, revises the design scheme of the system, and implements relevant control strategies. The interconnection and cooperation of different units in the production line can realize the adjustment of equipment and the monitoring of health state, carry out fault alarm and early warning, fault diagnosis, maintenance, etc. At the same time, the situation that cannot be handled will be timely fed back to the intelligent production execution process control system for re-optimization and change.

2.5.1 MES

In 1992, AMR (Advanced Manufacturing Research) in the United States proposed a three-layer enterprise integration model, which divided the enterprise into three layers: planning layer (MRPII/ERP), execution layer (MES), and control layer (process control system). The execution layer, also known as MES, is located between the upper and lower layers and plays the role of connecting and executing. MES provides operators/managers with the execution and tracking of plans and the current status of all resource elements (employees, equipment, materials, customer requirements, etc.). Many enterprises begin to realize that by taking data information out of the equipment layer, passing through the control and execution layers, and reaching the planning layer, enterprises can gain competitiveness by realizing vertical information integration through continuous information flow. The functions of MES include:

(1) As a bridge between production and management: The primary role of MES is to fill the gap between the upper production plan and the bottom industrial control. It is the bridge of information communication between management activities and production activities and the "information hub" between planning and control.

(2) Provide accurate real-time data: MES collects all kinds of data and status information of the whole process, from receiving orders to making final products to optimize management activities and make quick responses to production conditions that may change at any time. It emphasizes the accuracy and real-time of the data.

(3) Improved plant operations: MES improves equipment return on investment, as well as improves timely delivery, inventory turnover, gross profit, and cash flow performance (Fig. 2.10).

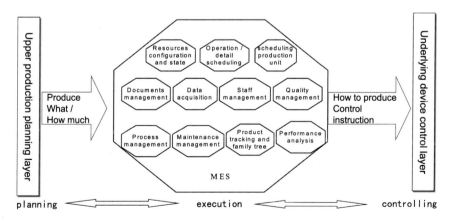

Fig. 2.10 The position and function of MES

MES plays a pivotal role in the IM system. In the intelligent workshop, production planning, equipment status, quality control, material distribution, production error prevention, operation guidance, and production statistics all need a lot of data collection to do this work well. Enterprises use MES to carry out production order management, production data collection and analysis, quality management, equipment management, production traceability, material management, and finally, to generate statistical production analysis and report management. A typical example is the virtual reality fusion application of the Haier workshop, which maps the collected data to the 3D model to display the real-time status. The staff can see each production line's real-time status and trend in the workshop from the office.

The typical functional structure of MES is shown in Fig. 2.11, which mainly includes:

1. Production modeling: includes enterprise object modeling, production process modeling, system infrastructure modeling, work instructions, engineering data maintenance, etc., to provide a model foundation for production simulation optimization.
2. Material management: realize the standardized logistics batch management, material receiving/returning/discarding, material loading confirmation, shortage Kanban, material tracing, and JIT pull warehouse management throughout the production from the supplier.
3. Production process management: realize task control (start, complete, decompose and merge), process monitoring (queuing time and cycle), staff training, and person-hour cost calculation, and can collect data independently or integrate with SCADA/PLC/DCS in real-time.
4. Quality management: Based on off-line and online SPC (Statistical Process Control), real-time integration with the test system, acquisition of critical data such as project data and defect data, and provision of professional analysis means such as abnormal early warning.

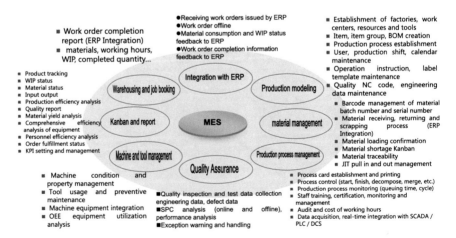

Fig. 2.11 Typical functional structure of MES

5.   Machine and tool management: Manage the condition and properties of equip-
      ment and tools such as reduction, opening, and stopping, and make fault predic-
      tion and cause analysis and tracking of equipment and tools to achieve the
      purpose of preventive maintenance and improve the utilization rate of the
      equipment.
6.   Kanban and report: provide accurate traceability from batch to batch, work in
      process, material conditions, the input and output, quality report, equipment
      integrated efficiency, material yield, and efficiency analysis, order status, KPI
      data display, such as interface include the trend report generator, query, analyzer
      tool such as, provide report generation, trend analysis, data query, fault recall,
      and other functions, have a variety of way on a visit to the real-time and historical
      data, reflect the production of the whole enterprise comprehensively.
7.   Finished warehousing and work reporting: provide the complete information
      of work orders, materials, working hours, WIP, completion number, and other
      relevant information to ERP.
8.   Integration with ERP: integration with mainstream ERP, APS, SCM, LabVIEW,
      RFID, PDA, and other third-party systems. Receive and manage the work order
      issued by ERP, and feedback the material consumption, work in process status
      and work order completion information to ERP.

Taking the powertrain workshop of an automobile company as an example,
Fig. 2.12 shows that the MES architecture scheme is generally divided into MES
server of the data center, MES device on-site, connection, and data acquisition of
on-site and automatic equipment. Through EAI middleware, MES communicated
with ERP to obtain primary vehicle data and reported it to ERP; Through OPC, MES
collects data from automation equipment and transmits production orders and related
instructions to production line PLC. MES also communicates with other systems such
as logistics execution systems through EAI middleware. MES system is configured

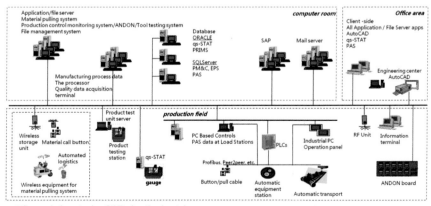

Fig. 2.12 Architecture scheme of MES in powertrain workshop of an automobile company

with one or more servers, which are used to collect real-time data of workshop automatic transport system and other automation equipment, including workshop vehicle crossing point information, operation status of automation equipment, alarm, failure, etc. MES needs to meet the following requirements: (1) High reliability; (2) High maintainability; (3) Monitorability; (4) Configurable and extensible; (5) Integrate the process control downward and support the operation and management upward; (6) Highly Shared information.

However, with the improvement of the complexity of enterprise manufacturing, products have changed from the traditional single variety and large quantity to the current multi-variety and small quantity, and the product life cycle has become shorter and shorter, the product structure has become more profound, and the design has become more and more complex. The traditional MES has been unable to meet these needs simultaneously.

Under the support of IM technology, MES is gradually transitioning to Manufacturing Operation Management (MOM). MOM includes not only MES but also EAM (enterprise asset management), JIT (just-in-time), QMS (quality management system), APS, EH&S (Environmental Health and Safety), etc. MOM is an integrated software platform that connects PLM software up and industrial control and automation systems down. MOM extends to the upstream and downstream industrial chain and is an integrated system that connects users and external resource providers and a connector between enterprises.

2.5.2 APS

APS (Advanced Planning and Scheduling) system is a kind based on constraint conditions, rules, the business model and algorithm, order to convert customer orders to the workshop, and intuitive graphical reflect the load condition of production resources and order schedule, rapid response to customer requirements of enterprise management information system. If the IM system is compared to humans, the human brain is equivalent to ERP/APS, and the central nervous system is equivalent to MES/PLC or DCS.

There are more and more contradictions in the operation of modern factories. For example, lean production pursues balanced production according to the beat, while production according to the order can easily cause imbalance. The product structure is becoming more and more complex, but the production planning and scheduling are required faster and faster; The production plan is exact and dynamically adjusted according to the field conditions. In order to solve these contradictions, high requirements are put forward for industrial software. The traditional ERP/MRPII/MES multi-stage decision-making method can no longer meet the needs of intelligent factories. Therefore, APS with a global model and overall optimal characteristics has been developed rapidly in recent years. The APS shall have the following characteristics:

1. Flatness and real-time response

The traditional planning system, from the strategic plan, business plan, sales operation plan, master plan, material generation plan, and material demand plan, is logically reasonable, but with too many layers and slow response. Therefore, the planning system of the APS system should be dynamic, flat, and combined, and the multi-stage calculation and decision-making should be integrated into one stage so that the optimal global solution can be obtained more quickly.

2. Multi-objective optimization calculation

APS is dynamic, the closed-loop system, it is a day to do the order of hours, computing capacity, and ability to simulate the scheduling, real-time meet four constraints: capacity, materials, employees, mold, whether long-term or short-term plan is to be optimized and enforceability, satisfy multiple targets at the same time, the shortest period of delivery, cost minimum, highest efficiency and sharing the shortest processing time, etc., and so that it satisfies all the requirements, and consider the overall supply, realize synchronous planning, enterprise resource constraints of the optimal planning. On the one hand, the vast computing workload requires an APS system based on memory computing mode and can respond quickly. On the other hand, software vendors of the APS system must adopt the applicable multi-objective model and the optimal solution algorithm according to different problems to build the software architecture.

The optimization algorithms of APS system mainly include: (1) mathematical programming methods, such as linear and mixed-integer programming and other mathematical modeling methods, which are more suitable for strategic planning, such as network location selection, procurement, and source search, which are somewhat simple and need to be solved accurately; (2) Heuristic algorithms, such as constraint theory or simulation, are more suitable for tactical planning or operational planning, such as production scheduling; (3) Artificial intelligence, such as expert system, artificial neural network, and genetic algorithm, is more suitable for problems with a large number of possible scheme choices; (4) Artificial intelligence based on complex System theory, such as dynamic adjustment algorithm based on MAS (Multi-agent System). Due to the limitations of computing power and the limited application of mathematical programming methods, most enterprises now use heuristic algorithms that are easy to develop, but artificial intelligence is gradually increasing.

3. Multiple scheduling operation modes

APS system scheduling operation modes usually include: (1) proactive scheduling, pre-preparing optimized scheduling for a group of orders; (2) Responsive scheduling, adapting to changes in changeable environments to ensure the feasibility of scheduling; (3) Interactive scheduling, using visual tools such as Gantt chart to drag the process plan manually. Gantt chart includes automatic programming of production planning, simulation of medium and long term production execution, automatic assignment of tasks to equipment, automatic balancing of equipment load, production preparation of materials and tools, automatic personnel scheduling and load balancing; (4) Advanced scheduling, intelligent optimal scheduling of the work-shop. Advanced scheduling can realize the automatic replacement of equipment, process routes, and arrange shifts to meet the delivery date; Order fulfillment can be tracked automatically to avoid order delays. It can also realize automatic scheduling equipment, full load of personnel to meet the utilization of resources. Modeling of advanced scheduling includes modeling of materials, process paths, production orders, shifts, resources, and rules. For example, rules need to be considered: order placing rules, sequence rules, selection rules, and resource allocation rules.

Interactive scheduling implemented by manually dragging a Gantt chart is the simplest scheduling optimization method, as shown in Fig. 2.13. For example, after receiving the customer order, the production capacity demand of each piece of equipment is calculated according to the number of ordered products and the production process route of the products. Reverse the schedule based on the completion date until the initial process is arranged; When it is found that the initial process time has fallen into the past time zone, it is not feasible. Therefore, a second attempt is made to delay the initial process to an idle capacity interval to fall into the future time zone. Accordingly, the downstream process is postponed and adjusted to ensure that the completion time of the final process meets the delivery requirements of the customer; By first arranging large tasks and tasks with the highest matching degree to the idle capacity interval, various operation optimization such as task insertion, combination and replacement can be carried out to reduce the scattered idle time and improve the utilization rate of equipment and resources.

4.   Precision and lean

Application of APS system of plant level, master the technology, equipment, employees, and energy and other kinds of data resources, can carry out rolling schedule, simulation and single processing, bottleneck analysis, and knowledge management, in order to cope with frequent changes to the single, customers and suppliers, variety change, downtime with uncertain events such as product quality fluctuation, and the material requirement time accurate to seconds, so can ask the

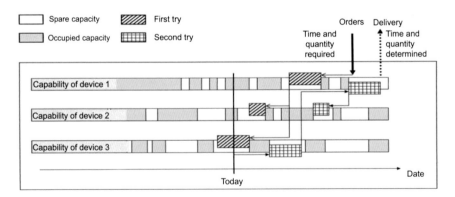

Fig. 2.13 Interactive scheduling based on trial and error

material supplier directly to send the material to production on time points corresponding to the location and products, from procurement and sorting, not accounting for inventories, even line edge libraries are not required, is the factory's production plan is fully integrated with external logistics. An SCM supply chain. In the future, based on the IoT technology, the APS system will analyze and optimize on the basis of device data acquisition. For example, if a device fails, the APS system will automatically eliminate the device during production scheduling.

Domestic enterprises have made remarkable progress in the construction of intelligent workshops. According to survey statistics, in the machinery industry, MES/APS achieved economic benefits of shortening product delivery cycle by 10–15%, increasing production efficiency by 15% and reducing operating cost by 10–15% on average.

[Case: Maoming Petrochemical Dynamic Production Scheduling]
Maoming Petrochemical Dynamic Refinery Production Scheduling system (Orion) enables production schedulers to anticipate production schedules and adjust scheduling in case of emergencies. The optimized scheduling results of PIMS were decomposed into ten-day scheduling and three-day scheduling, which ended the history of manual scheduling with blackboard in Maoming Petrochemical Company, improved the foreseeability of production scheduling and timely adjusted the scheduling in case of emergencies. Based on data model and balancing algorithm, the system realizes component balance, material balance and utility balance, assists integrated cost management, performance assessment and production management decision of petrochemical enterprises, and realizes intelligent, automatic production and management of petrochemical plant workshop.

2.6 Factory Layer

A factory usually consists of more than one workshop, whereas a large enterprise has more than one factory. As a smart factory, not only should the production process be automated, transparent, visible and lean, but also product detection, quality inspection and analysis, and production logistics should be closed-loop integrated with the production process. Information sharing, on-time distribution and collaborative work should be realized among multiple workshops in the same factory. The smart factory relies on the seamless integration of information systems, mainly including PLM, ERP, CRM, SCM and APS/MES five core systems, as shown in Fig. 2.14. A smart factory of a large enterprise needs to use ERP system to develop production planning for multiple workshops, and production scheduling is carried out by APS system /MES according to the production plans of each workshop. The granularity of APS system /MES scheduling can be days, hours or minutes and seconds.

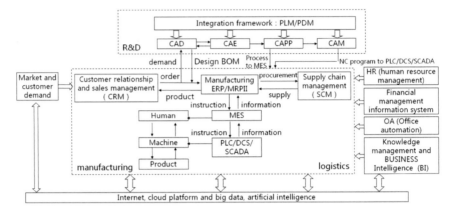

Fig. 2.14 Intelligent factory architecture and its core systems

A factory with automated production lines and industrial robots cannot be called an intelligent factory. The intelligent factory is based on the digital factory, which extends manufacturing automation to the highly integrated, flexible, and intelligent production system. It should contain six salient features:

1. The lean value chain system fully embody the concept of industrial engineering and lean production, by intelligent procurement, advanced manufacturing technology, intelligent logistics value chain system, can be achieved by order-driven, pull production, fast to adapt to the changing of the mold, many varieties of the mixed production line adopts the JIT/JIS (Just In Time/Just In Sequence) logistics, significantly reduce the WIP inventory and eliminate waste.

2. Digitalized and transparent operation management: Factory layer of PLM, SCM (supply chain management) and CRM (customer relationship management), ERP (enterprise resource planning), workshop layer of MES, APS, control layers of PLC/DCS software industry, such as to thoroughly mix the advanced manufacturing technology, information network technology, cloud computing, big data and the artificial intelligence technology, real-time display of the factory operation data and charts, display equipment running status, and can be found in the image recognition technology of video monitoring problem of automatic alarm, real-time insight into the factory operation, implement multiple workshops between collaboration and resource scheduling.

3. Decentralized control system: One of the critical goals of the smart factory is to provide customized products and services in a specific way according to the personalized needs of end customers. Only form composed of sensor network, control systems, robot control system, realizes automatic interconnection from machine to machine (M2M), through the network flat, based on centralized control and distributed computing model, through the intelligent device between the main body of independent consultation to solve emotional problems, improve manufacturing flexibility and ability to cope with market uncertainty and dynamics.

4. Flexible automated production lines: Intelligent factories can have a mix of production modes. A high degree of automation should be realized for the product line with few product varieties and large production batches, or even a black light factory should be established. For the small batch product line with multiple varieties or customization, the emphasis should be on realizing less humanization and man–machine coordination. Through AGV, rack manipulator, hanging conveyor chain, and other logistics equipment to achieve the process of material transfer, and the configuration of a material supermarket, shorten the logistics path; Extensive use of power equipment to reduce the labor intensity of workers.

5. Green and humanized factories: can timely collect equipment, production line, and workshop energy consumption, and analyze and optimize to achieve energy-efficient utilization; In the dangerous and polluted links, the priority should be given to robots instead of manual workers. Reducing noise, cutting cooling lubricant and other pollutants can realize waste recovery and reuse; Building a highly efficient, energy-saving, green, environmentally friendly, and comfortable humanized factory; Realizing green manufacturing.

6. Human–machine intelligent integration analysis system: realize the coordination and cooperation between human and intelligent machines, to expand, extend and partially replace the mental work of technical experts in the manufacturing process; It is composed of cloud computing, big data, and control system. It is capable of advanced analysis and modeling, self-learning, self-maintenance, and self-adaptation. It can be used for collection, modeling, simulation, analysis, reasoning, prediction, judgment, and planning. Make products and services intelligent and configurable.

The control system, value chain system, and analysis system with the above characteristics are combined with the cyber-physical system (CPS) to form an intelligent factory.

The smart factory is based on a digital factory. The construction of a digital factory is also an arduous task. According to the survey, 91% of industrial companies invest in digital factories, but only 6% think their factories are "fully digital".

[Case: Smart factories of Haier and Red Collar]
The demand for building intelligent factories in China is strong, and smart factories such as Haier, Midea, Gree, Red Collar, Dongguan Jinsheng, Shangpin, Sophia, etc., have emerged. Red Colla's entire enterprise is like a "3D printer" driven entirely by data. Red Collar supports independent customer design, production driven by customer demand, customer design and order processing process without the participation of designers, no manual conversion and paper transmission, real-time data sharing and transmission. Every employee obtains data from the Internet cloud, operates according to customer requirements, ensures the accurate delivery of orders from all over the world, and realizes the seamless connection between customer personalized needs and large-scale production and manufacturing with Internet technology. Haier Foshan washing machine Factory can realize the configuration, production,

and assembly according to orders. It adopts a highly flexible automatic unmanned production line, widely applies precision assembly robots, adopts an MES system for the whole order execution management system, and tracks the whole process through RFID, realizing the interconnection of M2M and M2H.

2.6.1   PLM

PLM (Product Lifecycle Management) mainly includes CAX software (CAD/CAPP/CAM/CAE and other tool software) and PDM (Product Data Management). It has the functions of document and version Management, workflow Management, project management, configuration/formula Management, etc. It is also the general name of information technology related to product innovation. From another perspective, PLM is the idea of managing product data information throughout the product lifecycle, from creation to use to final obsolescence. Before the emergence of the PLM concept, PDM mainly focused on the data and process management of the product development process. Under the PLM concept, PDM is extended based on department collaboration.

Product structure is a core concept of PLM. To adopt computer-aided enterprise production management, the first step is to enable the computer to read the structure of the product and the materials contained therein. In order to facilitate computer identification, the product structure must be described in a particular data format and document, which is BOM, or Bill of Material. Therefore, BOM is also called a product structure table or tree. In the process industry, this might be called a "recipe."

BOM is the product structure identified by the system. In the configuration management function of the PDM system, when the appropriate conditions are input, the system will filter out the parts that do not meet the requirements and quickly combine to form a variant product that meets the requirements of the rules. As shown in Fig. 2.15, input the retrieval conditions, such as cycling environment and price, into the bicycle BOM of redundant nodes, and the PDM system can delete the tire nodes that do not conform to the rules, and the remaining parts that meet the requirements, to quickly obtain the bicycle variant design scheme that meets the user's requirements.

BOM is the standard core document of ERP and PLM, so it becomes the interface between PLM and ERP, and it is the bridge between design and manufacturing. The product goes through three major stages: engineering design, process planning, and manufacturing, so there are three major BOM. First of all, product engineering design in the CAD system generates BOM from the engineering design perspective, which reflects the detail of design drawings. It is called engineering BOM, or EBOM. Then the process engineer takes EBOM as the basis for process design. According to the requirements of the production process, data such as process plan, assembly sequence, and working procedure are added to THE BOM to make it into the planned BOM, namely PBOM. Finally, according to the manufacturing sector has generated PBOM, steps for detailed design describes the task time, material ratio of this product,

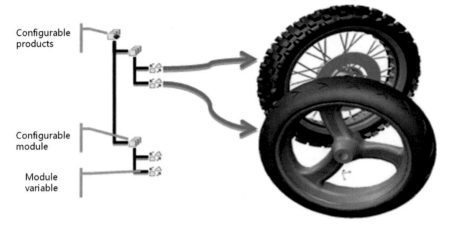

Fig. 2.15 Definition of product structure and configuration rules and selection of components by rules

as well as the equipment used, jig information, to produce the manufacturing BOM, namely MBOM, the BOM is not ready to direct production and is the critical basis of financial accounting cost. You can even create a service BOM, which expresses the product-service system to be delivered. Thus, changes to the BOM reflect the phases of the product life cycle and the synergies between the various departments, as shown in Figs. 2.16.

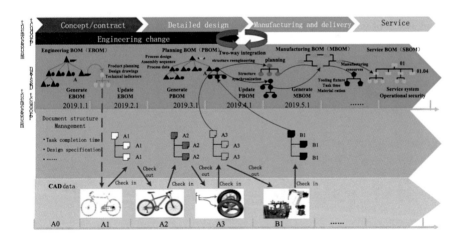

Fig. 2.16 BOM is a bridge connecting the design, manufacturing, production, operation and management of the enterprise

2.6.2   ERP

Enterprise Resource Planning (ERP) is a management information system that mainly faces the manufacturing industry and integrates three streams of enterprises: logistics, capital flow, and information flow. In addition to MRP II (including production control, supply and marketing management, and financial management), ERP also emphasizes the integration and integrated management of all enterprise resources, including human resources, and extends to non-production public institutions. Among them, the core functions of production control are master production plan, material demand plan, and capacity demand plan.

Traditional ERP reflects such a production model that emphasizes centralized control, characterized by a predetermined stable equilibrium state and rhythm, a fixed information transmission mechanism, and overall central production control, as shown in Fig. 2.17.

1.   **Master Production Schedule**

Master Production Schedule (MPS) specifies the product series in the business plan or production outline according to the customer contract and market forecast. It determines each specific final product (for example, a particular type of bicycle) in each specific period. MPS is the first critical planning level.

MPS is a critical link in the planning system and a junction between the top and the bottom. An influential MPS is an enterprise's commitment to the needs of its customers. It makes full use of the enterprise's resources, coordinates production and marketing, and achieves the business plan objectives expressed in the production plan outline. If there are no MPS and MRP is run directly according to the forecast and the demand of the customer order, the resulting plan will match precisely the market

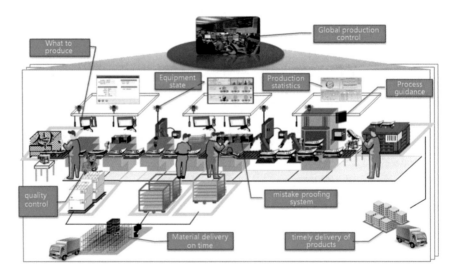

Fig. 2.17   Schematic diagram of traditional centralized control production mode

forecast and the actual order demand in terms of quantity and time. However, As the forecast and actual orders are unstable and unbalanced, the direct arrangement of production will sometimes lead to overtime work and failure to complete the task, and sometimes equipment and staff are idle, which will significantly impact the enterprise. Thus, Only after MPS balances supply and demand by time segment and absorbs the impact of uncertainty and fluctuation of demand can it be used as the input information of the following planning level: material requirement planning (MRP) to achieve balanced production.

The establishment and control of MPS are essential to the success of the ERP system. It is called a "master" production plan because it plays the role of "master control" in an ERP system.

The preparation steps of MPS are as follows:

(1) Choose a sufficiently long planning outlook period (more remarkable than the maximum cumulative lead period) and an appropriate planning time zone; (2) To count orders and make demand forecast; (3) Calculate the total demand according to the more significant value of the actual order demand and the predicted demand; (4) Draft MPS based on specific strategies (balanced strategy, following strategy or mixed strategy); (5) Calculate the estimated availability and production for each time zone; (6) Expected availability in time zone $0 =$ initial availability; (7) estimated availability for time zone $K + 1 =$ (estimated availability for time zone K) + (estimated availability by MPS for time zone $K + 1$) $-$ (total demand for time zone $K + 1$), $K = 0,1 \ldots$; (8) In the process of calculation, if the estimated allowable consumption in a time zone is negative, a production schedule quantity or procurement, etc. shall be arranged in that time zone, which is usually decided by a planner; (9) Evaluate the feasibility of the draft MPS with a crude capacity plan; (10) Simulation optimization and MPS optimal scheme confirmation.

The strategies of MPS include:

(1) Equilibrium strategy: The production plan is balanced and stable, and the production quantity does not change with demand fluctuation. The production quantity curve is horizontal. The production quantity in each period is the same, and only the total quantity meets the market demand. When sales are lower than production, inventory increases; when sales are higher than production, inventory decreases or sells out.

(2) Following strategy: or catch-up strategy refers to the production plan changes with the change of demand, and the production curve is the demand curve. This strategy has the fewest inventories but requires frequent adjustments in factors of production, including hiring and firing or changing of positions and the purchase or idling of critical equipment.

(3) Mixed strategy: It refers to that the production curve shows a ladder shape, which follows the demand for a long time but is balanced in a certain short time. There is a combinatorial explosion problem in the mixed strategy, and there is even an infinite number of schemes that require a better algorithm to find the optimal global scheme, while the commonly used trial-and-error method usually only gets the optimal scheme.

2.   **Material Requirements Planning**

Material Requirement Planning (MRP) refers to the enterprise management mode that takes each item for plan object, reversing schedule based on a completion date, distributing the order of the schedule according to the length of the lead time of each item.

Input information sources for MRP systems include MPS, parts orders outside the plant, quantity forecasts as independent requirements items, inventory records, and BOM.

MPS is the primary input source to the MRP system and provides quantitative information of "what products to make". MRP system should be decomposed layer by layer according to MPS items to get the demand of various parts.

As an independent requirement, it is only necessary to add the order quantity of such materials to the gross demand of related materials.

The inventory record document is composed of the inventory record of various materials, which provides the quantity information of "what materials are already available". The actual demand is the amount of stock of such material subtracted from the gross demand of the corresponding material. The inventory record file must be updated through various inventory transactions.

The BOM contains product structure information, or information of "how-to", which is used as a basis for requirements decomposition, as shown in Fig. 2.18.

An example of material requirements planning is shown in Fig. 2.19. A product X contains a part A, an A contains two C, and a C contains an outsourced part O. The calculation process is as follows:

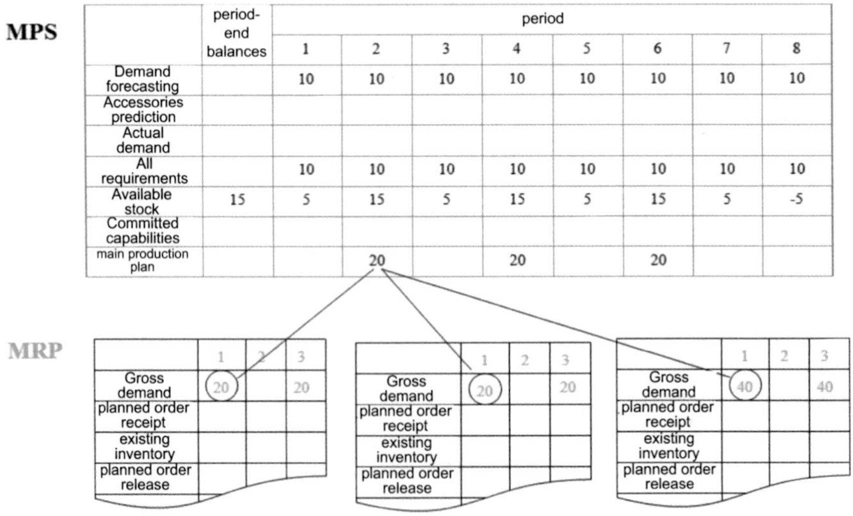

Fig. 2.18 An example of the relationship between MPS and MRP

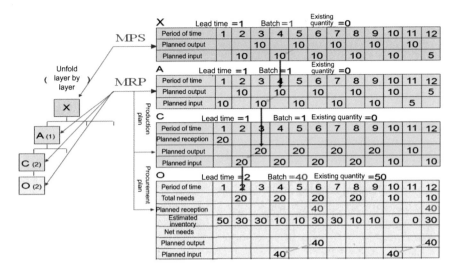

Fig. 2.19 Example of material requirements planning

In order to do this well, the capability requirements plan needs to be applied.

(1) It needs to start production in a time zone 4 after MPS decided to producing 10 X in time zone 5 because the lead time is 1 time zone;

(2) Since a product X contains one part A, 10 A must be produced at time zone 4, so production must start at time zone 3;

(3) Since an A contains two C, it can be calculated that 20 C should be produced at time zone 3. Similarly, production needs to start at time zone 2.

(4) We can calculate that 20 purchased parts O should be produced at time zone 2. Therefore, production should be started at time zone 0 because the lead time of purchased parts O is 2. However, due to the existing inventory of 50, the actual demand is $-30 (= 20$–$50)$, so there is no need for production or procurement.

At this point, since O is a purchased part, the MRP operation is completed, and the material production plan and purchase plan for some time in the future are obtained, as well as the inventory forecast and inventory status information about the future and suggestions for issuing, adjusting or canceling the plan, etc.

Often, the initial schedule may not be feasible, and the ERP/MRP system is challenging to determine the schedule's feasibility accurately. The frustrating thing is that when all the previous management steps are smooth, the final production process may not be satisfactory. ERP/MRP can adjust the time according to people's wishes without telling people that it is impossible to complete the task by this time. Therefore, it is necessary to compare the actual demand for materials with the production capacity. If it is found that the current master schedule is not feasible, managers may decide to increase production capacity (e.g., by purchasing bottleneck equipment, working overtime, outsourcing, etc.) or modify the schedule. In order to do this well, the capability requirements plan needs to be applied.

3.   Capacity Requirement Planning

Capacity Requirement Planning (CRP) is a project management method to calculate the required Capacity of MRP. Specifically, CRP is to accurately calculate the various resources required by each production stage and each work center, determine the human load, equipment load, and another resource load, find the bottleneck of capacity as early as possible, and balance the production capacity load.

The broad capacity requirements plan is divided into the simple and detailed capacity plans. The crude capacity plan referred to the main production plan after the main production plan, through the key work center production capacity and planned production capacity comparison, to determine whether the primary production plan is feasible. Detailed capacity planning means that after the closed-loop MRP obtains the demand for various materials through MRP calculation, it calculates the workload assigned to the work center in each period, determines whether it exceeds the maximum working capacity of the work center, and makes adjustments.

The object of CRP is capability, as shown in Fig. 2.20. CRP converts material demand into capability demand. Specifically, it converts MRP's planned and issued production orders into a load of each work center in each time zone, as shown in Fig. 2.21.

The logical process of capacity demand planning takes the order required to load, required capacity, and available capacity as input and gets the balance load/capacity through calculation and coordinated man–machine adjustment.

When considering the calculation method of capacity demand planning, it is necessary to convert the material demand of material demand planning into load hour, that is, to convert the material demand into capacity demand. The planned order of MRP and the work center and production calendar, as well as the work center shutdown and maintenance situation, finally determine the available capacity of each work center in each period. To put it simply, take a work center M1 and two kinds of materials as examples, and the calculation process of capacity requirement planning is shown in Fig. 2.22.

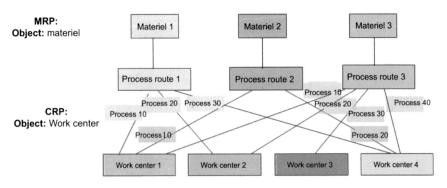

Fig. 2.20 Relationship between MRP and CRP

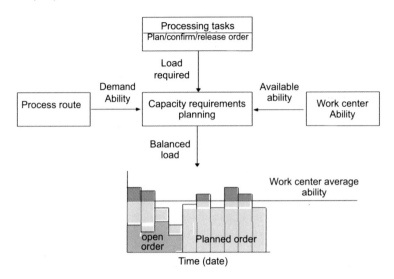

Fig. 2.21 Logical flow of capability requirements planning

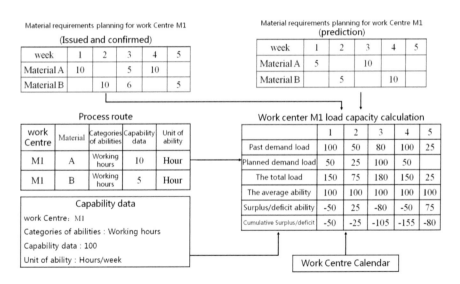

Fig. 2.22 Examples of calculating CRP

Capacity requirements planning is the process of determining short-term capacity requirements. Its main steps are as follows:

(1) Data collection

Required inputs include MRP plan, current workshop load, process route, work center capacity data, work center calendar, etc. Usually, in the specific calculation of capacity demand plan, according to the number of materials in the planned order issued by MRP and the period of demand, multiplied by the fixed hours in the respective process route, the capacity demand list can be converted into the total load demand, plus the working hours of the work center in the unfinished order in the workshop. Then a list of available capabilities of the work center is established based on the actual existing capabilities. Only with these data can we proceed to the next step.

(2)   Calculate and analyze the load

Assign all task orders to the relevant work center, then determine the load of the relevant work center and calculate the load of each relevant work center from the process route record of the task order. Then, the load situation of each work center is analyzed, residual/insufficient capacity is calculated, and the causes of various specific problems are identified so that the problems can be solved correctly.

(3)   Capacity/load adjustment

There are three ways to solve the problem of underload or overload capacity: adjust capacity, adjust load, and adjust both capacity and load. Measures to adjust capacity include adjustment and redistribution of Labour; Increasing the number of workers as needed; Arranging training; Arranging overtime; Rearranging process routes and placing part of orders to underloaded alternative work centers; Outsourcing. If you are overloaded for an extended period, consider outsourcing specific bottleneck jobs to suppliers. The measures to adjust the load include: parallel operation of one-piece flow (or single piece flow), in which some completed parts are passed to the next step for synchronous processing; Batch production, subdividing a batch of order into several small batches, and arranging production on the same machine simultaneously; Reduce the preparation lead time, standardize the preparation process, and reduce the preparation time; Adjust an order to advance or delay an order, or to complete part of order first, postpone the rest, or cancel some orders. Moving the production process forward or backward can be challenging due to priority requirements and the number of components available.

(4)   Confirm capacity demand plan

The output includes the workload report of the work center. After analysis and adjustment, the modified data will be re-entered into the relevant documents and records. After multiple adjustments, when the capacity and load reach balance, the capacity demand plan will be confirmed, and the task list will be formally issued.

After the above three clear and sequential steps of MPS, MRP, and CRP, the production and outsourcing plans were obtained, and their feasibility was evaluated. However, although the steps of this traditional enterprise resource planning are clear, the logic has the following problems:

(1)   Infinite capacity hypothesis

MRP's assumption of infinite capacity in BOM expansion makes the production plan defective. Although ERP software has a CRP module for capacity balance calculation, it can only passively check capacity, and cannot automatically balance capacity conflict when capacity conflict occurs, and can only be manually balanced by the planner based on experience. When the system and the product are more complex and the capability conflict is severe, manual adjustment can only solve the capability conflict of 1 or 2 bottleneck links but cannot solve the capability conflict of the whole enterprise.

(2)   Fixed lead time

The BOM development of MRP is carried out according to the fixed production lead time. Production lead time is generally composed of production preparation, processing, transfer, and waiting time. Waiting, preparation, and even processing time are all uncertain parameters in actual production. The current MRP system cannot handle the uncertainties of various times, so it can only be set as fixed time according to the statistical average value. The probability distribution for each time is different. For example, the wait time should be a function of the system load, which varies significantly with the system load. When the system production load is light, the waiting time may be short or even negligible, while when the system production load is heavy, the waiting time may belong. The processing time is affected by the upstream process and the quality problem of the process. When the quality problem needs to be checked or corrected, the processing time will fluctuate abnormally.

(3)   Optimization mechanism

There is no optimization mechanism for the operation logic of MPS/MRP. MPS, MRP, and CRP, are executed sequentially and lack mutual coordination and systematic optimization. Therefore, the production plan formulated often lacks feasibility. This serial check after the event, repeated correction of the processing method, for the enterprise brings excellent inconvenience, is a lot of production planner's nightmare. Although MRP has a closed-loop feedback mechanism to balance the capacity load by manually adjusting the master production plan in case of capacity conflict, this adjustment is quite difficult for complex systems, and the complexity of the adjustment is often beyond the scope of human intelligence. Planners lack the calculation means of the total balance of various departments and facilities, so they cannot realize the overall coordination and optimization of internal resources. Using the APS system and optimization algorithm to coordinate as a whole is the direction to solve the problem. However, the current level of the APS system is not yet a comprehensive optimization tool. It is difficult to consider multiple layers of constraints in the round, establish a complete model, and the solution is within a sufficient time to complete. Hence, only in local optimization solution can be got on the simplified model of the problem.

(4)   Uncertainty

The fixed lead time problem is a particular case of the uncertain parameter problem. Production and operation decisions involve several types of uncertainty. In the IM model, especially in the open cloud manufacturing mode, and business decision-makers with information is often not entirely transparent, the customer is not fixed, between the customer and customer relationship with various competition between producers, will inevitably lead to delay or get interactive information is challenging to obtain accurate, reliable and complete information. This kind of access to information asymmetry affects further cooperation intention and game action. Therefore, the uncertainty of the production service system in the IM model includes four kinds: fuzziness, the fuzziness of the qualitative index, and the interval estimation of the quantitative index in the process of resource optimization decision. Randomness, such as the change of downtime and time of production equipment; Grey, such as customer order quantity has not been precisely determined, only know the range of change; Uncertainty, such as production costs and benefits, is objectively determined, but subjectively unclear. Some information often has two or more kinds of uncertainty, and different combinations of uncertainty often need to be processed by different mathematical methods. There is no consensus on the mathematical processing of information with four kinds of uncertainty simultaneously. As a result, the uncertainty of the problem has received more attention in recent years.

Various attempts have been made to deal with uncertainty. Project Evaluation and Review Technique (PERT) is an early attempt to incorporate uncertainty into the project schedule. The three-point estimation method uses the most pessimistic time, the most optimistic time, and the most likely time to estimate the expected completion time of each active node in the PERT chart. Monte Carlo technique, also known as random sampling technique. It combines the solved problem with the probability model and obtains the approximate value of the problem by simulation and random sampling.

A mathematical programming model is used to find the best scheme under uncertain conditions. It can be divided into three types: stochastic programming, robust optimization, and dynamic programming. Stochastic programming is generally based on the assumption of a known probability distribution, so it has some limitations in practice. Robust optimization partly solves the problems brought on by fuzzy data. It is usually assumed that the distribution of the unknown but belongs to a probability distribution set. By constructing a suitable set, the probability of the original optimization problem can be transferred to consider the worst-case and random optimization problem. However, the probability in the process of collection construction is complicated, quickly leading to improper collection design calculation process is difficult to convergence. At the same time, robust optimization pays more attention to the optimization scheme in the worst case, so the final scheme is likely to be too conservative in the uncertain environment, and the cost of robust conservatism is often reduced production efficiency. However, the dynamic programming model is usually limited by dimension problems and has low efficiency solving. Therefore, how to maximize the interests of all parties and reduce risks in the decision-making of production and service operation is one of the critical problems in the application and implementation of IM mode.

(5)   The logical problem of the combination of similar terms

The guiding principle of MPS/MRP is that fewer varieties are better, and larger batches are better than the small ones. Therefore, the idea of "merger of similar items" in the MRP algorithm is always carried out: the merger of similar products (or parts and materials) of different orders, the merger of similar products within the same period (month, week, day); merger of similar products within the same period (month, week, day). Products are merged in MPS, parts are merged in MRP, and there is also a need to merge purchase orders and increase the size of the purchase. Unreasonable batch setting often results in the waste of capacity resources and the increase of inventory cost.

First, the merge of similar items makes it impossible to track orders. Due to the operation of "similar item merging" and "mass customization", customers' personalized order requirements flow into the ocean of mass production like streams, which are difficult to distinguish and track. When the production capacity is insufficient, and all the required parts cannot be produced in the expected time and quantity, it is difficult to determine which customer orders will be affected and make the corresponding processing according to the specific situation.

Secondly, due to uncertain events such as emergency orders, technical accidents, and quality accidents, the rationality basis of some "similar item merging" operations occurring in a certain regular order may disappear, and the merging order needs to be reversed or changed, which is difficult to operate due to the complexity of the system.

The first three problems are capacity constraints of production planning; The last two problems are the core problems of production planning model design.

The research direction is to replace the core module of ERP with an optimization model and integrate the functions of MPS, MRP, CRP, APS, and other systems into a model so that:

(1)   Capability constraints are considered when BOM is expanded.
(2)   The lead time is variable. The optimization model automatically determines the processing time of each process and the waiting time between processes according to the optimization criteria. They are functions of system load and uncertain events.
(3)   To make the parameters reflect the uncertainty of the objective world more truly, establish the uncertainty optimization model and solve it with a fast and accurate algorithm.
(4)   Complete the formulation of the master production plan, material demand plan, and capacity demand plan at the same time and optimize the product portfolio.

4.   Integration of ERP with SCM, WMS, APS, MES

The vertical integration interface of ERP, APS, and MES is shown in Fig. 2.23. Theoretically, the upper ERP makes plans to decide what to produce and how much to produce. The middle-level MES makes decisions about producing and issues control instructions to the control layer. As the coordination center, APS is responsible for scheduling and scheduling optimization.

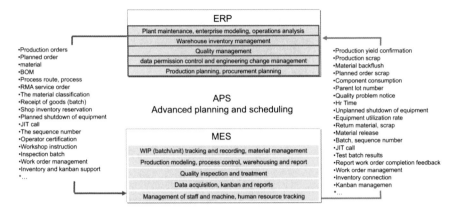

Fig. 2.23 Vertical integration interface of ERP, APS and MES

However, in practice, first of all, there are overlaps of functions. For example, the ERP system contains some functions of MES, such as workshop control of production control management module, manufacturing standard, and other functions, which were initially MES strengths. MES also includes some functions of ERP system, such as human resource management, production process control, quality inspection and processing, and other modules, which are better functions of ERP system.

Second, because of the ERP system and the APS system, MES the three types of the system software in different emphasis in the process of modeling, and generally implemented by different software companies, forming heterogeneous data formats, database, and integrated architecture, the joint operation of ERP and APS and MES system has brought a series of technical problems, which include system complexity, implementation cost. The data communication and integration of two systems usually need to be solved by middleware technology.

Data fusion and system integration of ERP, APS, and MES is an important research direction. If the system integration is considered in the planning and design stage, the ERP/APS/MES function will be more outstanding and powerful, producing the effect of $1 + 1 > 2$. After obtaining the underlying data support provided by MES, the ERP system can make accurate production plans according to accurate and real-time information and make proper scheduling on the production site according to real-time data to improve the just-in-time delivery rate and production resource utilization efficiency.

The integration framework of ERP, SCM, WMS, APS, MES, and other systems is shown in Fig. 2.24. APS system and MES combine to play a connecting role.

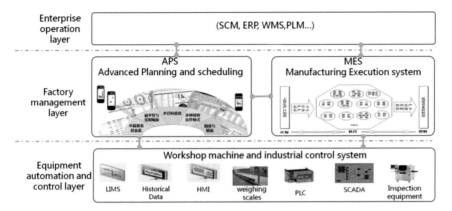

Fig. 2.24 Integration framework

2.6.3 SCM

Supply Chain Management (SCM) is a Management mode that connects suppliers, manufacturers, distributors, retailers, and end-users through sharing information. SCM focuses on end-to-end integration from raw materials to final consumption.

[Case: Cisco's inter-enterprise synergy]
Cisco orders 82% of its customers are placed online. A customer in Cisco's e-commerce sites to order products, a series of information will be automatically generated, Cisco ERP system audit order, then the SCM system to supplier purchasing information: a third-party manufacture circuit boards, another manufacturing shell, supply power distribution unit, and other general equipment, again by a factory assembly into finished products. The ERP system of each supplier factory was informed and prepared quickly, and the whole process was run automatically and in real-time. There was no storage, no inventory, and no paperwork. The essence of the SCM system is that upstream and downstream partners open up their information and let them share information that was previously considered confidential.

2.7 Coordination Layer

2.7.1 Internal Coordination

The department collaboration and cooperation problems are some of the biggest problems that the core management team of a large company can encounter. Internal collaboration enables all departments, project teams, and even enterprises and partners in the entire supply chain to share customer, design, production, and operation information. As shown in Fig. 2.25, the internal collaboration of an enterprise is to

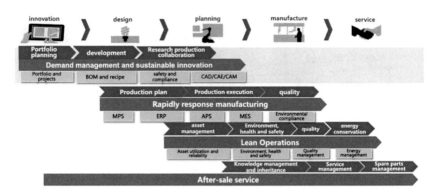

Fig. 2.25 An example of multi-process collaboration within an enterprise

change from the traditional serial working mode to the parallel working mode, and achieve optimization in the degree of structure and complexity, to minimize the time for new products to be launched, shorten the production cycle and quickly respond to customer needs.

Huawei CEO Guo Ping and Huang Weiwei put forward the concept of "cloud, rain, and ditch". "Cloud" refers to external industry changes, technological changes, market opportunities, internal core values, and other environmental factors. "Rain" refers to the business activities of various departments within an enterprise. "Gap" refers to cross-department and cross-field workflows, such as IPD, CRM, LTC (marketing and Supply chain management), ITR (problem management), and other cross-department process systems. Electricity can be generated only by turning clouds into rain, which collects and flows into trenches. Huawei's CEO uses this analogy to emphasize the importance of inter-departmental processes within the enterprise. The function of the process is to bring the power of everyone together to produce more incredible energy.

Take the R&D process as an example. Without a collaborative R&D process that flows smoothly and across departments like rain, Huawei would not be able to organize thousands of people to develop 4G and 5G wireless communication systems. But too many processes constrain thinking and stifle innovation. Therefore, the process should be continuously simplified and optimized, structured and normalized. There are also rules about how structured a process should be. When uncertainty is high, such as when disruptive technologies emerge and have a significant impact on the industry, the planning and R&D processes should be less structured and complex to increase flexibility on the contrary. When the external environment is relatively stable and the industry changes slowly, a more structured and strictly standardized process is needed to continuously optimize the existing business and products. Therefore, the R&D design process structure should be lower than that of the manufacturing process because the generation of new ideas and new schemes has high uncertainty of results.

2.7.2 Inter-enterprise Collaboration

In recent years, with the rapid development of the Internet and communication technology, fundamental changes have taken place in the world manufacturing paradigm, is not only the material resources but also the development and utilization of the global network of intellectual resources, break through the barriers of traditional physical space, realize the organic combination of physical and virtual, the zero distance product/service and market development; Based on the information technology platform, the unprecedented coordination between economies of scale and economies of scope is realized. It breaks the boundaries of traditional industries and enterprises and realizes the high-speed synchronization and integration of R&D, design, manufacturing, and service. In the new global manufacturing network platform, different enterprises can quickly participate in R&D, design, production, logistics and services, and other value-added activities.

As shown in Fig. 2.26, ultimately, inter-enterprise synergy evolves into interconnection synergy oriented at all stages of the life cycle of products and services, including the interconnection synergy of R&D, manufacturing, supply, logistics, marketing and service, and the construction of corresponding networked ecosystem.

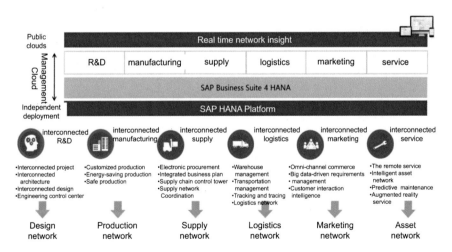

Fig. 2.26 SAP Industry 4.0 solution based on SAP HANA platform

2.8   Vertical Integration Through the Hierarchy

Vertical integration is the classic content of traditional manufacturing information. The German "Industry 4.0" strategic plan implementation recommendation points out that vertical integration is mainly about "integrating various IT systems at different levels, such as actuator and sensor, control, production management, manufacturing and execution, and enterprise planning". More specifically, vertical integration is the enterprise in different IT systems, manufacturing facilities, including NC machine tools, robots, and other digital production equipment to conduct a comprehensive integration, to create a highly integrated system within the enterprise to realize the R&D, planning, technology, production, service, each link between data flow automatically, realize the bottom of the top orders, the lower data upload at the top, for the future smart factory in personalization, digital and networked production support.

The process of cross-enterprise business collaboration is shown in Fig. 2.27, which includes horizontal inter-enterprise collaboration and inseparable inter-enterprise collaboration with vertical integration, forming a T-shaped integration architecture, which mainly includes six steps:

1.   Establish a product modularization platform. Get through the transformation of BOM from design to manufacturing, and realize the integration of PLM and ERP.

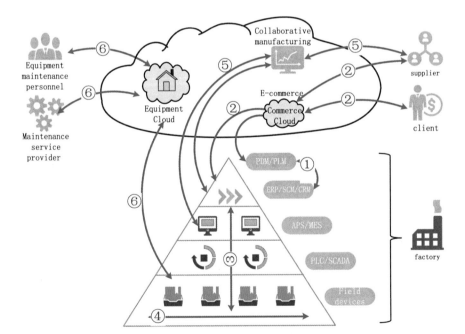

Fig. 2.27   Business collaboration processes across the enterprise

2. Provide online product customization tools, and connect with the production system to realize the integration of e-commerce platform with PLM and ERP, and perceive and predict the demand information.
3. Set up an open integration factory to realize vertical integration of flow, integration of ERP and APS/MES, PLC/SCADA, and equipment layer.
4. Establish data communication between machines and machines and between machines and applications to realize the general application of machine data.
5. Supply chain coordination, establish supplier coordination mechanism and realize the integration of ERP and SCM.
6. Set up the equipment cloud, analyze the big data obtained from the equipment, and realize predictive analysis, including equipment health management, energy consumption analysis, tool/fixture/tool analysis, etc.

Vertical integration through the control system level will achieve flat industrial control. In the traditional operation management mode, the tasks assigned by the ERP system must be assigned to the workshop by the managers and transferred to the industrial control system through MES or workshop managers to control the manufacturing equipment. Such a workflow cannot cope with the task changes brought about by the real-time customization of customer orders, nor can the orders be seen by customers, and MES cannot be directly connected to the manufacturing equipment. Based on service-oriented architecture (SOA) and vertical integration concept, make the ERP enterprise management and operation management of MES, field control layer of the industrial control system, interconnection, and mutual understanding, make manufacturing equipment in real-time to perform tasks assigned by the ERP system, to realize the visualization of the order and change at the last second.

German enterprises carried out an elaborate layout in advance many years ago in this respect. The country has a leading edge in vertical integration based on its substantial advantages in embedded technology and industrial software. In April 2007, Siemens company with $3.5 billion merger UGS companies in the United States, the latter of the world's leading CAD/CAM/PLM system into a bag, well-known manufacturers, Camstar MES companies in the United States in 2014 and m&a, at this point, Siemens formed R&D, management, production, equipment, controller and other hardware and software system comprehensive integration, on a global scale to achieve the most robust ability of vertical integration. As another essential supplier of "Industry 4.0", SAP also completed the acquisition of MES vendor Visiprise in 2008, bridging the gap between enterprise resource management and production management and achieving vertical integration of upstream information systems.

2.9   Practical Case: Production Execution Optimization Model of Chrysler Corporation

It was a bright morning in mid-April 2017. Waiting up, the first thing that Adams remembered was the production planning and control (PPC) meeting to be held later in the day. Adams was recently promoted to section manager and was looking after production planning in the stamping department, also known as the stamping shop. He reported to Davis, the functional head, new to the stamping shop. Davis had been transferred there from the weld shop just two months ago.

"In PPC meetings, they do not usually call someone at my level, so it must be something serious," Adams thought. The agenda discussed capacity constraints that became evident from the increased demand forecast for the upcoming months. The director of the production would chair the meeting. All divisional and functional heads would attend and contribute to the discussions.

Adams recollected a conversation he had with Davis the previous day:

Davis: Adams, have we calculated the load per press for next month?

Adams: Yeah, there is pretty much no change in this month's plan.

Davis: Sound good. We need to explain the calculations in tomorrow's PPC meeting. Management has specific queries regarding stamping shop planning. Please take a print of your work, and also paste them into a Powerpoint presentation. Hmm… Can you join the meeting?

1.   About Chrysler

Chrysler Corporation is the third-largest automobile manufacturer in the United States. The company also recognizes that its staff are its main strength and has adopted norms such as wearing fabric uniforms of the same color and quality and having lunch together in the same canteen, regardless of their level of seniority. As the primary organizational form, the team addresses a variety of operational issues, including on-the-job training, quality teams, and continuous improvement. That culture permeated Chrysler. The company focuses on three key indicators: product quality, safety, and cost. Chrysler is committed to continuous improvement by following the "plan-check" (PDCA) cycle promoted by Edwards Deming. The PDCA concept is cross-functional and job level.

The Chrysler plant is located south of Detroit, Michigan, consists of three vehicle plants (plants A, B, and C), each of which includes A welding, painting, and assembly plant. Factory A also has a stamping shop. Chrysler has ten models of cars it needs to produce, eight of them at its Michigan plant and two at other plants.

The stamping shop of Factory A consists of 6 punch presses, all numbered according to the age of the equipment: M1, M2, M3, M4, M5, and M6. M1 is the oldest machine, and M6 is the latest. The stamping plant is Chrysler's largest division. It supplies rammed body panels for three Michigan plants, some other Chrysler plants, and some joint ventures. As a result, stamping shop resources are widely used. The departments in the stamping workshop are the quality department, mold maintenance department, safety management department, production planning department, production scheduling, shift department, budget control department. Each department is headed by a department manager who reports to the functional heads.

2. Stamping process and production plan

Stamping production is mainly the use of stamping equipment and die to achieve the metal materials (plate) batch processing process. Coils come from many suppliers around the world. The coil is trimmed to the exact size with a shear blade or forming blade, and the blank is then fed into the forming die of the punching machine, where the blank is pressed into the body panel. Trim stamping and forming stamping action are collectively called "stamping". Cutting blades are often universal while forming blades and forming molds are usually dedicated. A forming mold is used for forming and stamping a specific part, and after a particular batch of a certain part is produced, the mold is removed and installed with another mold to produce another part. This kind of replacement is called "mold change", "mold shift", "mold switch", to meet the needs of personalized product mixed line production. Mold changing time is different for different presses. The larger the punch tonnage is, the longer the die-changing time is. The cost per punch is also different. For example, the large M2 costs RMB ¥4.55 per punch (stamping), while the small M1 costs ¥2.03 per punch. The cost of each shift is fixed. The cost of each shift refers to the cost incurred by a machine for each shift, including labor cost, oil, electricity, water and gas cost of facilities, etc. Table 2.2 shows that the cost is fixed at ¥105 and has nothing to do with the production of this shift. The stamping cost is linearly related to the number of stamping times, that is, the batch size.

The production batch size is driven by the size of the storage space for the parts and the pallet capacity for the parts. Pallets are used not only to hold parts but also to move them from one place to another. The typical batch size is 2.5 days of inventory. After the stamping process is completed, the plates are moved to the welding shop.

The PPC department was responsible for planning the production of vehicles in all plants following the demand presented by marketing. Production plans need to be synchronized with the plant's capacity requirements over some time, usually within a month, as shown in Table 2.3.

Press machine	Mold change time (s)	Cost per shift (RMB)	Single stamping cost (RMB)
M1	1125	105	2.03
M2	4500	105	4.55
M3	1425	105	3.64
M4	1075	105	2.73
M5	1075	105	2.73
M6	1500	105	4.06

Table 2.2 Average switching time and cost of punch presses

Table 2.3 Main production plan of 8 models in the first two weeks of July 2017

Car models	day															Fortnight
	1	2	3	4	5	6	7	8	9	10	11	12	13	14	15	
CH1002	53					53	59	53	52	53			53	53		429
CH1011	55					55	55	55	55	55			55	53		438
CH1030	79					79	79	80	79	79			79	79		633
CH2011	38					39	40	35	40	40			40	40		312
CH1333	35					30	30	26	30	30			35	35		251
CH1153	66					67	67	57	67	67			67	67		525
CH1043	9					9	9	9	9	9			10	10		74
CH2341	71					71	71	71	71	71			71	71		568

Davis and Adams presented the part-machine allocation and capacity calculations made for the coming month in the PPC meeting, as shown in Fig. 2.28. The director of the production expressed a concern:

"I appreciate the efforts you are exerting to achieve a sound plan. Much of your planning, however, rely on experience and intuition. It is good, but it can sometimes lead to personal biases. I would like you to develop an objective and scientific process that will work with intuition. Please focus on lowering the cost of production as much as possible, but without compromising quality..."

Model /Machine	Daily Demand	MACHINES						LHS (parts)	RHS (parts)	FLEXIBILITY HURDLE MATRIX					
		M1	M2	M3	M4	M5	M6			M1	M2	M3	M4	M5	M6
CH1002	53	7	2	6	2	3	0	20	20	0	0	6	0	0	0
CH1011	55	7	4	3	3	5	3	25	25	0	0	3	0	0	3
CH1030	79	3	6	2	10	0	3	24	24	0	0	0	0	0	3
CH2011	39	7	4	4	3	2	3	23	23	0	0	4	0	0	3
CH1333	31	8	4	2	0	3	3	20	20	0	0	2	0	0	3
CH1153	65	6	4	7	2	2	3	24	24	0	0	2	0	0	3
CH1043	9	0	0	2	0	0	0	2	2	0	0	0	0	0	0
CH2341	71	1	3	2	13	0	3	22	22	0	0	2	0	0	3
CM128374 ②	136	1	1	0	0	0	0	2	2	0	1	0	0	0	0
CM128361 ①	7	0	1	0	0	0	0	1	1	0	1	0	0	0	0
No. of shots ③ (S_j)		2111	1696	1474	2231	735	1020								
Average shots per minute ④		2.53	4.93	3.25	4.03	3.72	2.73								
Shot cost (RMB)		2.03	4.55	3.64	2.73	2.73	4.06								
Total stamping time needed (hr)		13.89	5.73	7.56	9.22	3.30	6.22								
Time per changeover (hr)		0.31	1.25	0.40	0.30	0.30	0.42								
Total changeover time needed (hr)		5.00	14.50	4.43	3.94	1.79	3.00								
Total stamping time + changeover time as shift multiples		2.52	2.70	1.60	1.75	0.68	1.23								
Shifts needed ⑤ (Y_j)		3	3	2	2	1	2								

Shots cost	29605.87
Shift cost	1365
Total cost (Z)	30970.87

Fig. 2.28 Parts—machine current allocation and capacity calculation table

Remark:

1. Total number of parts = parts on M1 + parts on M2 + parts on M3 + parts on M4 + parts on M5 + parts on M6.
2. Total = daily requirements per machine x parts.
3. Vehicles made at other Chrysler plants.
4. The average number of presses per minute is a measure of punch productivity. Data for the past three months have been taken into account. This does not include the replacement time of parts.
5. Minimum number of shifts required = total stamping number ÷ (average stamping number per minute × 60 min per hour × 7.5 h per shift). Any press can be changed up to 3 times a day.
6. The flexibility obstacle matrix shows the number of parts of each car model that can be stamped on each machine. For example, the "6" in the first line indicates that the six parts of the CH1002 model can be stamped on the punch.

Intuitively distributing parts to punch presses is a daily task—a senior assistant reporting to the production planner used to assign presses by hand in trial-and-error iterations. Now, inspired by the comments of the production director, the department head and production planner have decided to introduce a scientific method to optimize the allocation of production tasks and costs. As both men know, the complexity lies in that not all parts can be used interchangeably in any punch.

In order to make this production plan, they need to discuss and choose mathematical optimization models, think about the implementation process and methods, compare cost savings and the impact of this change on relevant personnel.

They tried to introduce an integer programming model, considering that it was constructed and solved in two steps. The first model contains all the relevant elements except the switch time. The second model introduces switching times, adding constraints, assumptions, and modifications to the model. The second model is an extension of the first.

They first considered the decision variables of the model, including the number of stamping parts allocated on each punch press of each model of car, is X_{ij}, the daily shift arranged for each punch is Y_j. Set X_{ij} is the number of parts for the car model i assigned to machine j. There are currently 10 models of cars and 6 machines; In other words, $i = 1,2,...,10, j = 1,2,...,6$. Hence, from X_{11} to X_{106}, there are 60 X_{ij}. Set Y_j is the number of shifts per day that the machine j should be running.

The objective function is to minimize the total cost, which has two components: the cost of stamping and the cost of shift switching. Table 2.2 shows the cost of stamping on different machines. For example, in M1, each stamping cost is ¥2.03, and it costs ¥105 to operate any machine for a shift. According to the decision variable, the expression of the total cost (Z) can be obtained:

$$Z = 0.29 S_1 + 0.65S_2 + 0.52S_3 + 0.39S_4 + 0.39S_5 + 0.58S_6 + 15(Y_1 + Y_2 + \cdots + Y_6)$$

In which, for all $j = 1,2 \ldots 6$. According to the daily demand, there are:

$$S_j = 53X_{1j} + 55X_{2j} + 79X_{3j} + 39X_{4j} + 31X_{5j} + 65X_{6j} + 9X_{7j} + 71X_{8j}$$
$$+ 136X_{9j} + 7X_{10j}$$

S_j is the total number of stamping parts assigned to machine j for all models.

There are five sets of constraints in the model, each of which comes from a different consideration: all parts of the car model must be allocated on six presses, and no part must be left in the unallocated state. That is to say, if we add the distribution of a particular part to all machines, we must get the total number of that part. A punch press can be changed at most three times a day in shifts, with mold changing once in each shift; The average production capacity of each punch press limits the average number of die stamping within 1 h; The maximum usable time per module is 7.5 h; Certain parts can only be punched on certain presses.

They used the Solver function of Excel to find the optimization scheme of this model and used the "simplex method" as the solution method. In the best solution obtained, the distribution of parts and punch scheduling are shown in Fig. 2.29. It is noted that M1 presses carry out 3 die change shifts, M5 presses carry out 2 times, while M2, M3, M4, and M6 presses carry out only 1 time each.

In Model 2 (Fig. 2.30), the production planner introduces mold change time into the model equation.

In Model 2, the production planner introduces mold change time into the model equation. After one batch of production, the mold group is replaced, and the next batch of production begins. The die change time is shown in Table 2.2. This cost is not directly expressed in monetary terms but at the cost of lost production time during the switchover, which is a non-value-added activity. It should be noted that all parts assigned to the punch for stamping usually need to wait while the die is

Model /Machine	Daily Demand	M1	M2	M3	M4	M5	M6	LHS (parts)	RHS (parts)	M1	M2	M3	M4	M5	M6		
CH1002	53	13	0	6	0	1	0	20	20	0	0	6	0	0	0	Shots cost	25384.03
CH1011	55	0	0	3	18	1	3	25	25	0	0	3	0	0	3	Shift cost	945
CH1030	79	6	0	0	0	15	3	24	24	0	0	0	0	0	3	Total cost (Z)	26329.03
CH2011	39	14	0	4	1	1	3	23	23	0	0	4	0	0	3		
CH3333	31	15	0	2	0	0	3	20	20	0	0	2	0	0	3		
CH1153	65	19	0	2	0	0	3	24	24	0	0	2	0	0	3		
CH1153	9	1	0	0	0	0	0	1	2	0	0	0	0	0	0		
CH2341	71	0	0	2	10	7	3	22	22	0	0	2	0	0	3		
CM128374	136	0	1	0	0	1	0	2	2	0	1	0	0	0	0		
CM128361	7	0	1	0	0	0	0	1	1	0	1	0	0	0	0		
No. of shots (Sj)		3418	143	973	1739	1965	1020										
Average shots per minute		2.53	4.93	3.25	4.03	3.72	2.73										
Shots cost (RMB)		2.03	4.55	3.64	2.73	2.73	4.06										
Total stamping time needed (hr)		22.4868	0.48311	4.98974	7.18595	8.81166	6.21951										
Total stamping time as shift multiples		2.99825	0.06441	0.6653	0.95813	1.17489	0.82927										
Shifts needed (Yj)		3	1	1	1	2	1										

Fig. 2.29 Calculation based on optimization model 1

Model /Machine	Daily Demand	M1	M2	M3	M4	M5	M6	LHS (parts)	RHS (parts)	M1	M2	M3	M4	M5	M6		
		MACHINES								FLEXIBILITY HURDLE MATRIX							
CH1002	53	1	0	6	13	0	0	20	20	0	0	6	0	0	0	Shots cost	25690.56
CH1011	55	0	0	3	19	0	3	25	25	0	0	3	0	0	3	Shift cost	1260
CH1030	79	20	0	0	1	0	3	24	24	0	0	0	0	0	3	Total cost (Z)	26950.56
CH2011	39	0	0	4	8	6	3	21	23	0	0	4	0	0	3		
CH1333	31	1	0	2	0	14	3	20	20	0	0	2	0	0	3		
CH1153	65	3	0	2	16	0	3	24	24	0	0	2	0	0	3		
CH1153	9	0	0	0	0	2	0	2	2	0	0	0	0	0	0		
CH2341	71	12	0	2	5	0	3	22	22	0	0	2	0	0	3		
CM128374	136	0	1	0	0	1	0	2	2	0	1	0	0	0	0		
CM128361	7	0	1	0	0	0	0	1	1	0	1	0	0	0	0		
No. of shots (Sj)		2711	143	973	3520	822	1020										
Average shots per minute		2.53	4.93	3.25	4.03	3.72	2.73										
Shots cost (RMB)		2.03	4.55	3.64	2.73	2.73	4.06										
Total stamping time needed (hr)		17.84	0.48	4.99	14.55	3.69	6.22										
Time per changeover (hr)		0.31	1.25	0.40	0.30	0.30	0.42										
Total changeover time needed (hr)		4.63	1.00	3.01	7.41	2.75	3.00										
Total stamping + changeover time as shift multiples		2.99	0.20	1.07	2.93	0.86	1.23										
Shifts needed (Yj)		3	1	2	3	1	2										

Fig. 2.30 Calculation based on optimization model 2

changed. By introducing die change times for punch presses for all parts, the average number of switches per part per day is considered based on the characteristic of "a typical batch size is 2.5 days of inventory".

We note that by accumulating all the punches, the optimization scheme requires fewer switches per day than the current scheme, from 13 to 12 based on optimization model 2, thus saving the fixed cost of one switch per day.

The optimized allocation of stamping parts across multiple presses resulted in a significant reduction in total stamping costs from ¥30,971 to ¥26,951, a savings of ¥4020 or 12.97%. Assuming that the punch operates 300 days per year (allowing downtime for maintenance, breakdowns, etc.), this translates into savings of ¥(574) × 300 = ¥1,205,400 per year. While that amount itself is small for a large organization like Chrysler, it must be noted that it represents only a tiny fraction of the cost savings generated by six punch presses.

Thinking Exercise

1. What is the most cost-effective allocation scheme for stamping machines?
2. How does the best solution compare to the current allocation?
3. In the production process, in addition to the cost of stamping and mold change, what other costs are involved in production? Please consider shifts, tests, and transportation.
4. Can other types of data help refine decisions?
5. Can production standardization help control some of the costs associated with this solution?
6. What is the interpersonal relationship with the stamping shop and what are the considerations when implementing a cost-effective scheme?

7. Try to build a mathematical optimization model and solve it by hand and compare with the results in the case.

Homework of This Chapter

1. What is the relationship between product family, product variable, product view, product sub-item and material? Please draw a description.
2. Please understand the position of various industrial software in the system from the perspective of strategic, tactical and operational levels.
3. Try to point out the differences between ERP, APS and MES systems, and illustrate with examples, which can be used in life, such as cooking.
4. How to understand the flatness and rapid response of the APS system? What are the problems with traditional planning systems and multi-stage decision-making processes? Please give an example.
5. Why is the MPS an impact absorber? What does it do? Please give an example.
6. In an inventory-oriented manufacturing enterprise, production planning should be formulated for one year if the initial inventory is 1000 units, the annual sales volume is 5000 units, and the ending inventory is 2000 units. So, what's the monthly productivity?
7. According to the sales forecast in the following Table 2.4, what are the monthly production volumes required by the master production plan under the balancing strategy, following strategy and mixed strategy respectively?
8. Try to express the three main production plan strategies in the above question.
9. Under what circumstances will MRP be rearranged?
10. Why is BOM an integrated interface between design and manufacturing?
11. What do you think is the most essential feature of SCM?
12. Try to analyze the information flow between a company's ERP, SCM and corresponding functional modules in its supplier's system.
13. Why does the 3D model have non-geometric parameters, such as material performance data, experimental data and supplier data? How does the 3D model relate to digital twins?
14. What are the main difficulties in Collaborative R&D by global suppliers?

Table 2.4 Monthly sales forecast

Beginning inventory = 200 Ending inventory = 200						
Month	1	2	3	4	5	6
Sales forecast	200	300	400	600	300	300

Reference

1. Ministry of Industry and Information Technology, Information and Industrialization Integration Management System Requirements (GBT 23001-2017).

Chapter 3
Life Cycle of Intelligent Manufacturing

3.1 Inspiration Case: The Tool Life Cycle Management of Huizhuan Company

Huizhuan Green Tools Co., LTD. is a cutting tool manufacturing company established in Guangzhou, mainly engaged in ultra-hard precision tools, ultrasonic machining tools, four-axis five-axis dividing plate, machine tool cooling system, and other products. It is the tool supplier of Apple Electronics Company and Benz Automotive Company.

In the production process of manufacturing enterprises, tool management is undoubtedly one of the essential factors affecting production efficiency. Whether tool management is reasonable and scientific largely determines the level of production efficiency. The original tool management mode of Huizhuan Company is shown in Fig. 3.1. The tool data is shown in Table 3.1, Huizhuan has a large output of cutting tools. A cutting tool usually has a short service life and needs frequent maintenance as a consumable. Based on the original model, the company has some problems in the process of tool maintenance quality control, such as lack of timely tool maintenance and low efficiency of resource elements.

1. Problems existing in maintenance quality control

    (1) Tool maintenance is not timely

         1) Machine capacity is limited, and the production plan cannot be arranged in time for cutting tool repairing and grinding. An unreasonable production plan leads to an unbalanced production load.
         2) The production department sometimes forgets orders, resulting in delayed cutting tool repairing and grinding.
         3) The feedback from the sales staff is not timely, resulting in the delay of tool repairing and grinding.
         4) The technical staff did not provide the grinding plan in time, resulting in the tool grinding being put on hold.

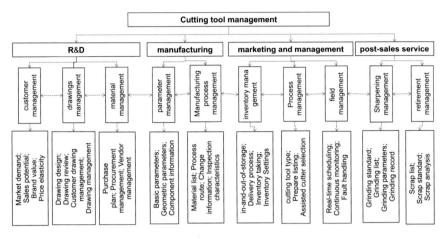

Fig. 3.1 The original tool management mode

Table 3.1 Main tool types and basic data

Serial number	The classes of cutting tool	Material	Production value (million RMB)	Cost	Mean life (h)
1	The hard cutting tool	Carbide	50	Middle	100
2	Superhard cutting tool	PCD/MCD/ND/PCBN/ceramic	60	High	1000
3	Thread cutting tools	High speed steel	60	Low	80
4	Ultrasonic cutter	Ultrasonic system and tool handle	100	Ultra high	20,000

(2) The efficiency of resource elements needs to be improved

1) The manual recording mode makes more personnel occupied in the tool circulation area.

2) Low OEE occupies a large amount of capital.

3) The lag of information management of tools leads to a low utilization rate of tools and CNC machine tools.

4) The process monitoring system of Keynes can be used to measure the tool wear offline. However, the system is not connected with the MES system and CNC machine tool in real-time, which results in the need to manually set the tool compensation amount on the CNC machine tool according to the tool wear.

5) The tool management has not formed the system management mode, lacks the data fusion, data analysis, data management, and standard system, lacks the optimized tool cutting technology standard and the use specification.

2. Management mode of the tool life cycle

The ideal model of the whole life cycle of the cutting tool management should include product planning, R&D, design, procurement, logistics, manufacturing, wear trace monitoring, real-time online diagnosis, fault prediction, recycling, repair, cost control, customer collaboration. It should have a complete set of systems to operate and control through the whole production cycle of the service to provide the needed tools for production systems timely and accurately, with high quality.

Given the existing problems, Huizhuan company developed a set of management systems to optimize and integrate the tools, personnel, equipment, and other resources in supply chain. Firstly, design and development of tool management database. Secondly, the relationship between tool life and tool parameters and machining parameters is analyzed through big data, and a tool life prediction model is established to accurately reflect the highly nonlinear relationship between tool life and its influencing factors. Finally, the management system of B/S mode is designed, and the functions of tool parameter management, tool assembly and disassembly management, loan management, purchase management, inventory management, online diagnosis, life prediction and management, and grinding production plan optimization are completed, to realize the tool life cycle management.

Its core idea is shown in Fig. 3.2a. Tool life cycle management is a process of comprehensive management of the health status of the tool and its use process. Each link of the tool life cycle should be systematically evaluated from the perspectives of reliability, quality, safety, cost, benefit, environment, system, and partner collaboration. Through the evaluation calculation, the wear amount and the remaining life of the tool can be estimated, and the grinding plan can be made in advance to improve the production balance rate and OEE. Secondly, it is necessary to recover, maintain or repair the coating of the tool before the tool wear is out of tolerance or the broken edge leads to the product quality problems or accidents, so as to ensure the customer's production will not be interrupted, increase the use times of the tool, extend its life cycle, reduce the cost of the tool and improve the cutting quality of the product. The overall logic framework of the tool lifecycle management platform is shown in Fig. 3.2b. The effect is enhanced through remote monitoring service and health prediction, and the efficiency is improved through intelligent scheduling and visualization support of digital mobile devices.

3. Hybrid intelligence based on big real-time data

(1) Surface fitting and prediction equation

When the tool wear and failure mechanism is simple and straightforward, and it is easy to extract and separate the natural influencing factors, using a curve or surface fitting can effectively use data sets to dig out the relationship between fault results

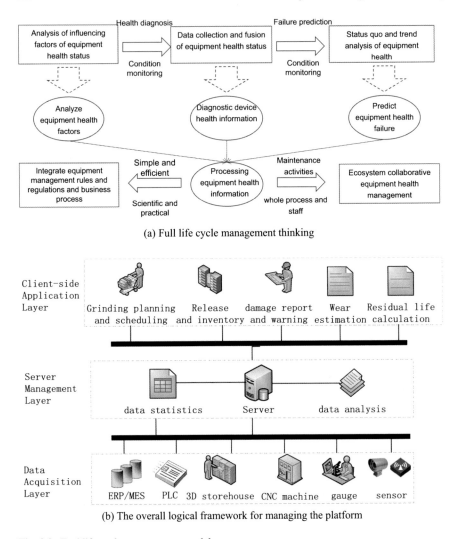

Fig. 3.2   Tool life cycle management model

and influencing factors, and obtain the fitting function equation. When the system has input, it can quickly make evaluation and prediction.

The reasons leading to short tool life include (1) Improper choice of tool type and material, tool rake Angle, cutting dosage, cutting depth, and workpiece surface roughness, resulting in insufficient blade strength. (2) Improper protection of cutting tools during production or transportation leads to blade breakage. (3) Improper temperature control causes the tool to be quenched and heated in use, and the blade material is subjected to thermal impact, resulting in thermal stress and thermal crack. (4) The rigidity of the machine-fixture-workpiece-cutter processing system is not enough, which causes vibration in the cutting process. (5) Incorrect machining mode leads

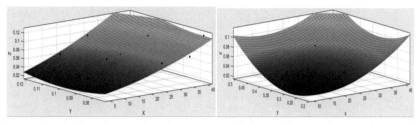

(a) Cutting depth and processing pieces   (b) Workpiece surface roughness and processing
temperature

Fig. 3.3 Fitting surface of the wear of milling cutter on back surface

to shorter tool life, such as improper selection of spindle speed, feed speed, cutting width, and other parameters.

For various factors affecting wear, real-time cutting parameters and wear measurement data of the tool are collected, 95% confidence is selected for each parameter combination classification, and quadratic polynomials fit a variety of surfaces. Each surface corresponds to a specific prediction equation group.

Figure 3.3a shows the relation surface between the cutting depth (Y-axis), the number of workpieces (X-axis), and the wear amount (Z-axis). It points out that when the cutting depth is selected at 0.10 mm, then the tool wear is the least. This rule suggests that the text should choose the optimized cutting depth when carrying out process design and workers are cutting. Figure 3.3b shows the relation surface of workpiece surface roughness, processing temperature, and wear amount. It indicates that the tool wear is the least when the workpiece surface roughness is 0.227 Ra. Therefore, raw material procurement or material pretreatment should be subject to corresponding standards. The higher the temperature, the faster the wear, so it is necessary to keep effective cooling during cutting.

The structure of wear assessment and residual life calculation model based on hybrid intelligence is shown in Fig. 3.4. The rule library generated from rule extraction can be used to supplement the predictive performance of "black box model" such as neural network. After extraction of real-time data by eigenvalues, data fusion operation is conducted with detected wear measurement value, model estimation value obtained from the prediction equation, predictive value of BP (Back Propagation) neural network, to obtain more accurate current wear evaluation value and calculate the remaining life. The predicted life value of the combined model is more accurate than that of the basic BP neural network and the fitted surface prediction model. The relative error is kept within 5%.

The tool's entire life cycle management process is shown in Fig. 3.5.

The life prediction model of each tool type is stored in the database. At the beginning of each new tool, real-time process parameters such as cutting parameters, cutting time, and cutting depth are collected through the network and recorded in the database. According to the tool life prediction model stored in the database, the maximum cutting time of the tool can be calculated before each cutting tool is

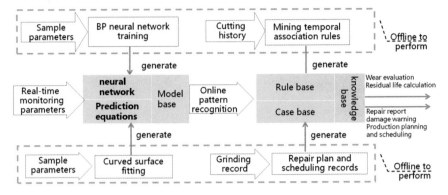

Fig. 3.4 Wear assessment and residual life calculation model based on hybrid intelligence

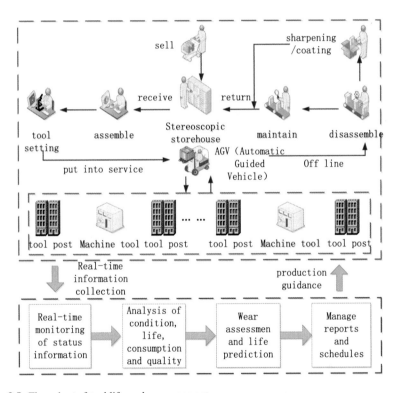

Fig. 3.5 Flow chart of tool life cycle management

used, which can provide early warning for tool repair and loss reporting. Combined with the rule base obtained by mining association rules from cutting history records and the case base obtained by grinding records, the optimized production plan and scheduling scheme are presented.

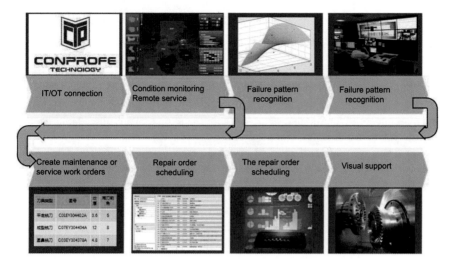

Fig. 3.6 Intelligent enhancement effect and Internet visualization to improve efficiency

4. Tooling lifecycle management process

The tool's whole life cycle management process is shown in Fig. 3.6. Functional modules include resource management, procurement, inventory management, online monitoring, wear assessment and residual life prediction, production planning and scheduling, knowledge management, and system management. Under the traditional management way, tool tracking card entry, statistics and sorting, on-site inspection and daily repetitive work occupy 80% of the workload, when using the whole life cycle management system, through the automatic upload on behalf of manual entry of data, as well as the AGV (automated guided vehicle) instead of manual handling, reduce the proportion of the above work to 20.5%, staff can put more time into production improvement and the R&D work. The company's service improvement plan is as follows: firstly, it provides customers with intelligent logistics and cutting process optimization services based on the hybrid intelligent system through intelligent enhancement effect; Secondly, it realizes visual management on-site through industrial Internet to improve efficiency. Finally, based on the complete production cycle management remote monitoring platform to provide technology, quality, efficiency, cost, and other comprehensive outsourcing services, then the international tool management service steps forward to level 5, the highest level.

Thinking Exercise

1. What changes have taken place in the key aspects of the company's product management? What is technical support? What has happened to the management model? What new business models can be created?
2. What benefits can the company achieve after the reform?

3.2   The Introduction

The life cycle of product and service is a chain set composed of a series of interrelated value-creating business links such as design, production, logistics, sales, and service. The similarity of this chapter and the second chapter is to discuss one dimension of IM. The difference is that the life cycle dimension of IM discusses the whole process of time-concept, dynamic and value-changing activities, while the spatial level of the IM system discusses the spatial depth structure.

The whole life cycle of product-oriented service includes five business links, namely design, production, logistics, sales, and service, covering the whole process from receiving customer demand to providing products and services. Compared with the traditional manufacturing process, the IM process focuses more on the intelligent application of each business link and the improvement of intelligence level.

Each activity in the product life cycle is interrelated and influences each other. The composition of the product life cycle is different in different industries.

1.   Design refers to the process of product construction, simulation, verification, optimization, and other R&D activities based on all constraints and selected technologies of the enterprise;
2.   Production refers to the process of creating necessary material data through labor;
3.   Logistics refers to the physical flow process of goods from the place of supply to the place of receiving;
4.   Sales refers to the business activities in which products are transferred from the enterprise to customers;
5.   Service refers to a series of activities and results between the provider and the customer, including maintenance, monitoring, etc.

In the five main links of the product life cycle, the difference between IM and traditional manufacturing is shown in Table 3.2.

This chapter tries to extract an implementation path of IM divided by life cycle stages from the perspective of the whole life cycle of products. Most small and medium-sized enterprises with comprehensive design, manufacturing, and sales departments can start from the R&D business link and gradually expand to the downstream link in stages to realize IM, as shown in Table 3.3.

3.3   Design and R&D

Design is to form an implementation plan through product and process planning, design, reasoning verification, and simulation optimization. The improvement of the integration of "intelligence + manufacturing" is achieved through changing from a two-dimensional to a three-dimensional structure, from internal collaboration with engineers to external collaboration with customers and partners, from experience-based reasoning to knowledge-based parameterized modeling and simulation. The

Table 3.2 Differences between intelligent manufacturing and traditional manufacturing

	Conventional manufacturing	Intelligent manufacturing	The impact of intelligent manufacturing
1. Product design and development	Conventional products; function oriented design; two-dimensional design, long cycle	Personalized design based on the combination of virtual and real design; customer-oriented design; three-dimensional digital design, short cycle	Life cycle oriented design concept; Change of design methods and means; Change of product function
2. Manufacturing	The processing is carried out according to the plan; Automatic processing and manual inspection; Highly centralized production organization; Man machine separation; Reducing material; metal and plastic	process is flexible and can be adjusted in real time; Intelligent processing and online real-time detection; Flexibility of production organization; Real time tracking of networked process; Additive manufacturing; Man machine integration and intelligent control; composite material increases	Change of labor object; Change of production mode; Change of production organization; Change of quality monitoring mode; Diversified processing methods; New materials and processes are emerging
3. Logistics and supply chain	Traditional logistics; simple business relationship between enterprises, lack of coordination	Integrated logistics; information collaboration; R & D collaboration; production collaboration	Forming a benign ecosystem; collaborative innovation; expanding the scope of management
4. Management and sales	Internal management; human management; transaction marketing; physical store sales	Extended to upstream and downstream enterprises; computer information management; machine and human interaction management; demand mining and sales forecasting based on big data; Omni channel marketing	Change of management object; change of management mode; change of management means; big data analysis; combination of online and offline

(continued)

Table 3.2   (continued)

	Conventional manufacturing	Intelligent manufacturing	The impact of intelligent manufacturing
5. Services	Product itself; after sales service	Product service system; Product life cycle	The scope of service object is expanded; the service mode is changed; the service responsibility is expanded

personalized, collaborative design based on the cloud, big data and knowledge base reflects the rapid increase of customized demand. The Digital model is an essential foundation of intelligent manufacturing we cannot ignore.

3.3.1   Development Trend

The starting point of value chain transformation is the R&D link. The mainline of future R&D should be carried out around the theme of innovation, and the following trends are presented:

1. Design of intelligent product: Product features continue to increase, and the presence of the mode "product as a service" makes the product R&D from predominantly mechanical design into a truly interdisciplinary systems engineering, and it not only includes the traditional machine, electricity, soft and other professionals but also includes the product connected, embedded services, new disciplines such as user experience.
2. Satisfaction with highly personalized customization requirements: Different customer layers, regions, and application objectives require product diversity. In the front end of R&D, supporting personalized customization should be considered, significantly reducing customization costs. In the future, enterprises will increase support for software customization to reduce the cost pressure brought by hardware diversification.
3. Rapid and continuous product improvement: Continuous improvement is pivotal to obtaining a sustainable competitive advantage. The future enterprise will extract products operating quality information through the Internet in real-time to realize the continuous improvement of arbitrary points in the cycle of life.

Table 3.3 "Intelligence + manufacturing" fusion depth model with single life cycle dimension

Level	R&D	Production	Logistics	Marketing and management	Service
L1	Design of 2D drawing; formulate relevant standards and specifications for product design	2D document based process design; two dimensional tooling design; master production plan and MRP	Information management of order, schedule and information tracking	Information management of sales plan and customer relationship; paper management mode	Paper document management of maintenance data and other elements
L2	2D and 3D design; EBOM management; product customization management; internal collaboration	3D aided process design; NC code aided generation; PBOM / MBOM management; MES; ERP; intelligent production line design	Using advanced Internet of things technologies such as barcode, RFID, sensor and GPS;	Customer demand forecast / actual demand drives production, procurement and logistics plan; R&D and manufacturing data link; integration of engineering change	Information management, content management, R&D and service are not integrated
L3	Full 3D structure design MBD; R&D and production and operation intelligent interconnection; parametric design and simulation optimization based on knowledge base; intelligent product design	3D process planning, tooling design and process simulation; equipment interconnection and real-time monitoring; intelligent operation guidance based on Internet of things and AR; integrated integration	Automatic operation, visual monitoring and optimized management of vehicles and routes in the transportation process; transportation integration of various strategies	Integration and collaboration of design and manufacturing data model based on 3D MBD; interconnection of development and service information; product quality engineering	Digital maintenance analysis; interactive electronic manual; maintenance engineering design; maintenance execution process management MRO; feedback optimization based on service system

(continued)

Table 3.3 (continued)

Level	R&D	Production	Logistics	Marketing and management	Service
L4	Requirements engineering; model-based system engineering design; verification based on virtual prototype; R&D innovation based on knowledge base	Digital plant design; production engineering; manufacturing supplier collaboration; equipment health management; whole process closed-loop and adaptive	Optimization of transportation path based on knowledge model	Intelligent product technology state management; intelligent interconnection quality management; accurate sales forecast based on knowledge model	MRO of maintenance execution process management; intelligent monitoring of product operation based on IOT; one-stop service; service based on knowledge base of cloud platform
L5	Implement cloud service of product design based on big data and knowledge base, and realize personalized and collaborative design of products	Manufacturing subcontract based on digital delivery specification; cloud manufacturing; real-time scheduling, correction and Optimization Based on Intelligent Algorithm	Lean management, visual intelligent logistics	More accurate sales forecast based on artificial intelligence; collaborative supply chain management among enterprises in the whole value chain; operation management and personalized marketing based on big data	Maintenance supplier digital maintenance and data delivery; customer collaboration; product health management; time service; intelligent personalized service by using customer service robot or big data intelligent analysis

4. Combination of virtuality and reality and Digital Twin: It is, in a narrow sense, a multi-scale simulation process making full use of physical models, updating of sensors, running history data, integrating multidisciplinary. It completes the mapping in the virtual space, reflecting the lifecycle of the corresponding entity. In a broad sense, an enterprise aiming at IM is a combination of virtuality and reality. The CEO of Foxconn believes that Foxconn is a six-flow enterprise consisting of a virtual flow of "technology, information, capital" and an actual flow of "people, process, and logistics".

3.3.2 Business Model

The business model includes three levels: Internal R&D, collaborative R&D in a wide range of value chains, and continuous optimization of R&D.

3.3.2.1 Internal R&D

The enterprise optimizes the R&D process through centering on intelligent products and personalized customization. The business model is as follows:

1. Customized demand engineering

Before product design, R&D personnel will make serialized product planning to meet customization requirements. That is personalized and customized design planning. Then marketing personnel provide customers with flexibly configured recommendation schemes based on the platform and provide visual product interaction experience using handheld terminals and augmented reality technology based on customized results. During the product development, the customer can real-time monitor the whole development process and make corresponding changes according to their needs.

Customer customization is a form of concurrent engineering. Personalization needs the support of the software industry, including CAD/CAE/CAPP/CAE/DFX/PDM/ERP systems. Taking the automobile industry as an example, it would not be possible to prepare materials according to the personalized production plan when the plan is issued. The supplier must prepare each vehicle's materials in advance to achieve the customization requirement. It means that before production starts, all the modules needed should be well-prepared and assembled according to the order requirement. It requires automating the flow of data and various industrial software to implement this process.

2.   Digital Mock-Up (DMU) and Model-based Definition (MBD)

The core of IM is digitization, and the core of digitization is modeling. MBD refers to an integrated 3D data Model to express complete product definition information, which becomes the only basis in the manufacturing process. MBD 3D digital product definition technology changes the product design fundamentally, which no longer needs 2D engineering drawings. Furthermore, the technology has a significant impact on the downstream of the process, including process planning and production. It enables the production links to be simulated, and has caused significant changes in digital manufacturing technology, opening the 3D digital era.

The use of models to reproduce product characteristics and the study of existing or designed products through experiments on models is called simulation, or computer-aided engineering (CAE). Figure 3.7a shows that an operational calculation model is established for the aircraft. It records and analyzes data dynamically to explore the motion characteristics of the aircraft to reduce the number of physical wind tunnels and improve design efficiency. Methods like finite element analysis are introduced to achieve the dynamic visualization analysis test to ensure the reliability of the design scheme. Figure 3.7b shows that the Yuchai machine has overcome subjective-judgment problems through the simulation analysis method. In the past, they depended on expert meetings without data. However, the discussions were always inconclusive. With the improvement, they cleared the long-term stubborn illness, which is the engine connecting rod fracture and cylinder hitting accident.

IM in product design is one of the essential roles of redefining the product model and data exchange standard, making intelligent product design between different users in different departments of the value chain can be complete, accurate, timely data exchange. By the consistency of the product model, data integration and extraction are more secure. For example, engineer A uses Siemens PLM NX software, engineer B use Dassault software Catia, engineer C use Autodesk Inventor. Therefore, it is not easy to communicate between software and engineers. However, with the birth of ISO 10303, three engineers, A, B, and C, understood each other's designs. Model-based 3D systems engineering in ISO 10303-242 is a valuable standard for manufacturers and suppliers in aerospace, automotive, and other industries. It

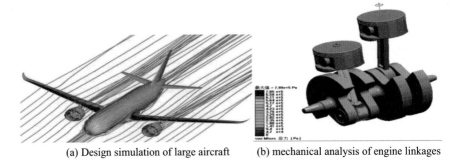

  (a) Design simulation of large aircraft   (b) mechanical analysis of engine linkages

Fig. 3.7   3D digital product design

includes PDM, design guidelines, associated definitions, 2D drawings, 3D products, and manufacturing information, etc. Reference to international product design standards can improve the efficiency of data exchange and use in intelligent product design, form a consistent product model, and ensure the security of information and data.

Digital Twins is the model of the fusion of virtual and reality, which is the most critical foundation of IM. For example, various sensors are installed on a bicycle to monitor the rider's speed, acceleration, physiological parameters, and other information. These parameters can be transmitted back to the digital product model during the cycling process as the boundary conditions required for simulation optimization. In the previous product simulation, many boundary conditions and parameters were imaginary. However, now based on this mode of virtual and reality fusion, the actual operating parameters of the product can be transmitted back to the data model to realize the combination of virtual and reality.

Digital twins can also be combined with engineering simulation technology. In the previous simulation work, boundary conditions and loads were all the ideal states assumed by the emulators themselves. Now, based on the application of the Internet of Things, the actual operation data of the product sensed by the sensors can be taken as the simulation conditions. For example, in offshore wind power equipment maintenance, if there are data abnormalities, the traditional method is to adjust or maintain the equipment according to experience. At the same time, the simulation is conducted based on real-time environment and state parameters of the equipment to analyze the effect of the measures taken, which is safer and more reliable.

The architecture design and virtual verification of intelligent products are carried out by using the model-based system engineering method based on the customization requirements of intelligent products. In the architecture design, the intelligent characteristics of products are mainly described through a behavior model to guide the implementation and verification of the intelligent characteristics in the future. Virtual validation supports the complete closed-loop process management of validation requirements, architecture design, virtual simulation, and problem correction.

Discrete manufacturing enterprise in terms of product R&D has been widely used CAD/CAM/CAE/CAPP/PDM/PLM software and systems, but many enterprises' application software level is not high. Enterprises can improve the application level of industrial software in the R&D field through the following ways:

Firstly, to shorten the product development cycle, MBD technology-based virtual digital prototype can be established to carry out multi-disciplinary system-level design, detail design, experimental verification, and simulation to support the multi-disciplinary collaborative design of machine, electricity, and software. Different from traditional product design, complex product system emphasizes more on intelligent characteristic-oriented design, embedded software development, and integrated management. Reduce physical tests through simulation; Depending on the implementation of standardization, serialization, and modularization, support mass customization or personalized customization; The simulation technology and test management are combined to improve the confidence of simulation results.

Secondly, based on the engineering application of DMU, Augmented Reality (AR) was introduced in the R&D process to realize hardware-in-the-loop and real-time mock-up to improve the level and efficiency of product development.

Finally, with the help of PDM, data can be shared through the data platform from the beginning of product design so that no matter how many layers are separated, relevant personnel can see the latest version of data in the background. Under the integrated framework of PDM, CAD/CAM and other computer-aided systems are used to transfer data to the machine tools directly without 2D drawings to realize the integration from design to manufacturing and shorten the launch-to-market time. Compared with discrete manufacturing enterprises, process manufacturing enterprises have generally started to use PLM systems to achieve process management and formula management, and LIMS (Laboratory information management system) has also been widely applied.

[Case: the birth of MBD standard]
In the development of the Boeing 777, the 3D design and 2D drawing coexist because the 3D model still could not express the material information, process information, testing information. This leads to much extra work and causes errors, such as 3D assembly drawings, and the 2D drawings for assembly workers are hard to understand. Therefore, ASME (American Society of Mechanical Engineers) formulated the MBD standard in 1996, making it possible to express all the standards, process, manufacturing, and testing on the 3D model without needing two-dimensional drawings.

3.3.2.2 Collaborative R&D in Wide Range of Value Chains

In the future, the collaborative business will break restrictions such as geographical location, resource shortage, and complex application deployment, realize R&D collaboration on a global scale, and support new R&D models such as mass innovation.

1. Collaborative R&D by global suppliers

Enterprises can realize efficient R&D collaboration with suppliers based on the global supplier collaborative environment. The main collaborative businesses include supplier technology interface management, delivery demand, plan management, R&D prototype collaboration, delivery data collaborative management, engineering coordination, AR-based R&D collaborative review, etc. The consistency of industrial software and efficient collaborative design mechanism is the foundation of collaborative R&D by global suppliers.

[Case: Airbus A380's delay]
Airbus is owned by Germany, France, Britain, and Spain, with Germany and France owning 22.5% of the shares to maintain the balance of power. Boeing is betting that medium-sized planes, which make direct flights easier, will go mainstream. Airbus

thinks the best way to improve traffic congestion is with bigger planes. The A380 was designed to be the largest long-range commercial airliner in history, with a standard range of 15,000 km and a capacity 20% higher than the 416-seat Boeing 747. In October 2006, Airbus announced a delay of at least two years in the delivery, which cost Airbus $6 billion. The main reason is the inconsistency between the development tools used in the design of A380 architecture and the software versions used in the assembly. Airbus employs about 41,000 people with offices and production facilities in 16 locations in four countries. Branches everywhere choose software without paying any attention to what version other companies are using. German engineers designed the aircraft's fuselage using Dassault's computer-aided design software CATIA V4, while engineers in Toulouse, France, designed the assembly process using a new version of CATIA V5. It was not until the final assembly stage at the French assembly plant that engineers from both countries realized that the software they were using was incompatible with each other, and when the engineers insisted on using their software and methods to solve the problem, things got worse. There are signs that earlier in the process of the design of the A380, as a result of the independent and loose coupling cooperation main body, some instructions, review, communication between countries increased management costs by 25%, many devices need to be moved from a factory to another factory and spend much time, it also needs to pay the high freight cost.

2. Collaborative R&D on the cloud platform

On the one hand, the accelerated development of new technologies needs to rely more and more on the joint R&D strength of the industry, universities, and research institutes and requires the joint participation of multiple institutions in the technology chain and enterprises in the industrial chain, which also puts forward higher requirements for collaborative R&D, especially for remote collaborative R&D. On the other hand, more and more enterprises choose alliance between giants to reduce costs and consolidate and strengthen their competitive advantages in the industry. For example, GAC and Nextel, BMW and Great Wall Motor, Volkswagen and Didi, South Railway and North Railway are merged into CRRC (China Railway Rolling Stock Corporation), etc. Enterprise merger is not only the merging of financial statements but also the integration of internal resources, including R&D resources. Since the merged enterprise still works in different places, the need for collaborative R&D in different places gradually appears.

For the enterprise alliance or the participating members of mass innovation, a collaborative R&D cloud platform is established to provide customized R&D tools and application environment on demand and support each collaboration subject to carry out efficient collaboration based on the collaborative cloud environment.

According to the specific requirements of the remote collaborative application scenarios of member enterprises in collaborative R&D, the application of cloud technology can deploy the R&D cloud data center based on collaborative R&D application. In the cloud data center, IT system infrastructure is deployed in the cloud-based on cloud technology, realizing the virtualization, resource sharing, and elastic expansion of computing and storage networks. Cloud data centers can effectively solve the problems of designing data interaction, conflict resolution, and real-time response between R&D teams in different places. At the same time, the deployment structure of the private cloud can also effectively ensure the security of stored data based on making full use of the existing computing resources of enterprises.

3.3.2.3 Continuous Optimization of R&D

Through the IoT and other technologies, operation and maintenance links are closely combined with the R&D links to promote the continuous and rapid innovation of R&D. The new product development team relies on resources, information, and support from critical internal suppliers provided by other parts of the enterprise. Its main models include:

1. Concurrent Engineering and Integrated Product Development (IPD)

Concurrent engineering, including DFM (Design for Manufacturing), the DFA (Design for Assembly), DFE (Design for Environment), DFSC (Design for Supply Chain), are collectively referred to as DFX, prompting decision-makers to take into account manufacturability, assimilability, suppliers network and logistics costs, process design, scrap recycling and downstream links such as green environmental protection.

IPD is a significant practical result of concurrent engineering. It optimizes the R&D process as follows:

(1) Inter-departmental and inter-system collaboration. Use cross-functional product development teams to effectively communicate, coordinate and make decisions to get products to market as soon as possible. In industries with rapid technological progress, cross-departmental teams are generally adopted by the majority of enterprises. Experts in various fields form joint forces, and various functional departments step back in the background to provide resources and support for these cross-departmental teams.

(2) Parallel development mode. The idea of parallelism is applied to the development activities, and many subsequent activities that need separate serial activities are carried out in advance through reasonable interface design to solve the problem of project division and cooperation proposed before.

The product package is a core concept in R&D process optimization. The product package is the sum of tangible products, intangible services, and influences, including performance, function, price, packaging, quality standards, pre-sales service, after-sales service, warranty, corporate image, corporate brand recognition, etc. The product is only a critical component, and it is a physical entity that can be sold independently, as shown in Fig. 3.8. It is necessary to carry out the design, development, and verification of a complete product package in product design. Traditionally, the design validation of the service system is usually placed at the end of the development phase, while in the IPD model, these factors are placed in the product package.

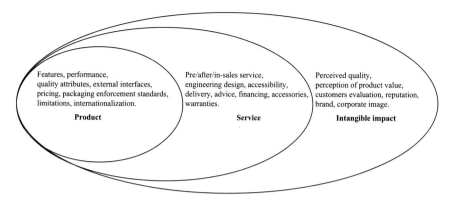

Features, performance, quality attributes, external interfaces, pricing, packaging enforcement standards, limitations, internationalization.

Product

Pre/after/in-sales service, engineering design, accessibility, delivery, advice, financing, accessories, warranties.

Service

Perceived quality, perception of product value, customers evaluation, reputation, brand, corporate image.

Intangible impact

Fig. 3.8 Product package concept diagram

They are designed and validated together to shorten the product development cycle and improve customer satisfaction.

By extending products to the concept of product packages, IPD is, in a broad sense, an end-to-end product management system and can even extend to corporate management systems.

(3) Structured process. Referring to the product development process of IPD mode, the optimized product development process is divided into six stages: concept stage, planning stage, development stage, verification stage, release stage, and operation and maintenance stage. During the whole R&D process, the decision review points should be stipulated, including the development work to be done in each stage, the corresponding sequence, and relevant necessary documents. This solves the structured process problem.

2. Data-driven R&D optimization

Through the IoT, extensive data analysis, and other technologies, the production and product operation links are connected, and the product R&D is continuously optimized through real-time monitoring of production quality and operation effect. R&D personnel can also use considerable data analysis to predict product failure problems and solve product defects through improved design.

[Case: Guiding the design through data analysis]
A lighting manufacturing enterprise used the industrial design network collaborative service platform for the lighting industry cluster in Guzhen Town, Zhongshan, to collect data. It found that the total output value of household lighting accounted for about 1/3 of the total market output value, and the annual growth rate was 20%. The decision of the family to buy the lighting mainly rests with the young wife, who accounts for about 40% of the survey population. The requirements for lighting quality are fashion 34%, innovation 21%, simplicity 16%, interest 15% and color 14% respectively. Other factors account for a small proportion. Brighter white was

chosen by 60% of consumers, while 25% of the younger generation opted for cool colors such as black. Attach importance to the style and neglect price; focus on the living room, and overlook kitchen and bathroom. Therefore, the designer decided: focus on the living room and neglect kitchen style and just keep their lighting styles consistent and put forward the purchase of the living room lighting with kitchen lighting as a gift; The target customers are young women, then designers design a warm and light color, fashionable and novel modelling, high-grade lighting. Finally, the enterprise improved the ability of independent developers to avoid the dilemma of a price war to survive.

3.4 Production and Manufacturing

When the product design prototype is complete, the next step to consider is manufacturing. Production philosophical depth rises from information management to each element and the centralized control of the process, finally achieve procurement, production planning, and scheduling, production operation, warehousing logistics, and completed the whole process of closed-loop adaptive feedback, data-driven continuous optimization, implementation is based on intelligent algorithm and constraint conditions of real-time scheduling optimization with global manufacturing.

3.4.1 Development Trend

The manufacturing link is the key to resource utilization and cost consumption. How to manifest excellence in the manufacturing link, realize continuous operation and intelligent production of factories, and become the critical direction of the future transformation of production-oriented enterprises. The future trends are as follows:

1. Combination of Additive and Subtractive and 3D printing: "Additive" refers to additive manufacturing, and "Subtractive" refers to traditional cutting technology. Over the past decade, additive manufacturing (AM), known as 3D printing, has been an innovative technology to build three-dimensional objects by adding layer upon layer of a given material. It puts a significant change in production process theory where the product has to be designed, developed, manufactured, and distributed. This new technology is based on designing the product prototype using CAD software and printing the model in 3D using a high-quality plastic material. The potential application of AM in the modern automotive industry results in new prototypes and products characterized by the specification of designs, which are cleaner, lighter, and safer with shorter lead times and lower costs. Challenges that can face this technology is sophisticated customization where advanced 3D printing would be a feasible way for

realization. That needs much development in the printing technology to make them fast and accurate. The real advantage of AM for the automotive sector is its ability to break existing performance trade-offs in two fundamental ways: First, AM reduces the capital required to achieve economies of scale. Second, it increases flexibility and reduces the prototype manufacturing time.

LSP (Laser Shock Peening) can be applied to enhance the fatigue life of 3DP parts to improve their fatigue life, thus pushing 3DP technology to the stage of mature application.

2. Combination of black and white and composite materials: "black" refers to the hard metals represented by iron or iron-based alloys, "white" refers to non-metallic materials such as white ceramics, and polymer materials such as soft materials. The combination of black and white refers to the composite material formed by various engineering materials. The 787 is the world's first composite aircraft with a structural weight of 50%, which means that more than 90% of the parts and structural parts are made of composite materials. It uses only 15 L of petrol for 100 km for five people. The development history of the aero-engine is the development history of composite materials. The core component of an aero-engine is the "blade", whose manufacture accounts for more than 30% of the workload of the entire engine. Not only are the turbine blades subjected to temperatures over 1700 °C, but they are also subjected to pressures equal to those at the base of the Three Gorges Dam. Scientists found that with the increase of the content of aluminium, titanium, tungsten, and molybdenum in the super-alloy, the material performance continued to improve, but the thermal processing performance declined. After adding high-temperature resistant cobalt, the life and reliability of the material can be improved by order of magnitude. The addition of rhenium nickel alloy single crystal blades made of rhenium with a melting point of 3180 °C enables the engine to resist deformation and cracking for a long time under the conditions of fierce heating and cooling as well as strong mechanical impact and vibration.

3. Digital/modular process planning: Production units will continue to optimize the industrial structure in the future, and the supply chain collaboration system will constantly change. From the perspective of assembly, digital and modular methods will be used in process planning and design to provide a basis for flexible production and supply chain system adjustment. Component manufacturers will use digital means in process planning to link R&D with the main contractor. In terms of service, mixed reality (MR) technology is adopted to realize information synchronization and digital twins between digital virtual space and real physical space, and between production logistics site and background management and monitoring, so as to facilitate users' perception and interaction. Digitalization of products, process design, tooling design, manufacturing process, logistics planning, processing and assembly, inspection and packaging, and other links are unified in the intelligent platform.

4. Remote monitoring and continuous operation: After the construction of the intelligent workshop is completed, vertical integration of equipment, workshop, and the enterprise shall be realized, and based on this, capabilities of remote

monitoring of plant operation and equipment health management shall be built, so as to realize the continuous operation of the factory and reduce the impact of production efficiency caused by equipment shutdown.

5. Intelligent production: will be combined closely with the production process and plant operation management, namely based on IT/OT interconnected implement integration of production and operational controls, constantly introducing new intelligent production equipment, and the corresponding network infrastructure construction and logistics facilities, support production planning, process design, production execution, equipment operation, material distribution process, such as integrated management.

6. Intelligent enterprise: It is an organism based on digital perception and uses advanced intelligent technology to enable the enterprise to realize hearing, vision, openness, independent learning, and independent evolution in a multi-directional and complex ecological environment.

[Case: the "Smart River" of Guodian Dadu River]

Guodian Dadu River is a large-scale basin hydropower development company integrating hydropower development, construction, and operation management. The energy industry is facing more and more stringent environmental requirements, hydropower enterprises are facing fierce market competition in energy competition, the era of pooling resources and pooling scale has ended, and the contradiction of relative excess power capacity is prominent. Based on automatic contains anticipation, autonomous decision-making, autonomous evolution of the concept of automatic management, in order to "accurate prediction, intelligent control, scientific decision-making" as the core of the multidimensional variable scale forecast control integration platform has been built in Dadu river set control center put into operation, realizes the Dadu river downstream of the cascade hydropower stations group of remote control, unified scheduling and intelligent human–machine collaborative operation.

Dadu river wisdom scheduling construction hydrologic meteorological forecast as the core, build high precision hydrologic meteorological forecast system, to strengthen all kinds of weather, water, extensive market data, and analysis, change from past experience with the qualitative forecast, the forecast is given priority to the traditional forecast model, solve the forecast area general, large time span, and the problems of low prediction precision. Large amounts of data are needed to improve the accuracy of water-situation weather prediction. Supported by the data of the National Weather Service of the United States, the European National Climate Center, and other world authorities, and combined with the self-built 105 telemetry stations and the automatic water-situation measurement and reporting system covering the whole basin, Dadu River has realized the function of the fixed-point, time-fixed and quantitative forecast. In 2015, the company took the lead in signing the basin information sharing agreement with five major development and operation entities of the Dadu River trunk stream. Through the independent development of the basin information sharing platform, the online collection and data sharing of the whole basin water situation data can be realized, and the real-time balance and optimization of the dam's water storage and power generation can be achieved according to the predicted data. By mining and applying big data information, the accuracy of water situation prediction of 9 controlling sections of the Dadu River trunk stream is guaranteed to reach more than 90%. The data shows that Pubugou and its downstream hydropower stations have added 3.77 billion kilowatt-hours of electricity generation and 910 million yuan of electricity sales revenue. We reduced thermal coal consumption by 1.301 million tons, soot emission by 884,000 tons, and greenhouse gases by 3.388 million tons. Four billion cubic meters of flood storage, the maximum peak clipping rate of 40%; Labor productivity increased substantially, saving 58 million yuan in costs.

3.4.2 Business Model

3.4.2.1 Digital Factory Planning and Simulation

Digital factory planning means that the producer considers how to build a digital factory to produce the products defined in the first stage, including process planning and design, factory planning and design, and introduces digital, modular, and other advanced technologies and methods in the planning and design process, to improve the planning efficiency and quality. The steps to establish the digital factory include: (1) digitizing the attributes of each piece of equipment used in the factory according to the IEC standard attribute base; (2) Establishing the association relationship between each device, which is divided into a constituent relationship and functional relationship. For example, describe the composition of PLC and the matching relationship of process parameters such as current and voltage; (3) Add the geographical location information of the equipment to the digital factory database to define the IP level and area, such as whether it is an explosion protection zone; (4) Establish the way of information exchange between tools and database in the whole product life cycle. The information in the digital factory database will be used and exchanged by various tools in the whole product life cycle.

By combining the digital twins of the product and the production line, the new production process can be simulated, tested, and optimized before the actual launch, with the following business support mainly realized:

1. Parallel collaboration between R&D and manufacturing: Based on maturity, process personnel can carry out the derivative design of digital process, design of special-purpose large tools, material quota, process division, and other work in advance. For example, before the final design of a part is finalized, purchase a mold based on the determined maximum size of the part.
2. Process planning and design: pay attention to the integrated management of digital planning, design, process, and resources, and pay attention to the standardized management of process and the establishment and application of

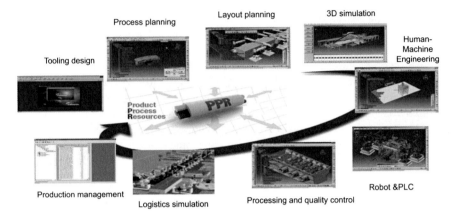

Fig. 3.9 Integrated manufacturing planning system functions

process knowledge base. For the final assembly unit, the idea of modularization is introduced, and the process staff carries out the modular process design so as to realize the seamless connection between modular production and design.

3. Factory design and simulation: The factory planners and designers adopt digital means to comprehensively plan, design, and simulate intelligent equipment, logistics distribution channels, network infrastructure, and staff operation in the factory.

4. Modular and flexible tooling design: oriented to intelligent production line, modular and flexible tooling will be realized in the future. That is, the tooling supporting environment can be rapidly established based on modularization for the mixed assembly lines.

Take the automobile industry as an example. The functions of the integrated manufacturing planning system are shown in Fig. 3.9. The basic elements of the Product manufacturing Process are abstracted into Product, Process, and Manufacturing Resource, namely the PPR model proposed by DELMIA Systems. The actual process is the result of the coupling effect of the three elements. The functions of the complete body-in-white welding design platform include product design, process design, fixture design, and robot programming, as shown in Fig. 3.10.

As shown in Fig. 3.11, man–machine engineering software can be used to simulate and evaluate the situations of front-line production employees in various manufacturing or maintenance environments in a 3D virtual scene, so as to achieve the following goals: verify man–machine tasks; Analyze the accessibility and operability of assembly workers; Evaluate and predict people's work efficiency; Attitude/line of sight analysis; Optimize workshop layout. The following values are obtained: whether the early feedback station design is correct; In 3D environment, the manual assembly task is confirmed completely. Used to generate assembly instructions or for operator training.

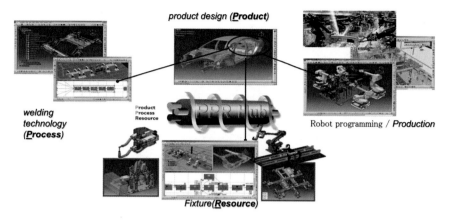

Fig. 3.10 Complete body-in-white assembly and welding design platform

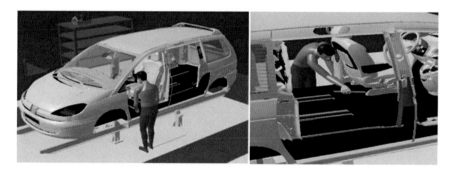

Fig. 3.11 Virtual manufacturing: Manual assembly station simulation (software used: DELMIA product of DSO System)

In the process of digital factory planning, strive to achieve the following transformation and upgrading:

1. The intelligent production line will develop from linear static to modular dynamic production line;
2. From functionally limited MES to covering the whole process of the value chain through integration with ERP, MES, and APS systems;
3. Change from single type of work to better man–machine collaboration and man–machine integration;
4. From failing to meet the requirements of personalized customization to meet the requirements of customization.

Taking the Application of Dassault Digital factory simulation solution DELMIA (Digital Enterprise Lean Manufacturing Interactive Application) in automobile enterprises as an example, as shown in Table 3.4, a large number of automobile OEMs realize the following strategic goals through Digital factory planning: accelerating

Table 3.4 Value of digital manufacturing planning system to the transformation of traditional automobile OEMS

No.	Item	Value	Main problems solved	Data case
1	Plant design and optimization	(1) Improve line design and scheduling, eliminate bottlenecks, increase manufacturing capacity by 15–20%; (2) Reduce investment by 20%; (3) Reduce inventory and production time by 20–60%	(1) Virtual factory modeling (2) 3D factory layout (3) Simulation and analysis of production logistics system (4) Simulation and analysis of production scheduling system (5) Production line scheduling (6) production layout analysis and Optimization: analyze various layouts according to distance, frequency and cost	Volkswagen, Ford, DC, GM, BMW, Toyota, etc. "Volvo has gradually optimized the plant layout and saved millions of unnecessary capital expenditure," Volvo said
2	Welding	(1) Shorten the time of welding project by 25%; (2) increase the output by 10%; (3) reduce the project investment by 5%; (4) reduce the engineering change order (ECO) by 20%; (5) shorten the production time by 10%; (6) reduce the machine equipment and cost by 10%	(1) Welding production line planning (2) welding production line design (3) detailed design of welding station (4) body welding and testing (5) report and Archive	Audi, VW, BMW, GM, Ford, Chrysler, Mazda, PSA etc
3	Painting	Management, simulation and analysis of coating quality and color distribution in the whole coating process	(1) Coating quality analysis; (2) coating color distribution analysis; (3) Painting robot programming	Audi, VW, BMW, GM, Ford, Chrysler, Mazda

(continued)

Table 3.4 (continued)

No.	Item	Value	Main problems solved	Data case
4	Final assembly	(1) assembly time reduced by 20%; (2) ECO reduced by 20%; (3) production time reduced by 15%; (4) the labor and prototype cost reduced by 5%; (5) Production increased by 10%; (6) the labor productivity increased by 10%	(1) General assembly line planning (2) assembly line design (3) detailed design of assembly line station (4) Performance analysis of general assembly line (5) report and Archive	Audi, VW, BMW, GM, Ford, Chrysler, Mazda, PSA etc

the NPI (New Product Introduction); Shorten the production line preparation period, accelerate the mass production and market; Enhance production flexibility and improve total production efficiency; Improve production quality; Improve production line capacity; Shorten production line shutdown time; Reduce the cost of change; Implement changes promptly; Reduce engineering and operational costs; Unified definition and exchange of manufacturing processes; Facilitate enterprises to implement industry norms; Introduce new products and effectively deliver them to manufacturers and suppliers to improve the capital utilization rate of enterprises, as shown in Table 3.4.

3.4.2.2 Continuous Optimization of Production Execution and Data Driving

When the planning of the digital factory is completed, IM will enter a stage of substantial operation: production execution and optimization. Integrating traditional industrial automation technology and IT has formed a typical vertical architecture, 5-tier enterprise vertical architecture. In a five-tier architecture, requests for data are either event-driven or cyclic in response to requests from the device or software system at the next level, and the next level always acts as a server or responder. For example, an HMI system may request its status to the PLC system or issue a new production formula or BOM. The data acquisition device converts the electrical signal of the sensor into digital form, then assigns a timestamp by PLC, and then sends the information to MES to further provide related services.

Under the new IM mode, production execution and continuous optimization focus on the following business support:

1. Process/production coordination. Process and production execution link to realize highly digitalization and integration synergy, namely, on the one hand, technology provides material to the production quota, digital process design, data changes, etc., on the other hand, the production provides feedback of execution status, changes of execution state to process.
2. Integrated management and control of production operations: The traditional MES focuses on managing production execution process tracking and time quota. In the future, enterprises, workshops, equipment, and other vertical enterprise resource level, production business level, and quality, cost management level will be integrated based on CPS (Cyber-physical Systems). Different roles such as equipment manager, material manager, workshop director, factory director, general manager of enterprise production, the person in charge of quality department and user side can realize real-time visual monitoring on different dimensions of production operation.
3. Intelligent production guidance based on IoT and AR: From the perspective of improving workers' workshop execution efficiency, future workshop workers will obtain more comprehensive information, express information in a more intuitive way, and process information more quickly. For example, workers

can get the pushed production execution information through the handheld terminal in real-time, including product design data, 3D process rules, tooling and resource equipment information, etc. In addition, through wearable devices or handheld terminals, workers can scan the position or part identification code in real-time, display the parts to be assembled and simulate the assembly process based on AR technology, and also see the real-time status of the parts processed by specific smart devices.

4. Equipment health management: From the perspective of maximizing the operating efficiency of equipment, future equipment managers can realize the health management of equipment based on the IoT, big data, and AR technology, which mainly includes: real-time monitoring and health assessment of equipment state based on the IoT; Predicting future failures based on big data; Based on the AR guidance equipment maintenance implementation.

5. Process/production line optimization based on quality big data: the quality data and real-time status data collected in the equipment production process, parts processing and assembly process can be used for auxiliary analysis, and the process and production line design department can optimize and improve accordingly; Process optimization uses knowledge management and optimization technology to achieve the optimal matching of process routes and parameters to achieve the production objectives of high output, low power consumption, and high efficiency.

[Case: Xugong's small orders and Intelligent turntable production line]

The customer service center of Xugong handles crane orders from more than 300 sales points around the world every day. According to the previous production efficiency of the workshop, they have abandoned many small orders with complicated configurations but a small quantity, because if the order does not reach a certain scale, the production will lose money, but the abandoned order will lead to the loss of customers. The turntable is the core of the crane to carry gravity, weighing up to 6 tons. Xugong's 14,000-square-metre workshop produces only 20 turntables a day and is often half-filled with backlogged workpieces. The turntable has 18 production processes, and each process is divided into 5 or 6 working steps. After the completion of one process, the turntable must be transported to the following process for further processing as soon as possible. It usually takes 20 to 30 min for the turntable to be transported to the next station. Xugong has revolutionized this inefficient, traditional manufacturing process through the development of intelligent production lines. The intelligent railway logistics vehicle carrying the flexible workpiece tray will dock with the welding workstation automatically. This process used to be done manually, and the docking accuracy is difficult to control. Now, the intelligent operation is realized entirely by machines. Two hydraulic bearings will lift the 6-ton turntable workpiece up to the level of the welding workpiece before it enters the welding station by the logistics vehicle carrying the work tray. The ability to maintain an absolute level of work determines the quality of the weld. The engineers got the technical idea from the strings. Next to the hydraulic bearings, they attached steel wires that could be stretched with the hydraulic bearings. The encoders connected with the wires fed back the tension data in real-time 360,000 times per second, constantly sending adjustment instructions to the hydraulic system to ensure that the lifting was always horizontal. The turntable rises steadily under the precise control of the hydraulic system, and the real-time feedback of high-frequency data ensures the level. After the turntable lift was in place, the logistics car began to move. Because the inertia of motion is considerable, it is difficult to stop accurately within 1 mm. Engineers are also inspired by the principle that piano pedals can adjust the volume of sound by using three petal-shaped sensors to control the speed of the logistics car. When the logistics car touches the first sensor, the logistics car starts to slow down. Brake when the second block is triggered; The third trigger stops completely. Finally, the logistics truck and the 6 tons of heavy industrial parts loaded reach the welding station precisely, and the height is also accurate. Through this intelligent turntable production line, a workshop produces 40 units per day and can undertake small orders of various complex configurations.

3.4.2.3 Global Manufacturing Synergy

The future global manufacturing collaboration will rely on the Internet and other technologies to eliminate regional barriers and further improve the management standardization level of manufacturing subcontracting and collaborative efficiency of manufacturing subcontracting. From transforming the manufacturing business model, realize the maximum and efficient utilization of manufacturing resources based on the cloud platform through IoT, cloud computing, and other technologies.

1. Global manufacturing subcontracting coordination: The total manufacturer shall control the complete process of manufacturing subcontracting requirements, planning, execution, delivery, and status control; The subcontractor will have easy access to subcontracting requirements, deliver results according to plans and specifications, and interact with changes in real-time.
2. Cloud manufacturing: Through the cloud manufacturing platform, the traditional business model of equipment resources bound to production capacity is changed, the resource integration advantages of manufacturer alliance are given full play, resource redundancy, and idleness are reduced, and the on-demand supply of resources is realized.

[Case: Global co-manufacturing of Airbus A380]
Airbus has a procurement network covering 30 countries worldwide, with more than 1500 suppliers joining in, among which 40% are American suppliers, and the rest are European, Asian, African, and Australian suppliers. Airbus buys everything from raw materials to aircraft structures and landing gear. Airbus has established a procurement portal, Sup@irWorld, to ensure timely communication of procurement information throughout the supply chain. In Europe, Airbus is headquartered in Toulouse, France, and the A380 is assembled in the main assembly plant in Toulouse, France.

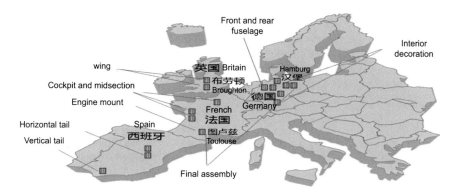

Fig. 3.12 Distributed specialized collaborative manufacturing of Airbus A380 passenger aircraft

The front and rear sections of the fuselage and the aircraft interior were designed and manufactured in Germany and shipped from Hamburg, Germany; Wings are designed, manufactured, assembled at Broughton in the UK, and shipped to France; The cockpit, middle fuselage, and fuselage and wing connection were completed in France; Both tailfins were designed and manufactured in Spain. These countries worked closely together to produce the world's largest airliner, the A380, as shown in Fig. 3.12.

3.5 Logistics

3.5.1 Development Trend

The future trend is to develop along the path of "digitalization, networking, and intelligence":

More and more manufacturing enterprises pay attention to the automation of logistics and production automation. Automatic 3D warehouse, automatic guided vehicle (AGV), and intelligent hanging system have been widely used. In the logistics center of manufacturing enterprises and logistics enterprises, the application of intelligent sorting systems, stacking robots, and automatic roller table systems are increasingly popular. WMS (Warehouse Management System) and TMS (Transport Management System) are also widely concerned by manufacturing or logistics enterprises.

In the era of IM, the demand for personalized customization puts forward many brand-new requirements for intelligent logistics systems. In the car industry, for example, when the number of components explodes, configurations add up to 10 to the 32nd power, meaning that two identical cars will not be rolled off a production line for a month or more. In order to support this mode of production, it is required that the intelligent logistics system in the IM system should change to the direction of digitalization, networking, flexibility, and intelligence of the whole process.

The improvement of the philosophical depth of logistics starts from the information management of order, schedule, and information tracking, to transportation integration through a variety of strategies, and then to the optimization of the transportation path based on the knowledge model, and finally to realize the lean management and intelligent visual logistics. The future trend is to develop along the path of "digitalization, networking, and intelligence":

1. Digitalization of the whole process

In the framework of IM in the future, intelligent logistics systems can intelligently connect and integrate all internal and external logistics processes and realize full transparency and discrete real-time control of the logistics network. The foundation of realizing this goal lies in digitization. Only when the whole process is digitized can the logistics system have intelligent functions.

2. Networking

All kinds of devices in the intelligent logistics system are no longer running in isolation. They are intelligently connected together through the IoT or Internet technology, forming a comprehensive network structure, which enables rapid information exchange and autonomous decision-making. Such a network structure not only ensures the high efficiency and transparency of the whole system but also maximizes the role of each device.

3. Intelligence and self-organization

Self-organization is an essential embodiment of intelligence. Connected intelligently through logistics facilities, they have decentralized, autonomous decision-making capacity. They can be not only the executor of tasks but also the initiator of tasks. Unlike the previous forms of production logistics, intelligent production logistics system can not only obtain information from intelligent equipment but also intelligently control production logistics activities, such as transportation, handling, loading and unloading, packaging, etc. Moreover, the intellectualization of logistics route planning in the production system is the foundation to ensure the effective connection of the IM process. At the same time, to ensure the smooth flow of the logistics process, the logistics process of all kinds of interference intelligent discrimination and tradeoff decision. Finally, the intelligent production system mode is intuitively realized through the intelligent management platform. The key issues in the system mode, such as intelligent path planning and interference management, are integrated to realize the intelligence and efficiency of production logistics.

3.5.2 Business Model

3.5.2.1 Lean Logistics

The three typical Lean Logistics modes are a sequential introduction, sequential establishment, and Set Parts Supply (SPS).

The sequential introduction refers to parts sequence being determined by suppliers outside the assembly plant and introduced to the location and immediate use of parts, according to the product assembly instructions. Satellite suppliers will order the required parts and thus maximize the elimination of line side shelf inventory and handling waste and improve the efficiency of logistics.

In this mode, those parts of large volume, high price, heavyweight, packing inconvenience, abundant type or color will have their inventory costs and logistics costs significantly reduced, but it requires the supplier's distance be shorter, the supplier's production transportation period is less than the assembly cycle, at the same time also requires accurate production order, supplier capability and higher quality of parts, because the parts are delivered and directly used by the assembly, and there will be no quality check.

Sequential establishment refers to parts sequence being determined in the assembly factory. After the supplier sends the parts to the assembly plant, because the parts are sorted by the supplier and are not consistent with the order of production requirements, they need to be converted to the order of production in the sorting field to keep consistent with the order of production line requirements, and sent to the line side in the specified time. The model is suitable for parts that require further processing prior to online assembly and conform to the downstream characteristics after processing.

SPS refers to a form of logistics in which components are delivered to a plant for inspection, stored in a temporary area called a Progress Lane, and the small and medium-sized components required for a single product are combined in special containers to supply the production line according to production schedule instructions (ANDON, Kanban). In other words, it is to package the parts of the same car at different stations to form a whole set of supply, so as to prevent leakage and wrong loading, reduce the demand for line area, and flexibly respond to product changes.

3.5.2.2 Smart Logistics

The intelligent logistics model organically applies IT, Internet technology, and IoT technology in production and logistics, and the model turns to personalized customization for the object. The new model reduces the dependence on manual operation; Production logistics operation through machine self-perception, learning reasoning, intelligent decision-making, self-solving problems in production logistics; Devices are interconnected with the MES system as the central core. Hardware and software collaborate, information is synchronized in real-time, and resources are shared.

Intelligent logistics should have the following typical characteristics:

1. Decentralization: the future logistics equipment will not be controlled and scheduled by the central control system but will be autonomously and decentralized;

2. Autonomy: The intelligent device has the ability to decide the work route by itself. For example, the Air-carry robot of Midea Group, Ande Zhilian, analyzes the navigation by laser scanning imaging.

3. High flexibility: flexibility of logistics refers to the flexible organization and implementation of logistics operations in order to realize the trend of "multi-variety, small-batch, multi-batch and short-cycle" that logistics operations adapt to consumer demands. On the basis of automation, the corresponding logistics system is required to have higher flexibility. Under the new production mode of customization, it is the logistics system of manufacturing enterprises that are impacted first. In order to meet the requirements of customization and rapid response, the material distribution mode needs to be more flexible and auto-mated and has the ability to make quick responses according to orders. The technical path is to use the IoT technology to achieve full network coverage, obtain accurate data automatically in real-time, and adapt to changes through data-driven processes. The rigid production system connected by a conveyor line is gradually transformed into a flexible IM system supported by robots. A flexible logistics system includes not only the flexible requirements of flow but also the flexible requirements of hardware and layout and takes into account the possibility of layout adjustment and system adjustment according to produc-tion demands in the future. Truck, pallet, turnover box, for example, become the basic intelligence unit. Each transport vehicle is independent. According to the location and status of one or more orders, they can match in the form of a many-to-many turnover box chooses suitable tray and interaction with other production equipment and transporter, bypass obstacles on the road.

[Case: pizzeria's delivery service]

Just enter the customer's phone number, and the pizzeria's computer displays all the important information about the customer—name, address, home location, nearby landmarks, and even the customer's history of pizza purchases. If the other branch is closer to the customer, the order is automatically forwarded to that branch. With a click of the mouse, the computer will not only display the customer's information but also indicate the customer's location on a local map, showing the street the driver needs to find; Automatically calculate the optimal path, display all delivery trucks traveling toward the store on the path, and automatically assign the delivery tasks according to the plan with the fastest delivery and the lowest fuel cost; If there is a traffic jam at a certain point on the path, the delivery plan and path will change automatically. If the time delay exceeds the threshold, the customer will be notified. Meanwhile, customer orders are displayed on computer screens that can be read by chefs in the kitchen, and all delivery vans are equipped with mobile phones so that if something goes wrong, the driver can always talk to the store or customer to fix the problem, find the quickest way to avoid reordering a pizza.

4. Intellectualization: For example, The Optimus Prime and Bumblebee system not only realize the intelligent operation of planar storage but also improve the efficiency of 3D space use through the design of grasping devices; In the future,

the 3D packing algorithm based on the heuristic algorithm and mathematical programming model will optimize the placement of goods, improve the space utilization of 3D shelves, containers, transport vehicles and so on, and reduce logistics costs. For example, according to the experience of the newly discovered knowledge, employees can input heuristic rules "for the rest of the items sorted by volume, bulky items into the first", "one face of the same item in the first stacked merged into a big item add" on the man–machine interface, these programs are added to the knowledge base of expert system, the heuristic algorithm will automatically apply all the rules, improve loading efficiency.

[Case: Bosch's self-organized production]
In the Bosch logistics center, each artifact is recorded in electronic tags (RFID). At every production link, the reader will automatically read the relevant information, feedback to the control center. Most of the production activities can achieve self-organization. For example, where a workpiece is located, what machine tools are processing, processing time, how much material inventory, whether to fill materials, etc. This information is directly integrated with the production management software seamlessly so that all the data in the production process can be efficient, real-time flow and visual presentation on the network, which can easily and effectively solve the problems encountered in the production process. When the new system came into operation, the factory's inventory was reduced by 30%, and production efficiency increased by 10%, resulting in savings of tens of millions of Euros. At Bosch's Beijing plant, all machine tools are connected. A computer manages all of the NC machine tools, assembly stored in a central server, the status of the machine tool, open, shutdown, operation, processing products, how many pieces of processing, the fault information, the utilization rate of machine tools, etc., all information are displayed automatically and accurately, then it realizes the production process of transparency and the organization.

3.6 Sales

Intelligent sales management takes customer demand as the core and uses big data, cloud computing and other technologies to analyze and predict sales data and behaviors, so as to drive the optimization and adjustment of production planning, storage, procurement, supplier management and other businesses.

3.6.1 Development Trend

In the background of IM, sales management and R&D, manufacturing, service and other links will be closely integrated, and reflect the characteristics of intelligence.

Different from traditional Business Intelligence (BI), the trend of intelligent sales management is as follows:

1. Visualization: Sales management will realize completely transparent monitoring at the vertical enterprise resource level, end-to-end value chain level and horizontal supply chain level.
2. Insight: Based on visual realization, the monitored sales status will be analyzed in the future based on big data analysis and other capabilities, so as to establish the correlation between data, find out the root causes and correlation characteristics of sales problems, and assist leaders in making decisions.
3. Optimization: Based on vision and insight, the company will be supported to optimize the sales management business system in terms of organizational structure and management process in the future, such as adding a new big data management department and supporting businesses to build the core knowledge assets of the company.

Specific business models include customer relationship management, collaborative sales, full-channel operation and whole-process O2O, and sales intelligence based on big data.

3.6.2 Business Model

3.6.2.1 Customer Relationship Management

Customer Relationship Management (CRM) is defined as an enterprise that uses relevant information technology and Internet technology to coordinate the interaction between the enterprise and customers, and to provide customers with innovative and personalized customer interaction and services. Its ultimate goal is to attract new customers, retain old customers, enhance customer stickiness and promote sales.

Driven by digital technology, the RFM model is widely used. R (Recency) refers to the time of the last consumption, and the more recent the customer is, the better the customer should be. The data shows that if consumers can get them to buy, they will keep buying; F (Frequency) refers to the number of purchases made by customers within a limited period, and the most frequent customers usually have the highest loyalty. M (Monetary) refers to the amount of money spent, and according to Pareto's law, 80% of a company's revenue comes from 20% of its customers.

The RFM model is an important tool and means to measure customer value and customer profitability. Combined with these three dimensions, if each dimension is divided into 5 levels, then customers can be divided into $5 \times 5 \times 5 = 125$ categories for data analysis, and then targeted marketing strategies can be developed. An example of marketing analysis of a baby products company is shown in Fig. 3.13.

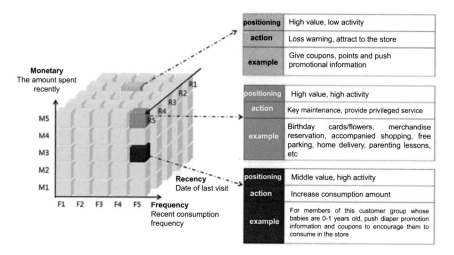

Fig. 3.13 Example of RFM model and marketing strategy

3.6.2.2 Cooperative Sales

The first change of sales management brought by the Internet is to promote collaboration among sales team members and between the sales department and the production and service department. Specifically, first, CRM, OA and e-commerce platforms are connected to provide integrated sales management functions to meet the needs of mobile scenarios. Second, use social tools, real-time communication, strengthen sales team collaboration; Third, customer-centric, customer information and product information integration, real-time connection between products and technical experts, improve the success rate of sales. Foreign Siemens global knowledge management system, SALESFORCE, domestic Fxiaoke and Ali DingDing are all representative suppliers of cooperative sales system.

The basic functions such as field statistics, sharing reports, and "PK assistant" function intuitively show the business competition between sales teams. Those functions of attendance clock, field management, document survey, instant call, relying on Internet telephony, can satisfy the daily demand, and save a lot of telephone expenses for the user.

[Case: 7–11's online order]
Any one of 7–11's more than 8,000 locations in Japan is an order point. There, customers can order using a touch screen, and the items can be delivered to any 7–11 store. Multiple orders can be consolidated and sent to the same store. 7–11 solves the problem of not having a credit card or not being willing to provide a credit card number online by accepting cash or credit card payments for items ordered on the touch screen. A similar example occurs in the sharing economy. After renting a car, the user drives to another city and can return it locally instead of driving back to the starting city. This model is very popular for the convenience of the user. People can

find the bike nearest to them through GPS. They can unlock, rent a car and pay by scanning the QR code on their mobile phone. At the same time, the bicycle can be returned quickly by scanning the code in the designated area or even any location. This convenient advantage of returning the bicycle has contributed to a new sharing sales mode.

3.6.2.3 Full-Channel Operation and Whole-Process O2O

The second change of the Internet on sales management is to open up the full-channel sales system, and finally promote channel coordination in the form of O2O (Online to Offline) to promote the change of sales process, including pricing, order processing, delivery, return, etc., and improve sales efficiency. Many companies allow customers to browse their products in real time. A furniture store, for example, uses 48 "WEB cameras" to display its inventory online and receive customer feedback, resulting in a significant increase in inventory turnovers.

[Case: Suning's full-channel sales system]
The combination of online, offline and logistics creates a new retail impact on pure e-commerce and pure offline. Jd.com, Suning and other e-commerce giants have put forward new concepts related to retail, and Hema Fresh, unmanned supermarket and other new formats have been launched. In order to realize online + offline transactions based on consumer experience, Suning launched a full-channel smart retail scheme. Consumers can shop online through Suning Store APP, and deliver goods within 3 km within half an hour. In the offline world, QR codes are set up beside the shelves, which are automatically followed by unmanned delivery vehicles. In addition, they can also go to the exclusive shopping cart settlement channel by themselves, so as to realize the shopping scene of self-service checkouts and mobile payment. In the community, users mainly focus on "three meals a day", featuring fresh, fruits, vegetables, cooked food and other categories; In the core business districts and crowded areas, differentiated products and service combinations such as fruit, daily distribution, hot drinks, medical (near hospitals), and sports (near sports centers) will be provided according to user group portraits to realize "thousands of shops have thousands of styles". Suning's full-channel sales scale increased by 44.55% in the first half of the fiscal year of 2018 through the construction of stores based on Internet and data and the strengthening of management quality control.

3.6.2.4 Intelligent Sales Based on Big Data

Enterprise customers, including end users, distributors and partners, are the most important enterprise resources. Satisfying the needs of customers through big data analysis is a reflection of the transformation from product-centered to customer-centered, and a sign of the transition from networking to intellectualization of sales business model.

[Case: Toyota: Extracting profits from data]
Toyota's sales intelligence based on big data covers the intelligence to customers. Toyota Motor Sales corporation (TMS) has 25,000 employees and more than 1350 franchised dealers in the United States. TMS uses data mining tools to mine dealers' data fusion, and analyzes the relationship between car owners' age, income, occupation and other attributes and the purchased car models. Therefore, targeted marketing can be conducted. Once it canceled a plan to build a new dealership in a city because it is found that increasing the inventory of an existing dealership in that city will increase sales; And so on. Where it used to take more than 200 h for an analyst to produce a cost report, which could only be done once a year, it can now be done almost in real time, alerting customers to delays, showing changes in costs due to incorrect delivery routes, and so on. According to the Automotive News Data Center, Toyota increased U.S. car production by 40%, but added only 3% to its workforce.

Through big data analysis, the retail industry can predict the sales volume, determine the inventory level and distribution plan; Potential hot routes can be identified in the airline industry; In the broadcasting industry, it is possible to predict the best programs to be broadcast in prime time and how for commercials to get the maximum return; In marketing you can analyze consumer demographics and predict which consumers will respond to ads. In banking, bad debt levels, credit card fraud, new customer spending and customer reaction to products can be predicted.

[Case: British Telecom's Consumer Behavior]
British Telecom (BT) is a large telecommunications company. Its 1.5 M users make 90 M calls a day. The company offers 4,500 products and services. It wanted to find the best way to reach individual consumers, and the solution was to build a warehouse of customer data. The company uses MPP (Massively Parallel Processing) neural computing technologies. The data warehouse has 16 GB of memory. Through this system, companies can identify individual products, product systems and consumer buying characteristics. One application is to determine which consumers are likely to be taken away by competitors and to predict trends in products with large sales. The market sector used to analyze data from 6 to 12 months ago, but now the information is near real-time.

3.7 Services

Service is a process of statistical analysis of product operation and maintenance, feedback to relevant departments by customer satisfaction survey and usage tracking, in order to maintain customer relations, improve products, so as to achieve vertical customer requirements mining, and then horizontal customer expansion. The improvement of service intelligence is the transformation of service mode from traditional informatization to feedback optimization based on service system, to service quality improvement based on cloud platform and customer service knowledge base or product fault knowledge base, and finally to intelligent and personalized service realization by using customer service robot or big data intelligent analysis.

3.7.1 Development Trend

The transformation from "R&D as the center" or "production as the center" to "service as the center" has become a development trend of manufacturing enterprises. Through service transformation, on the one hand, by providing customers with more value-added services, to create a new competitive advantage; On the other hand, through the innovation of service business model, to bring new profit growth point. Based on the IoT technology, after the products are sold, manufacturers can still perceive customers through the networking ability of the products, collect product usage data, provide prediction and warning services, and further excavate user service demands.

For example, Sany Heavy Industry Group conducted data analysis on the operation status of 200,000 units of construction machinery it sold, took the initiative to provide maintenance services for customers, and reduced its spare parts inventory. The formation of macroeconomic indicators called "Excavator Index" provides a reference for our government decision-making; Using its digital networking ability to realize remote visual control of excavators, it has completed the partial cleaning work inside the ruins of the Fukushima nuclear power plant in Japan, and won international praise. Another example is Haier's U + platform, which provides services for intelligent products sold out and gathers a large number of third-party resources to further meet customer needs and form a service ecosystem.

The main trends of customer service in the context of IM are as follows:

1. High agility of services. High agility of services: Customer services can achieve high agility with the help of new technologies and new business models such as internet-based operations and augmented reality-based service support. Agile service is to take the customer as the center, products for intelligence operations, introducing maintenance engineering and other advanced methods, using digital, virtualization, IoT, big data, such as technology, building planning and operating environment of the agile service, planning, operation and optimization of interconnected closed-loop service system, implement products reliable and efficient operation, the profit model.

2. Innovation of service business model: intelligent interconnectivity of products makes it possible to service business model innovation, enterprises can bring a new profit growth point by providing value-added services, such as establishing breakdown maintenance, preventive maintenance, predictive maintenance such as multi-stage maintenance mechanism, provide one-station service,

timing services, remote software upgrade services, products, operating hosting service, multi-level service mode; Enterprises can also break the traditional sales strategy of simply selling products and switch to the product-as-service model, adopting new business models such as customer operation data analysis service, performance assurance plan, global service collaboration, and product service system.

3. Service systematization: The transformation of service mode requires the essential support provided by the perfect customer service business system and application system. Therefore, enterprises need to gradually establish a complete digital customer service system of planning, delivery, execution, optimization, and global coordination through digital means in the future.

3.7.2 Business Model

3.7.2.1 Service Collaboration and Product Service System

1. Cooperation between development and maintenance: A digital maintenance planning analysis and design system need to be built to provide accurate input of operation steps, maintenance resources, and technical capability requirements for maintenance execution. In the future, the maintainability analysis will be closely integrated with the product and process development process. On the one hand, through correct input, the maintainability of the product will be guaranteed during the development stage. On the other hand, the research results can be effectively managed and easily traced.

2. Synergy between service and cultural construction: Culture represents products' unique taste, and culture and service constitute the extension layer of products. Products and services should have practical value and express a kind of cultural connotation. When purchasing products and services, consumers should not only obtain their functions and utility but also take their intrinsic cultural-added value. Cultural-added value has become a key factor influencing consumers.

[Case: Jiujiang Winery's R&D and cultural construction]
Jiujiang Winery is located in a wine town in Foshan with a strong Lingnan wine culture. It has a history of more than 200 years. At present, it has more than 20 invention patents, invested more than 30 million yuan in R&D every year, and participated in the formulation of the National Standard of Fermented soya bean Liquor.

Jiujiang "double steam" wine with Xi river water, rice, and soybeans as raw materials, after distillation, storage, also need to be put into fat pork for a month of immersion time. It is Jiujiang double steam's most core process characteristic. Fat pork contains many trace elements, such as fatty acids, which are helpful for soybean fermentation. At the same time, the fat pork can absorb impurities in the wine, so that the wine tastes smoother, with a unique smell of smooth. Jiujiang plans to take advantage of Guangdong's rich fruit, combine wine and fruit, launch more flavors of rice and fruit wine, to provide more choices for young people. In 2015, Jiujiang established the only rice wine research institute in China. Traditional wine-making mainly relies on uncontrollable natural conditions. However, with the continuous improvement of wine-making technology in recent years, the environment for wine-making has also reached a controllable level, including humidity, temperature, probiotics cultivation, and control of trace elements.

In addition to the R&D process, Jiujiang also attaches great importance to the enterprise operation of digital management. At the beginning of e-commerce, Jiujiang established its own O2O platform and realized product tracking and management through the management mode of "one bottle, one code". In addition, through the two-dimensional code, we can also enter the enterprise WeChat public platform, to achieve the interaction between enterprises and consumers. Some valuable customers have been settled down and formed a database. Enterprises will make use of big data technology to deeply explore the value of consumers.

In the aspect of cultural construction, Jiujiang established the first wine museum in China in 2009. Inside the museum, consumers can make their own wine. In recent years, the company has acquired some surrounding plants and land and plans to build a characteristic winery along the Xijiang River that integrates brewing technology, tourism, and wine collection, to expand the content further and business model of the existing museum. In addition, Jiujiang also cooperates with Hong Kong TVB to shoot the TV series "Jiujiang 12 workshop". Its content is to tell about the predecessor of Jiujiang Winery, 12 winemaking workshop twists, and the history of development. Jiujiang winery advocates healthy wine first, then healthy drinking. Abandoning the bad custom of unrestrained drinking, learning from the experience of western countries, studying the collocation of wine and food, the atmosphere of drinking, the temperature of drinking and the equipment of drinking, etc., has formed a scientific and quantitative "sommelier culture".

3. Integration of products and services: Enterprises shift from selling products to delivering measurable business results and provide a product service system formed by the integration of products and services. For example, seed supplier Monsanto has acquired a satellite company that can increase yields by 10 to 20% by scanning farmland and gathering valuable information about soil composition, temperature, and humidity to make the right decisions for farmers. Komatsu, a construction machinery firm, uses satellites to connect mining equipment for remote monitoring. Komatsu uses predictive tools to help customers with equipment maintenance, ensuring that equipment is readily available for mine use, resulting in increased profitability and customer loyalty. Rolls-Royce installs sensors in aircraft engines that collect engine data for predictive maintenance and improved product design. The use of big data solutions can help engineers diagnose engines and suggest repairs, as well as manage parts. The essential purpose of GE's industrial Internet transformation is to transform its business model from manufacturing to service. The most significant users of GE industrial Internet Predix platform, Haier, provide customers products and take

advantage of the existing R&D, logistics, manufacturing, and other resources to provide clients with customized product solutions and services.

Moving to this new business model requires a new pricing model that accurately evaluates the value created, a platform that manages new feedback cycle processes and data flows, gathers sensor data through cloud architecture, develops predictive and optimization algorithms, and innovates across the process.

[Case: remote monitoring of shield machine of China Communications Construction Group (CCCG)]
Digital management and real-time analysis of construction machinery can predict engineering accidents and reduce the loss of life and property. Shield machine sensors include soil pressure, displacement, temperature, flow rate, and other types of sensors. Years ago, poor tool monitoring led to a significant collapse during tunneling in Shanghai. In February 2018, due to inadequate monitoring data and the neglect of early warning, the sudden flooding at the subway construction site of Jihuaxi Road in Foshan city resulted in deformation and damage of tunnel segments, causing the collapse of more than 30 m of ground, and personnel fail to evacuate in time, resulting in a severe accident. Lu GuangMing, the head of CCCG, said that they had improved the entire lifecycle management system and related management system through the real-time data acquisition and data analysis of the shield machine cutter and soil pressure on the high frequency, the analysis results are shown on the remote real-time video monitoring system with a large screen. In the field of underground space of each project, set the shield machine sensor and HD video network, data induction ability reach millimeter level, through the Internet card, equipment operation data, and material flow and other data can be directly transferred to the group headquarters every 3–5 s, which used to be like installing clairvoyance. They can see each construction project in headquarters, all engineering personnel can be evacuated before various accidents such as tunnel collapse, they can change the method of route and process timely, prevent the occurrence of collapse in advance. Also, with the automation of the equipment, the workload that used to take 2–3 years for a bid section can now be completed in six months (Fig. 3.14).

3.7.2.2 Global Service Collaboration and Service Optimization

The Internet has changed the traditional service of regional distribution, and high bandwidth network, high-definition video, the application of virtual reality will implement the remote or the opposite effect, such as remote medical treatment, can initially concentrate in the central city, scarce medical resources cover more remote rural areas, reduce the area emergency patients mortality; Moreover, the Internet has made it easier for low-income people to access and use it. For example, in the financial field, the former banking financial services cannot be offered to the public because of its high service cost and limited capacity, while Internet finance enables enterprises to reduce IT costs and better control credit risks, so they can provide financial services for a wide range of low-income people.

(a) The front end of the shield tunneling machine being assembled and debugged

(b) the cutter plate of the shield tunneling machine being dismantled and maintained in the construction section of Foshan

(c) Control room in tunneling engineering

(d) Real-time data display and analysis of knife plate sensor

Fig. 3.14 Data acquisition and visualization of shield tunneling machine

Global service collaboration refers to the organization and collaboration system, MRO (Maintenance, Repair, and Operations) of global service built by manufacturers using the Internet; Maintenance, repair, and operation of supplier network, etc., provide network service information support. Its business contents include:

1. Service monitoring and quick response: after delivery of products in the manufacturer service management department will build a one-to-many service monitoring environment, based on the IoT technology for smart products and services in the process of real-time monitoring, and the problems in the process of monitoring to respond in a timely manner, provide one-stop service, through the collected data to optimize the maintenance plan and other projects, form a closed-loop service cycle of the security system.

2. Augmented reality and virtual reality supporting: provide customers with less paper service information support, customer and field service personnel to use a handheld terminal or wearable devices, real-time access to products' operating status, maintenance technical information, service planning process closely integrated with the process of practical maintenance, real-time access to products based on augmented reality operation information; Develop computer-based Training (CBT) facilities and digital Training materials to support Training personnel to conduct virtual Training based on augmented reality technology.

3. Spare parts prediction and predictive maintenance: provide customers with spare parts prediction and suggestions, help customers predict spare parts purchase and production requirements in advance through service monitoring and actual spare parts consumption, and manufacturing enterprises can arrange spare parts purchase and production plans in advance; After the fault is found by the operation monitoring platform, the technicians can repair the problem remotely or guide the on-site service personnel to carry out the maintenance work through the network. By analyzing the data of product operation and combining it with the fault knowledge base, we can predict the possible faults in advance and take actions in advance to prevent the occurrence of faults.

4. Service optimization based on the knowledge base and maintenance big data: with the help of cloud platform, mobile client, knowledge model and intelligent customer service robot, and other technologies, the customer knowledge is mined in a multi-dimensional way to establish a fault knowledge base and provide basic support for fault diagnosis and product health management; With the data collected in the service process and the data in the maintenance process, big data analysis can be carried out to optimize the design, process, service planning, and other engineering design links, so as to provide support for the next generation of product innovation and provide customers with intelligent services and personalized services.

3.8 End-to-End Integration Covering the Entire Product Life Cycle

End-to-end integration refers to the realization of the automatic flow of data between enterprises and the integration and cooperation among enterprises centering on products through the engineering activities of product R&D, production, service and other product life cycle of different companies cooperating for customer needs.

End-to-end (or point-to-point) integration of operating systems covering the full product life cycle will break down the barriers between existing management systems and integrate management into one. The original management systems (such as PLM, SCM, CRM, etc.) cannot be connected to each other due to different information formats and other reasons. For example, SCM information is often unable to be directly transmitted to ERP systems, but needs to be input manually, which makes resource management of ERP systems lose timeliness. Based on service oriented architecture concepts such as the original can't connect link management system, the application of different functional units and associated services, through well-defined interfaces and contracts, in a unified and general way to interact, eliminate obstruction, repetition, conflict in the management system and arbitrary, business management will be more smoothly and quickly.

The three core layers in the overall architecture of IM are: the enterprise operation and collaboration layer, factory optimization and execution layer, and factory connection and automation layer. Each layer supports six different intelligent indicators respectively, covering the full life cycle from R&D to delivery, as shown in Fig. 3.15.

The original purpose of end-to-end integration proposed by Germany's "Industry 4.0" is to enable SMEs, as "invisible champions", to overcome the disadvantages of enterprise size and participate in global competition by building a close industrial chain and ecological circle. End-to-end integration may be a breakthrough for Chinese manufacturers. It is believed that on the basis of China's mature industrial system and driven by the "Internet plus" strategy, Chinese enterprises will lead the world in end-to-end integration and lead smart manufacturing to achieve breakthroughs.

Haier's Internet factory enables users to have direct conversations with the production line through the Internet. Users' personalized needs can be reported to the production line for the first time, and users can remotely view the whole process of the production of their purchased products. In various links such as customization selection, order flow, production process, logistics and transportation, automatic and orderly flow of data is realized. Users will no longer be passive waiters, but participants and supervisors in the whole process, so as to meet the best user experience and realize zero-distance interaction between the manufacturing end and the consumer end. There are also many Internet companies in China, such as xiaomi, 360 and so on, which take the whole product life cycle as the main line and realize social cooperation.

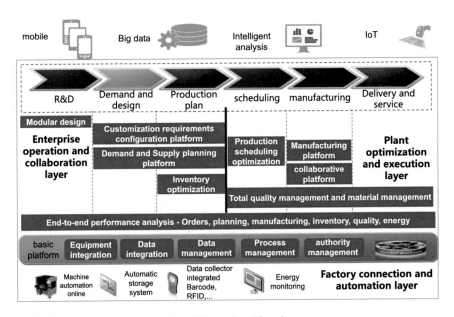

Fig. 3.15 Business logic process view of the product lifecycle

Future factory will change the current "the sparrow is small, but all-sided", to make full use of social resources development, including shared orders, personnel, materials, equipment, blur the boundary between the plants, break the closed and independent production mode, build into the open service platform, through the information system to suppliers, outsourcing cooperation factory, customers and internal each department closely linked into smooth "ditch", the data flowed smoothly. The factory not only produces products, but also produces data, which will become the core competitiveness of the enterprise, as shown in Fig. 3.16.

[Case: Product life cycle management of Shenyang Machine Tool Co., LTD.]
Shenyang Machine Tool Co., Ltd. (SM) has been a pioneer in the domestic machine tool industry since its birth. However, due to the lack of core technology for a long time, the products produced by SM can only target the middle and low-end market, which is not only less profitable, but also vulnerable to the impact of the economic environment. Guan Xiyou, the company's chairman, found that the reason for the low control accuracy of machine tool products was not hardware, but mainly software. If the lack of advanced control software and scientific parameters, no matter how good the machine parts will be assembled after the same noise as the tractor; Once the machine is disassembled and reassembled, the parameters should be adjusted if the natural frequencies of the various joint surfaces are different. SM's I5 system was successfully developed in 2014. The so-called I5 refers to Industry, Information, Internet, Integrate and Intelligent. It also refers to industrialization, information, networking, integration and intelligence. The I5 system operates on a touch-screen and can do most of its functions with just two button presses. The I5 system loads

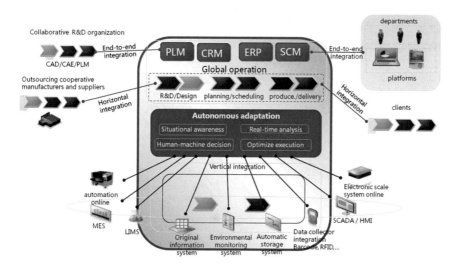

Fig. 3.16   Cross-system integration and end-to-end integration

an "expert system" that is embedded in a CNC machine tool to make programming intelligent, that is, to convert a 3D model directly into a 3D surface processing program.

SM allows users to use CNC machine tools at low cost through direct financing, financial leasing, U2U startup payment and other modes. Such services are connected by the financial module of the i5 system. During use, the system tracks the whole process of each machine tool part and component used by users, and records its batch, supplier and use situation; Based on this data, an accurate spare parts plan can be provided to ensure timely delivery, minimize the downtime, and make the machine keep running healthily; Based on the accumulated production data and experimental data of the i5 system, the service life of core components was modeled. The i5 system's intelligent full life cycle management system creates a manufacturing profile for each machine, enabling remote diagnosis, maintenance, and repair of problems by experts, even if they are not local, simply by connecting to the cloud. That way, by the time a component's expected life expires or something goes wrong, it may have already departed for its destination. Ideally, by the time the user's part is expected to be damaged the next day, the spare has arrived at the site the day before.

70% of the i5 system machine tool orders are leased by enterprise customers. The average charge for an hour of machine tool startup is only 10 yuan, which greatly reduces the one-time investment of enterprises. The sales growth of this product exceeded 300% in 2015 and 2016. Based on the big data of iSESOL (I—Smart Engineering & Services On Line), the trading parties can even generate various forms of resource pricing, such as sharing according to time, according to artifacts, according to value, etc., which will promote the occurrence of new transactions and make the ecology become more active.

The cloud ecosystem is the future of manufacturing. SM added iCAM, a cloud-based computer-aided manufacturing system, to the I5 system, and placed part of the machine tools in the cloud for users to call at will, so as to achieve mass customization. Anyone can log on to the platform via the Internet to design and manufacture personalized products directly online. Users only need to use software modeling, while processing programs can be generated automatically. In order to interact with users more effectively, shenyang machine tool introduced a user interface based on WeChat applet. For example, users can customize simple souvenirs with their zodiac and name online using WeChat applet. The influx of individual users will generate enormous traffic and demand.

Based on the generated big data, the machine tool production capacity can be visualized. The workshop Wisdom Information Management System (WIS) mounted on the I5 system can connect the machine tools of all factories, complete the production scheduling through mobile phones in different places, check the production status, and realize the online operation analysis. ICAM on an i5 system can intelligently match requirements to the nearest available idle resources. Processing tasks can be assigned to the most reasonable machine tools via the Internet, consumption is calculated through sensors, revenue is generated based on consumption in the form of startup fees, and the final product is generated by mail. When iCAM can no longer satisfy the requirement of configuration of supply and demand, the company developed the iSESOL data cloud platform as a cloud platform, focusing on the needs of the user information and the supply of information productivity, through the matching algorithm and online broker, the shenyang machine tool company as a platform operator to reap benefits.

When the machine tools reach the end of life, SM, based on data tracking, the first time to intervene in the recycling, so that the waste into treasure, the launch of second-hand machine tools, so that users sell the maximum salvage value. To that end, Shenyang Machine Tool has set up a "remanufacturing division" that takes old machine tools from the front end of the market and rejuvenates them with I5 technology. The company launched its industrial operating system, i5OS, in 2017. By the end of 2017, the total operation volume of i5 intelligent machine tools had exceeded 22,000. In the first quarter of 2018, the company realized operating revenue of 1.639 billion yuan, 85.42% year-on-year growth.

Q&A: What parts of the product life cycle does the company's new machine tool system cover? What benefits has it produced?

3.9 Practical Cases: Product Service System—Based on the Cases of Four Enterprises

The four companies in this case are all headquartered in Guangzhou, with mainly domestic customers and a small number of overseas customers. They are business-related companies, which focus on the production of Galanz microwave ovens and other household appliances. Kingfa provides modified plastics. Borche provides injection molding machines and Echom provides industrial design and manufacturing services. The enterprise information of these companies in 2015 is shown in Table 3.5.

1. Profile of Galanz Group

Galanz group is the world's number one manufacturer of microwave ovens, with research and manufacturing centers in Foshan and Zhongshan, Guangdong province, and 13 subsidiaries in its headquarters in China. Galanz positions itself as the "global manufacturing center of microwave ovens".

Galanz provides customised product looks or functions according to different geographical features such as temperature, humidity or special customer requirements. The company provides "professional, user-friendly and high quality" services while increasing product diversity. The general manager of the company put forward the guiding ideology of "service is the engine of the company". The company has invested heavily in building a service support network and standard workflow to help customers choose products and carry out product repair and maintenance. Safe operation and product-related knowledge learning are also realized in this category. In case of product failure, service personnel need to arrive at the site within two days to solve the problem and provide one-stop service.

Table 3.5 The case of the enterprises

	Galanz	Kingfa	Borche	Echom
The ownership of	Private	Private	Private	Private
Number of employees	40,000	2000	600	3000
Sales (RMB billion)	18	7.5	0.35	1.5
Date of Establishment (year)	30	16	10	17
industry	Home appliance	Chemical	Machinery manufacturing	Design
Product diversification	High+++	High	Middle	Middle
Degree of modularity	High	Low	High	High
product	White home appliance (microwave oven and air conditioner)	Modified plastics (5 product series, more than 2,000 specifications)	Injection molding machines.6 product types, about 60 specifications	Design, manufacture and planning of machinery, structure and mould
process	Discrete assembly	Continuous flow	Discrete assembly	discrete
Supply chain location	Assembly plant	Material supplier	Equipment supplier	Component supplier

Customers provide Galanz with market information by providing feedback and filling in questionnaires to support operational decisions. Through the establishment of data warehouse, Galanz has collected and accumulated a large amount of customer information and product use information. Advanced data mining tools are very useful in the feasibility analysis and scheme design improvement of products (such as electric steam furnace) and the analysis and research of market trends.

The general manager of the company said, "There is a lot of communication between the marketing, purchasing and R&D staff in the company, and they form cross-departmental teams for product development." R&D and production engineers are also involved in the provision of services by providing technical training to service personnel. Suppliers provide Galanz with new materials, new technologies and their application information, but are not directly involved in product development.

2. Introduction to Kingfa Science & Technology Company

Kingfa is a high-tech listed company specializing in the research, development, production and sales of high-performance modified plastics. It is the largest modified plastic manufacturer in China.

Kingfa's pre-sales staff select the appropriate plastic for the customer. Technicians help customers to improve the manufacturing process and solve the problems encountered in the actual production. After-sales service personnel also arrive at the customer's workplace to provide guidance on the injection molding process, as well as suggestions for product design improvements. There are clear rules for return visit control management. If customers have special needs, Kingfa provides customized products by changing the formula and ensures that customized products can be produced by standard machines and processes. To meet the needs of more than 60 application areas, the company's 5 production lines produce more than 2000 different formulations of plastic products.

Kingfa has a good relationship with collaborative suppliers such as Echom Co., LTD., whose high-end products need to cooperate with Kingfa because of its excellent plastic properties. In collaboration, Kingfa uses fluid dynamics analysis software to share information in real time with collaborative suppliers or customers such as Echom, Borche, etc., for collaborative design, such as discussing mold specifications. The company improves the efficiency of supply chain management according to customers' purchasing and channel management practices. Furthermore, clients provide Kingfa with market and demand information, which greatly reduces the distance between Kingfa and the final market.

Channels are available to facilitate communication within Kingfa. For example, the R&D department works with the marketing staff to provide services. The general manager stressed the importance of supply chain synergy, noting that "we have developed several domestic standards in collaboration with our upstream and downstream partners". Regular supplier meetings are held every three months. The working group established by Kingfa consists of suppliers, cooperative suppliers, internal employees and external experts. The working group is responsible for developing new products and solving customers' problems.

3.   Introduction of Borche Intelligent Company

Borche Intelligent Equipment Co., LTD. (Borche for short) ranks in the top three in China's plastic machinery industry in terms of its comprehensive strength and brand index. Borche has established 40 4S stores in China. The clamping force of injection molding machine ranges from 10 to 4500 tons, and the theoretical injection volume is 25–150,000 g, which can meet various personalized needs of different customers and provide "tailored" and "one-stop service". Services include helping customers select suitable injection molding machines, designing workshops and auxiliary facilities, monitoring production processes, initializing machines and correcting errors, and providing 24-h technical support. For example, before sales, it will provide drawings for the reconstruction of the customer plant, water, electricity and gas, etc., and the cross-functional team will follow up for two days, which is not only a product, mold, process, but also a system project. Because customers are not familiar with the selection, purchase, installation and construction of injection molding machines, they need a comprehensive service provider to help them assess the risks and progress. The machines, auxiliary facilities, material purchasing plan and workshop design plan provided by the company constitute the "Diamond Service" integrated solution to cover the customer's complete value creation process. In Borche, the probability of returning and replacing machines due to selection error is less than 1%, while the probability of returning and replacing machines in some peer competitor companies is close to 40%.

Design engineers and marketing staff visit customers every month, with a clear return visit plan, at least 6 times a year. Front-line service personnel and technical center staff spend one-third of their working time on field visits. Every month, the company will also organize open days for suppliers and customers, invite customers to listen to customers' requirements, hold talks and trainings with experts, and transfer customers' feedback requirements to suppliers. Compared with its main competitor, Ningbo Haitian Machinery Co., LTD., product customization service is the company's advantage. If a customer specifies parameters, such as a new requirement inspired by a new injection molding machine at an international exposition, Borche redesigns the standard injection molding machine to meet the customer's needs. But due to a lack of complete product package requirements definition and description framework, it tends to equate customer complaints and opinions about the products directly with customer demand, has formed six product platforms, more than 120 models of large product array and the output of 7000 units a year, including non-standard accounted for about 30%, which leads to Borche products price being 10% higher than domestic competing goods.

Customers provide Borche with new product concepts based on what they see in their marketing and overseas exhibitions. The customer asked Borche to disassemble and reverse analyze foreign competitive products in order to expand Borche's technical knowledge. Moreover, customers are directly involved in the design and production process of Borche. For example, one customer helped Borche solve a key problem in its product design by applying its own patents. After the product is launched, customers can also provide feedback and suggestions for improvement.

Weekly cross-functional and cross-project meetings to support front-line services and evaluate product development projects. Borche works with supply chain partners to solve customer problems. When selecting a supplier, Borche evaluates whether it can be one of the supplier's top two customers. If not, it will consider switching suppliers, because only when it becomes an important customer will it be valued and be able to handle emergency orders. Through formal meetings with suppliers, some suppliers would provide suggestions and information on new technology improvements. Suppliers also participate in the internal operation of Borche, including product design, manufacturing and mutual service to customers. But if the supplier enters into the design stage, it will face the difficult problem of product concept confidentiality.

4.   Introduction of Echom Technology Company

Guangzhou Echom Sci.&Tech. Co., Ltd. (Echom for short) is a famous industrial design group in China, covering TV, automobile, IT and other industries, serving nearly 300 customers around the world. Echom takes industrial Design as the core and DMS (Design, Manufacturing and Service) as the concept, which breaks through the complete industrial chain including industrial design, structure design, mold design, injection molding manufacturing, spraying, sheet metal, complete machine and other links, forming a service mode combining design and manufacturing.

Through a team composed of different professionals, the company is stationed in the customer's factory to select appropriate products for customers and provide after-sales services (such as technical support), so as to promote the customer's products to the market as soon as possible. Moreover, Echom integrates mechanical design, structural design, mold design, manufacturing and supply chain planning services. For example, it provides customers not only with the body of mold, but also with the plastic parts inside and outside the car. The general manager told the research team that "we will participate in the customer's product design and product series planning according to the customer's brand awareness, product function and market strategy".

Customers are the main source of innovation and improvement ideas for Echom. Customer cooperation with the company's market research, evaluation feedback to the company to bring market information and product reliability information. Echom classifies customers according to their abilities. Category 1 customers are directly involved in the product development of Echom. Through such synergy, the company identifies new market trends through customer needs and preferences, and the knowledge acquired can be applied to other customers.

"The walls between departments are very thin," commented the company's manufacturing manager. Product development is done by cross-functional teams. Echom helps customers collaborate with collaborative suppliers and manage the entire supply chain purposefully. Suppliers are also involved in the internal operations of Echom Company. Companies build win–win relationships with their suppliers, which improves transparency throughout the supply chain.

Thinking Exercise

1. What changes have these manufacturers made? What trends reflect the development of enterprises and intelligence?
2. In the process of product service system innovation, what kind of customer collaboration method does the enterprise adopt?
3. What benefits can customer collaboration bring to the enterprise?

Homework of this Chapter

1. What stages can product design and development be divided into? What is the significance of the digital 3D model for IM?
2. Why can digital manufacturing planning bring benefits to enterprises?
3. Why couldn't Xugong handle small orders before? What measures have been adopted to enable it to accept small orders?
4. What are the drivers of global manufacturing synergy? What's the difficulty?
5. What is the difference between SPS and Sequential introduction and Sequential establishment?
6. What systems and measures does the stores mentioned in this chapter use to improve customer satisfaction and achieve green logistics?

7. How is the intelligent production of Bosch realized? Where is it reflected?
8. What do stores collaboration and the sharing economy have in common?
9. In O2O, what is the focus of online and offline tasks? How do they work together?
10. Analyze the importance of data fusion.What do you think is the more valuable data that Toyota hasn't used yet?
11. Illustrate the application of data mining technology in various industries.
12. Why did the Jiujiang winery build a museum and promote "sommelier culture"?
13. What are the benefits of remote equipment monitoring?
14. What kinds of information systems are used for each stage of the product life cycle? What is the connection between the information systems?
15. What parts of the product life cycle does the SM company's new machine tool management system (I5 system) cover?What benefits has it produced?
16. What are the level classification and characteristics of product design in intelligent manufacturing capability maturity?
17. what are the level classification and characteristics of procurement in intelligent manufacturing capability maturity?
18. What is the position of the industrial Internet in the intelligent manufacturing system architecture?
19. What are the key difficulties in achieving equipment interconnection in intelligent manufacturing?

Chapter 4
Intelligent Features of Intelligent Manufacturing

4.1 Inspiration Case: The Supply Chain Intelligence Service Platform of Guangzhou Pharmaceutical Co., LTD.

Founded in 1951, Guangzhou Pharmaceutical Co., LTD. (GPC) is the largest sino-foreign pharmaceutical distribution enterprise in China, a leading enterprise of "Great Southern Medicine". The business of GFC is constantly developing, and the number of subsidiaries distributed in different places is constantly increasing. The business fields and coverage areas are constantly expanding.

1. Platform planning

With the rapid expansion of business, its cross-organizational collaborative and group control aspects of the demand is higher and higher, in the sales, distribution, procurement, warehousing, logistics, retail, medicine, electronic supervision of all links such as fine management is increasingly urgent demand, but many of the original information system fails to properly meet and solve the problem of new requirements. The function and main technical route of the original information system have inherent limitations under the application trend of mobile Internet, cloud computing, big data and the IoT. As a result, the original information planning and the Information System gradually are difficult to continue to efficiently support the company's business and management innovation and maintain the match between OT and IT.

Based on the competitive environment and current situation of the company, the company put forward the "bridge of health, benefit of the public" as the company mission, "integrity, stability, innovation, mutual benefits" as the core values, and to achieve the vision of "the best service provider of Chinese pharmaceutical supply chain with optimum service, the most widely network, comprehensive range, excellent management, talent excellent".

At that time, the company's existing information system cannot support the efficient operation of the marketing headquarters and subsidiaries. It is shown in: there is a lack of an effective information integration platform at the level of the whole group, and the various operating conditions and customer network of each subsidiary is not clear; B2B based e-commerce platform has not been established effectively; The existing CRM system is not perfect, customer basic data is separated from the system, lack of customer analysis and other high-end application modules; The information platform of logistics distribution network and system can not support its operation efficiently; The acquisition of subsidiaries' information construction is very difficult; Existing decision support function is weak, the production efficiency and utilization ratio of report is low; The human resource management is backward; the strength of financial management is insufficient.

In order to solve the above problems, it is necessary to realize comprehensive informatization through five aspects, namely, group-oriented structure, operation centralization, network terminals, service localization and management standardization.

The informatization development of GPC has reached the stage of deep integration, and the construction focus has shifted from system implementation to system integration and application improvement. The project team put forward the information guidelines of "strategic foresight, flexibility and stability, comprehensive promotion, value creation". The goal of informatization is to build a smart service platform for the Chinese pharmaceutical supply chain that integrates the four goals of "group management and control, full chain collaboration, service innovation, wisdom cohesion" into one, as shown in Fig. 4.1.

2.   Enterprise Application Integration (EAI) approach

If the company wants to become "the best service provider of Chinese pharmaceutical supply chain", it needs to integrate closely with the information of upstream and downstream enterprises to provide fast and efficient services, which requires the integration of enterprise layer. Before application integration, GPC had problems such as "information island", repeated storage of information and mixed system integration technology. The existing information system uses Application Programming Interface (API) to open the database to other applications and uses Web service technology for one-to-one data connection through the Shared Directory.

The ultimate goal of enterprise application integration is to provide enterprises with a complete solution that can integrate different applications together quickly and easily. Because the application system inside the enterprise presents a variety of different hierarchies, according to the characteristics of application integration at different levels, enterprise application integration can generally be divided into data layer integration, application layer integration, business layer integration and enterprise layer integration.

Because Enterprise Integration to solve the problems of the enterprise facing the industry chain extension, but also meet the needs of the company in industry chain partners, should introduce the concept of Enterprise Integration, and work together with upstream and downstream companies in the industry chain, on the basis of the established system, with the aid of the ESB (Enterprise Service Bus), build an inter enterprise integration system aimed at establishing a complete industrial chain. Due to the establishment of the inter-enterprise integration system, the companies in the

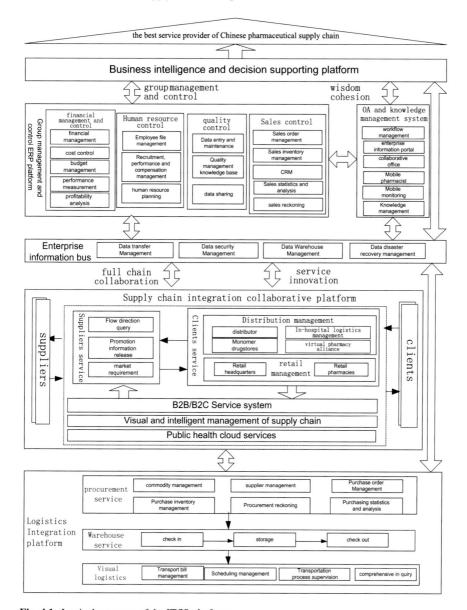

Fig. 4.1 Logical structure of the IT/IS platform

same industrial chain form a virtual company through the integrated network to form a community of interests and make concerted efforts to promote the development of the industrial chain.

An ESB is a way to implement EAI in an SOA architecture in a particular context. Currently, a number of Web services technology-focused providers have

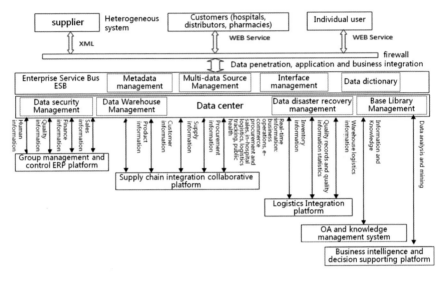

Fig. 4.2 Association of the enterprise service bus with each platform

provided enterprise application integration platforms based on Web services technologies, including WebMethods, IBM WebSphere, etc. IBM WebSphere has long been the industry leader in the ESB market, with stable and reliable characteristics; WebMethods has the characteristics of integration concept, single platform, expansion of virtual enterprise concept and flexible deployment, which is suitable for the rapid expansion of GPC. Therefore, it is recommended to use this platform.

The data exchange relationship between the planned application platform and the enterprise service bus is shown in Fig. 4.2.

Taken together, a set of reproducible efficient and rapid integration processes is formed in ERP of group management and control. The company can complete the information, management and business integration of a newly acquired enterprise within 45 days. In terms of the construction of supply chain integration collaborative platform, GPC independently developed software such as "hospital inventory Management platform" and "Medical centralized procurement platform" supporting mobile applications to realize the integration of hospital procurement and inventory services. In terms of the construction of logistics integration platform, it took the lead in improving the cold chain management system of drug circulation in China, and realized the real-time monitoring of temperature and humidity in the whole process from purchase to distribution.

3. IT organizational structure

Based on the business characteristics and IT organization history of GPC, IT is more appropriate to adopt the mixed structure of "centralization and decentralization". In other words, IT headquarters centralizes IT planning, control and standards, but disperses IT support resources to the business team to actively provide services for business personnel, so as to give play to the scale effect of IT resources on the basis of reflecting IT centralized control, so as to strengthen the strategic role of IT organization. Decentralization of IT personnel to various business teams will degrade the skills of IT personnel in the team. Therefore, the IT business department is responsible for making training and learning plans to promote the professional development of IT personnel.

4.   New forms
(1)   Supply chain integration collaborative platform for rapid collaboration

   1)   virtual pharmacy alliance: because the new GSP (Good Supply Practice) definitely requires drug must be unified procurement. Therefore, a virtual pharmacy alliance module is designed to strengthen the control of retail terminals, and organize a large number of monomer drugstores to form virtual pharmacy chains. It has the following functions: member management; Collaborative procurement; Information sharing; Professional management consulting including category planning, facade design, decoration design that licensed pharmacists are not good at. At present, there are about 3 thousand independent stores in China. These independent stores have a low management level. Adopting this mode can improve their management level and reduce operation cost.

   2)   Public health cloud services: The system provides cloud services related to the health industry to manufacturers, distributors and the whole society. It mainly has the following functions: drug trade supervision; Medical and health records of residents; Performance assessment of medical management personnel; Information sharing of patient testing in local hospitals; Remote diagnosis and treatment; Chronic disease management. Attach importance to the implementation of the new medical reform, actively participate in and assist local governments in medical reform, and build a public health system with the cooperation of local governments based on the concept of collaborative and integrated intelligent management of supply chain.

(2)   Logistics integration platform

   1)   Visual logistics management: RFID tag technology and GPS technology and other IoT technology are adopted to realize the integration with the cold chain information system and realize the following functions: information query of goods in transit; Real-time temperature monitoring and geographic location tracking; Multi-warehouse intelligent coordination and distribution; Data monitoring; Emergency deployment. The supply chain system can visually manage the entire logistics process from the manufacturer to the logistics distribution center, to the hospital pharmacy and finally to the patient, and uniformly allocate resources to ensure that the whole process of drugs is in a low-temperature and refrigerated environment, so as to separate logistics and commercial flow and improve the circulation safety of drugs.

2) In-hospital logistics management system: it is specially used for the management of hospital pharmacy. Connect the hospital information system with the ERP of GPC to support the information pharmacy trusteeship. It has the following functions: order publishing; Order sourcing organization and distribution; Wireless scan receiving; PDA shelf location recommendation; Warehouse delivery, warehouse removal and inventory taking; Information exchange. The procurement staff of the pharmacy in the hospital pharmacy can directly convert the systematic drug demand into purchase order, and issue to GPC through the system. GPC organizes the supply of goods according to the order, and uniformly distributes it to the designated pharmacy and pharmacy of the hospital. The hospital pharmacy administrator uses wireless scanning equipment to scan the receipt one by one. At the same time, the receipt confirmation and the import of electronic invoices were completed. Meanwhile, the order information, drug information and logistics distribution information interacted between the hospital and the company to ensure the completeness of information and improve the acceptance efficiency.

(3) OA and knowledge management system to enhance innovative service capacity.

1) Mobile pharmacist: With the help of wireless PDA and mobile pharmacist system, pharmacists can grasp the medication status of patients in real time and timely deal with various conditions. Through the improvement of hospital procurement and logistics management, hospital pharmacists are gradually liberated from the traditional warehouse operation and put into the real pharmaceutical service. Pharmacists can use the system at any time and anywhere to the hospital patients medication and emergency treatment. At the same time, the data processing of drug use is timely fed back to the drug store, based on analysis and processing, managers can more accurately predict drug use demand, make accurate procurement plans, make reasonable reserves, reduce inventory overstock, avoid out of stock, reduce procurement management cost, and improve the level of drug procurement management. In 2014, the "smart pharmacy" project of The Second People's Hospital of Guangdong Province, built by GFC, with the function of mobile pharmacists as the core, became the national industry benchmark. After the transformation, the pharmacy staff is reduced by 10%, the drug storage is reduced by 50%, the cost is reduced by 10%, the error rate is reduced to zero, and the efficiency is increased by 30%.

2) Mobile monitoring: Including the monitoring of salesmen and the monitoring of various customer services. Drug salesmen carry mobile monitoring locators, and with the help of GPS, the mobile paths of visiting customers can be displayed remotely and sent back to the system every day. Therefore, sales managers can visually see the work records of sales staff, monitor the frequency and path of customer visits, combine the company's overall sales activity data and historical data, give sales staff

suggestions on coverage areas and personalized sales strategies, and play a better supervisory role and improve efficiency. Doctors and nurses wear mobile monitoring locators that record the experience of administering medication to patients and prevent medical errors caused by forgetting or repeating the medication. Customer service systems are integrated with B2B systems, including general practitioner education, etc., on the site.

With the support of the new platform, GFC has achieved sales volume of 37.823 billion yuan by 2016, with a year-on-year growth of 12.81%, ranking among the top five in the industry in China. Its member enterprises have expanded to 22, becoming one of the largest pharmaceutical wholesale and distribution networks in China, providing more than 51,200 pharmaceutical products to over 19,200 customers, more than 5,700 suppliers and 98% of the people in 31 provinces, municipalities and autonomous regions.

Thinking Exercise

1. What steps can the IT planning process of GPC be divided into? What stage results are produced respectively?
2. What complex environment will the information system of Guangzhou Pharmaceutical Co., LTD face?
3. Why should such informatization guidelines be formulated?
4. Why do the company set these four goals?
5. What is the relationship between the information systems of GPC?
6. What emerging business forms can be generated based on the planned platform of GPC?

4.2 The Introduction

Among the three dimensions of IM standard architecture, the dimension of intelligent features (or intelligent functions) has the characteristic of intelligence depth obviously. The intelligent depth dimension includes five levels of intelligent requirements such as resource elements, connectivity, system integration, information fusion and emerging business forms.

4.3 Five Levels of Intelligent Features

1. Resource elements refer to the level at which an enterprise digitizes the resources or tools it needs to use in production. It includes strategy and organization, employees, design and construction drawings, product design and process documents, physical entities such as raw materials, manufacturing equipment, production workshops and factories, and energy sources such as electricity and gas.

2. Connectivity refers to the interconnection between equipment, equipment and control systems, people and machines, employees and enterprises through wired, wireless and other communication technologies.
3. System integration refers to the bottom-up integration of intelligent equipment to intelligent production units, intelligent production lines, digital workshops, intelligent factories and even IM systems through the integration of raw materials, spare parts, energy, equipment and other manufacturing resources through information technologies such as two-dimensional code, radio frequency identification and software.
4. Information fusion refers to the realization of collaborative information sharing on the premise of ensuring information security through the use of cloud computing, big data, blockchain and other new-generation information and communication technologies on the basis of connectivity and system integration.
5. Emerging industries refer to the level of value chain integration among enterprises, including personalized customization, remote operation, maintenance and service, network collaborative manufacturing and other new manufacturing modes.

Intelligent feature dimension is the comprehensive embodiment of intelligent technology, intelligent infrastructure construction and intelligent results, and the interpretation of Cyber Physical integration. It completes the whole process of perception, communication, execution and decision-making, and guides enterprises to develop mode innovation by using digital, networked and intelligent technologies.

4.4   Resource Elements

4.4.1   Strategy and Organization

In order to promote IM, enterprises need to adopt appropriate development strategy and organizational structure in the first step. In the process of implementing the IM strategy, China should fully draw on the advanced experience of developed countries, including the following aspects:

1. Leading enterprises drive small and medium-sized enterprises

The IM strategy of Germany, the United States and Japan is initiated and promoted by the industrial alliance composed of domestic leading enterprises. China's development of IM should also give full play to the leading role of leading enterprises and form a benign ecosystem where large enterprises lead the majority of small and medium-sized enterprises.

The ecology of IM in China is quietly changing. The emergence of a group of small enterprises engaged in IM technology services and providing solutions in China's IM industry is very important for the promotion of IM in China. Without such a group of small enterprises, a good ecological environment for IM cannot be formed.

On the other hand, there should be a number of large enterprises, they build their platform to serve the other small and medium-sized enterprises at the same time, such as Haier COSMOPlat platform, Foxconn's Internet industry, Huawei cloud platform and 5G—ACIA IM alliance, Midea group's "new generation of man–machine data link" strategy and industrial ecology Internet platform is not only used for its own business, but also for more small and medium-sized enterprises to provide the fundamental of IM services.

2. Central Research Institute

Another key point for large enterprises to lead innovation is the summary and promotion of Central Research Institute. Central Research Institute refers to the R&D system consisting of "Central Research Institute + Business division's R&D institutions". Central research institute is engaged in basic prospective studies and key generic technology research, and business division's R&D institutions engaged in the needs of the market demand of new product development. Both cooperate by reasonable operation mechanism, combined together to form enterprise development system with clear hierarchy, reasonable division of labor, close connection, which not only reduces the redundant construction of innovation resources, but also make innovations to be used to the greatest extent. GE Global Research focuses on technology research that is strategically important for the company's long-term development. In addition, GE sets up R&D units in infrastructure, energy, healthcare, transportation and other business divisions, which mainly carry out product R&D activities to meet the current market demand.

Draw lessons from the large manufacturing enterprise innovations which have the characteristics of focus, prospective, sustainability. The practice of the Central Research Institute of Midea group, and other enterprises has also proved that this kind of application patterns can improve the efficiency of R&D and innovation ability, then is conducive to solving a series of IM key common new technology.

3. Large enterprises of different types form IM industry alliance

Through the establishment of IM comprehensive industrial alliance, large enterprises become the main force of strategic implementation. There are several types of large enterprises and they should share different jobs:

(1) Leading manufacturer

Leading manufacturing enterprises such as Foxconn, Gree, Haier, Midea group, have very strong industrial strength and manufacturing experience and supply chain integration ability. They play the market leading and demonstration leading role on the ecological building, but there are also some bottlenecks including closed organization mechanism which is not flexible, lack of information platform technology experience, integrated talent reserve insufficiency, higher risk of transformation and development. Therefore, leading enterprises in the manufacturing industry should focus on accumulating industry knowledge and experience, and at the same time improve their capabilities in digitization, software and modularization. By deepening cooperation with other innovation subjects, they should explore platform-based solutions and services, so as to form a reusable and high-value platform innovation mode.

[case: the Internet industry of Midea group]
In March 2018, Midea group said during the annual strategy conference, after nearly three years of production standardization and digital transformation, Midea hasbasically realized the global collaborative production platform with the goal of "one Midea, one system and one standard", developed software to support the operation of the whole value chain. On this basis, Midea works with the world's leading robot enterprise, KUKA, to open up the R&D end, production equipment end, supply chain end, business end, logistics end and user end, in order to achieve "zero" inventory production, 100% logistics tracking management and "single" customized C2M. Midea Industrial Internet has been able to support the production and operation of more than 10,000 products in more than 40 bases around the world, and can provide flexible solutions for multi-layer processing and complex processing.

(2)   Internet giants

Internet giants such as Alibaba, Tencent, Baidu, etc., have a platform technology architecture, operational experience, business model innovation mechanism, ecological building capacity, big data and the advantages of artificial intelligence technology, but there are no industrial genes in them, they lack professional knowledge and industry business, lack industry professionals, these short boards are increasingly apparent in the process of promoting the Internet industry. Therefore, Internet companies should focus on developing industrial Internet platform architecture technology, giving full play to the advantages of large data, such as artificial intelligence technology, by deepening the cooperation with manufacturers, develop professional manufacturing knowledge, business and application scenario of cross-industry solutions, build perfect ecological, provide enterprises with high added value platform products and services.

(3)   ICT leading enterprises

ICT (Information Communications Technology) leaders, such as China mobile, China telecom, China unicom and equipment suppliers Huawei, their core advantage is the ability to Information and communication technologies, but they tend to have only part of the key enabling technology supporting platform development, the lack of a holistic view and platform operating experience, and their ecological building ability is weak. Therefore, ICT enterprises should focus on key enabling technologies such as equipment access, network transmission, simulation modeling and industrial technology software, so as to provide technical support and solutions for digitalization, modularization, platformization of manufacturing resources and online trading of manufacturing capabilities.

(4)   Universities and research institutes

Colleges and universities and research institutes with a position of neutrality, public welfare services, authoritative features, is to carry out the strategic, fundamental, common frontier technology research and improve the bridge and link, optimize the environment of public services but subject to its rigid system mechanism, the traditional service mode and market capacity is not strong so that it is often difficult to give full play to its advantages. Therefore, colleges and universities and research institutes should strengthen cooperation with the depth of the platform the innovation main body, and thorough going efforts to promote the process of "public entrepreneurship, peoples innovation", build a whole process service system based on the data evaluation, application of diagnosis, consulting training, admissibility rating, security assurance, promote the public service ability of multi platform sharing, cross-industry collaboration and massive subject collaboration, empower platform enterprise, guide enterprises precise application, optimize industry accurate services, to support the government to accurately adapt measures, promote open platform value of ecological construction.

4.4.2 Employees

To succeed in the industrial revolution of intelligent manufacturing, talent cultivation is the primary guarantee.

In the change of human–machine interaction paradigm, the machine should adapt to the needs of humans, rather than humans to adapt to the machine. How to recognize and adapt to the complicated process and the changeable working environment with high technical content of equipment and tools has also become a subject to be studied. IM requires cultural cooperation, and companies need employees who can master industrial technologies and understand artificial intelligence and data analysis systems. IM is the cross integration of various technologies. Its own development cannot be separated from a large number of professional and technical personnel, and the new industrial ecology it generates can absorb a large number of labor forces. At present, it is necessary to focus on the cultivation of classified talents and cultivate professionals in various fields from the aspects of scientific research, technological breakthroughs and engineering applications.

In terms of skill requirements, the implementors of IM should have a high level in integration of manufacturing and automation technology, information and communication technology, and software development and application. Each individual field for IM is very limited. professional students want to find a way to broaden the professional view, expand the disciplinary boundaries, because in the flat, self-organizing, flexible organization coordination, each employee and team need to work together with external organizations, being in dynamic links, they need to understand the work content and technical methods of others to complete the work, as shown in Fig. 4.3. For example, Huawei requires human resource managers to master deep business technologies so as to find excellent "warriors" for the enterprise. In 2017, China's Ministry of Education proposed the concept of "New Engineering", which aims to provide effective guidance for engineering and technical personnel training in the era of IM. World economic BBS in early 2016 issued a "work in the future—the

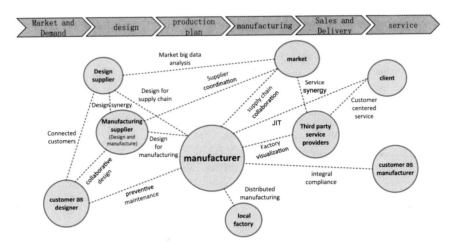

Fig. 4.3 Employees in a work organization covering the entire value network

employee skills and labor strategy for the fourth industry", gave reference opinions on the demand for the ability in the new era of IM, put forward the key personal ability including the cognitive ability, systemic skill and the capability to solve complex problems.

In terms of job changes, the IM system has significantly reduced the demand for employees in some positions. For example, JD plans to use AI technology and robots to reduce the number of employees from 160,000 to less than 80,000 in the next decade, but field operators must master new operation methods, test and maintenance methods.

According to Gartner (2018), the   interview with global 460 CEOs and senior manager shows that 54% of the companies suggest that its digital business goal is to transform, and all CEOs have said that talent professionalism is the biggest obstacle in the business process of digital implementation. The number of CEOs who made digital talent development the top three of their priorities rose from 16% to 28%.

4.4.3 Production Equipment

IM equipment (or device) is a general term for all kinds of production equipment with the functions of perception, decision-making and execution. It is the next generation of production equipment to achieve efficient, high quality, energy saving, environmental protection and safe and reliable production. Covers the entire industry from the key intelligent common basic technology (such as high-grade sensor, hydraulic parts and other key parts and components) to the measurement and control device and components (perception functions components such as camera, decision-making functions components such as numerical control system, perform functions components such as intelligent instrument and robot), and then to IM complete sets of equipment, a total of three aspects. IM equipment through the overall integration of technology to complete the perception, decision-making, perform of the integration of the work.

Under the mode of IM, the production equipment is undergoing the transformation of "part defines machine → software becomes part → software defines machine". What we're looking for in a production facility is a few pieces of code, a different version of software, and its modification can change the machine's function. Software can be used to drive production equipment, which will make the equipment more convenient, more technical content, more intelligent, because it encapsulates industrial knowledge, establishes an automatic data flow rules system. Software defines devices, products, and manufacturing, which is fundamental to IM.

In the way of industrial software embedding, from the past, the computer was placed on the side of the equipment, and the equipment was controlled through the interface. Later, it gradually developed into a special control cabinet to place the industrial computer. In the future, the computer and equipment gradually converge, fuse and integrate, and the industrial software is directly embedded into the system. The relationship between industrial software and hardware production equipment has undergone the micro, integration, dynamic and intelligent technological evolution of "separation → convergence → fusion → integration".

"Industrial 4.0" projects in Germany, constantly upgrading "Cyber physical system", make it have the "independent thinking" "smart factory", the discrete production equipment for Cyber physical system gains intelligence, and by that time, cloud computing and big data, which embody Internet thinking, is just an object to be used in the manufacturing industry, the German equipment manufacturing industry relative to the United States will highlight a competitive advantage.

At present, in addition to high-end CNC machine tools, industrial robots and 3D printing equipment, the common shortcomings of China's production equipment mainly lie in the following aspects:

1. Intelligent sensing and data acquisition equipment

Foreign brands monopolize the domestic market for high-end CNC equipment, as well as the production of real-time sensing and data acquisition technology. Domestic brands lack unified data interface standards and communication protocols. Due to the lack of application of OPC UA architecture standard, large international automation manufacturers have their own industrial bus and communication protocol, which blocks the data acquisition interface at the bottom of the digital control equipment.

2. Intelligent detection and assembly equipment

The appearance detection and structure detection of products still rely on manual work. Image recognition intelligent equipment and voice recognition detection equipment can not meet the requirements of manufacturing. For example, China's automobile detection and assembly equipment industry is still in its infancy, with different detection standards, which restricts the improvement of automobile manufacturing quality. Automobile inspection and assembly equipment will be digitized, miniaturized, networked, integrated and intelligent.

3.   Intelligent warehousing and logistics equipment

The complete applications of intelligent warehousing and logistics equipment, such as Bluetooth beacons, intelligent 3Dwarehouse manufacturing and automatic warehouse station connection, lag far behind the international advanced level. Taking Bluetooth beacons as an example, compared with traditional GPS positioning, Bluetooth beacons have advantages such as low power consumption, accurate positioning and low cost, as well as the "letter" function of sending information and the "mark" function of indicating location. The function of medium range communication of Bluetooth beacon breaks through the limitation of user types and has a broad application prospect in the manufacturing industry. Industrial enterprises can use beacon technology to improve the capability of asset monitoring and asset utilization. Warehouse operators can use it to track materials moving through the assembly line; Marketers can use location data to send offers to people passing through a store. However, there are three enterprise standards in the current Bluetooth beacon market, namely, Apple's iBeacon, Google's Eddystone and Radius Network's AltBeacon, but Chinese enterprises have not been seen.

4.4.4 Energy

The previous two industrial revolutions promoted technological progress through energy transformation to complete industrial upgrading. Therefore, the optimization of new energy and energy management is China's preparation for the new industrial revolution. Sustainability is one of the four goals of IM. Energy savings are a sign of sustainability. During the 12th Five-Year Plan period, energy consumption per unit of GDP was reduced by 18.4%. According to the comprehensive work plan on energy conservation and emission reduction released by the State Council during the 13th Five-Year Plan period, by 2020, the energy consumption per unit of GDP per 10,000 yuan will be reduced by 15% compared with 2015, among which electricity saving is the key for industrial enterprises. Digital, networked and intelligent technologies provide many ways for energy saving:

1.   To save energy by substituting digital technology for traditional experimental means

For example, a 30-min wind tunnel test of a 30-ton fighter jet would cost several million yuan and consume the same amount of power as a medium-sized Chinese city, while an hour wind tunnel test of a Boeing 787 would consume 8.8 megawatt hours, which translates into 3000 tons of coal. If digital technology is adopted, the virtual prototype can replace part of the physical wind tunnel for virtual wind tunnel test, which can significantly save energy. In the 787 program, Mike Bair, Boeing senior vice President, said, "In the 767 program, we have conducted physical wind tunnel tests on more than 50 different wing configurations. In the 787 program, only a dozen airfoils have been tested through physical wind tunnels, but they are of better quality than the 767. "This is because the 787 project has completed a large number of Computational Fluid Dynamics (CFD) analysis based on the shape of a digital aircraft, which has been fully tested and has reduced the number of wind tunnels blowing in physical objects.

For another example, the traditional development of a gun requires continuous testing, firing 2000 test rounds to finalize the gun. The Redstone Test Center uses digital technology artillery model, based on Virtual Proving Ground (VPG) technology to replace part of the artillery live-fire test, finally do verification experiment with only 20 rounds of a recoilless gun, greatly saving the energy and material losses and protect the environment.

2. Automation of electrical equipment

For example, it takes about half an hour to turn off the air-conditioning power in a building after work, floor by floor, wasting energy during that time. After installing the unified control system with automatic switching function, it will automatically shut down after the end of the work day. This simple way can save energy.

At present, in the production mode with fewer people, some machines are often forgotten to shut down or not close enough, resulting in idling of the machines and waste. Common measures are based on the correlation between the machine chain control, so that the unnecessary machine stops in time.

[Case: Automatic equipment connection of Guangdong Feng Aluminum Company]
Feng Aluminum was plagued by high power cost of electrolytic aluminum products, which accounted for 40–45% of the manufacturing cost. Engineers use digital meter for network monitoring millisecond energy consumption data and draw curve of energy consumption, after careful inquiry curve, it was found that workers often forget to close the oven door, then adopt the corresponding management measures, and according to the lean production and fool-proofing design, design the equipment having a linkage, make them close or open automatically, in order to reduce workers operating errors, eventually reduce the energy waste.

3. Network management of electrical equipment

The scheme is suitable for remote electrical monitoring of multi-workshop factories, buildings, homes, electric cars and trains. By dividing the power consumption system and installing measuring devices at the dividing points of different power consumption units, the internal power system of the enterprise is monitored, analyzed and managed. Integrated information such as current, voltage, power consumption and status information of various electric equipment can be displayed on the display screen or mobile phone to monitor and manage electric equipment in a visual way. When temperature, electric current and other data are abnormal, alarm will be automatically issued and the person in charge will be notified via SMS. It is an important way to realize sustainability to timely collect the energy consumption of equipment and production line, compare the change of energy consumption curve, determine effective energy saving measures and realize energy efficient utilization.

[Case: cleaning Dust cover]

Feng Aluminum installed digital electricity meters for the 6 extruders with the largest energy consumption, continuously carried out network monitoring, and constantly tried to adopt various technical or management reform measures. Finally, it is found that one day after cleaning the dust cover of one of the machines, the energy consumption curve shows that the machine saves 20% energy. Therefore, it was determined that regular cleaning of the dust cover is an effective energy saving measure. Without the separate installation of smart meters and comparative analysis, it is not possible to find this rule, thus achieving energy savings.

4.   Intelligent management of electrical equipment

The realization of intelligentization based on digitization, automation and network can further save energy.

[Case: Google uses artificial intelligence to save energy]

Each data center has some unique characteristics, such as climate, weather, building structure of each center, type of software system, etc., so it is difficult to establish a general optimization model. Deepmind, acquired by Google, has applied alphago-related technology to energy optimization in data centres. By developing a kind of artificial intelligence software, considering the data center fan, refrigeration system and windows, outside weather and people's usage, and so on, the statistics about 120 variables, through continuous monitoring found that cooling system is not adjusted in real time with demand fluctuations, then it keeps its data center "Cold Aisle" at around 27 °C to conserve energy, and uses outside air to cool its data centers instead of using energy-hungry cooling systems, and then through the adaptive control algorithm, the power consumption for cooling was reduced by up to 40%, the overall power efficiency increased by 15%, within a few years Google will save hundreds of millions of dollars in electricity bills.

5.   Move energy-intensive equipment to low-cost areas

For example, due to the large data industry in Guizhou has the congenital advantage, such as rich energy, pleasant climate, geological stability, etc., the national big data comprehensive experimental area in guizhou, and three major network operators, and multiple state ministries and Huawei, Ali, Tencent and other several big giant data centers also are located in Guiyang cave (as shown in Fig. 4.4a), they are storing the core data of enterprises. Due to constant temperature and humidity in the cave, PUE (Power Usage Effectiveness) is the highest, and air conditioning has the lowest energy consumption. Ren Zhengfei, President of Huawei, said: "The data center located in Guizhou will save hundreds of millions of yuan in electricity bills one year after it is completed and operational." Apple chose a location in Guian New

(a) The Guizhou data center of Tencent   (b) the cluster of airplanes and airborne conference rooms in the cave of Bahnhof

Fig. 4.4 Data center in the cave. **a** The Guizhou data center of Tencent, **b** the cluster of airplanes and airborne conference rooms in the cave of Bahnhof

District to build the iCloud National master data center. CEO of Data center operators Bahnhof company, said the efficient operation of data center is a perfect business model, and the company's most data centers will have all its waste heat released into the central heating system of local communities, one of its data center buildings in Stockholm, Sweden, a mountainous area in the south, as shown in Fig. 4.4b.

At present, most industrial enterprises with high energy consumption are far from enough in terms of electricity monitoring and energy saving management. The application of enterprise power consumption monitoring and management system is not universal. In terms of energy saving management, most industrial enterprises still adopt extensive management, which does not start from the overall situation of enterprise power consumption and lacks scientific and effective management means.

6.   Diversified, clean, intelligent and global new energy transformation

"Made in China 2025" states that "we will continue to take sustainable development as an important focus in building China into a manufacturing power, strengthen the promotion and application of energy conservation and environmental protection technologies, processes and equipment, and comprehensively promote clean production". In the future energy consumption structure, renewable energy and electric energy will replace traditional fossil energy and gradually increase the proportion of electric energy in energy consumption. The direction of the world's energy transformation is diversified, clean, intelligent and globalized. It includes the following five interconnected transformations: the transformation to renewable energy; Converting buildings into micro-power plants that collect renewable energy; Use of hydrogen and other storage technologies to store intermittent energy; Building energy sharing networks across continents; Zero vehicle emissions [1].

4.5   Connectivity

Connectivity is the amplifier of the industrial revolution.

The steam engine was invented in the first industrial revolution, but it was not until the advent of a network of railroads that breed large industrial cities, making the existence of trains a necessity; Electricity was invented in the second industrial revolution, but only through the power network, can realize the long-distance transmission of electricity, so that electricity really play a role; In addition, transportation networks, such as airplanes and ships, and telephone networks, further expanded the range of products for sale, all of which led to large-scale industrial production. Computers were invented in the third industrial revolution, which improved the efficiency of information processing. However, the emergence of the Internet promoted the application of mobile social networking, e-commerce, O2O, sharing economy, network collaborative manufacturing, etc., and sublimated the efficiency revolution into the transformation of economic life paradigm.

The three industrial revolutions have been accompanied by major advances in connectivity, namely, the railway network, the power network and the Internet. At present, driven by new information technologies, the objects of connectivity are still changing. From the interconnection between people to the interconnection between people and things and between things. The interoperable content extends from text, voice and image to video and business flow; The communication medium has expanded from the traditional telephone and telegraph communication network to the new generation communication network and Internet platform which can provide multimedia service. It is this 3D enhancement of connectivity that has promoted the evolution of the Internet into the Internet of everything. Based on the new-generation communication network, Internet and IoT platform, it will provide more in-depth connectivity services for more complex and diverse objects.

4.5.1   Network Environment

The most important challenges to implementing IM are the ubiquitous, fast Internet infrastructure and the standard interface of Cyber-Physical Systems. The interconnectivity of equipment is the key to the transformation from traditional factory to smart factory. The traditional connection takes the network as the main body and the application is tightly coupled on the network. New information and communication technologies decouple, platform, intelligent and cloud, forming a multi-layer connection system, and the platform gradually becomes the main body of the connection. The enterprise shall first establish wired or wireless factory networks to realize the automatic issuance of production orders and the automatic collection of equipment and production line information; To form an integrated workshop networking environment, to solve the networking problems between equipment of different communication protocols and between PLC, CNC, robots, instrumentation/sensors and industrial control/IT systems; The video monitoring system is used to monitor, identify and alarm the environment and human behavior in the workshop. In addition, the factory shall achieve an intelligent level in the control of temperature, humidity, cleanliness and industrial safety, including the safety of industrial automation systems, the safety of the production environment and the safety of personnel.

As shown in Fig. 4.5, realizing the interconnection and real-time control of production equipment, sensors, control system and management system based on industrial Ethernet, so as to achieve safety and energy saving, will be the core technology of intelligent factory. Specifically, from the scene of the I/O, sensors, PLC system and the workshop level HMI, the production equipment of the SCADA collected data through wireless or wired to the database server, realize self-discipline coordination M2M, and make the network operation technology (OT) and plant level of SCM, ERP, MES, CAD/CAM/CAPP/PDM, information technology (IT) system interconnection. With the exponential growth of the underlying data volume and the penetration of high-bandwidth applications such as virtual reality into the industrial field, the connection between enterprises and the factories of cloud applications has put forward higher requirements in terms of reliability, bandwidth, delay and jitter performance, business form and other aspects, and it is necessary to make innovative attempts. With the advance of 3GPP 5G standard and the approach of commercialization, 5G will be widely used in the industrial field.

Through the above identification technology, network deployment and data acquisition platform construction of each device, a typical information data acquisition system is shown in Fig. 4.6, which should include the following contents:

1. Equipment: The equipment layer mainly includes the main equipment of the power production line. The device has a standardized network interface and the function of data transmission and sharing.
2. Data acquisition: OPC technology is adopted to realize intelligent interconnection and interworking of devices, connecting CNC devices with different

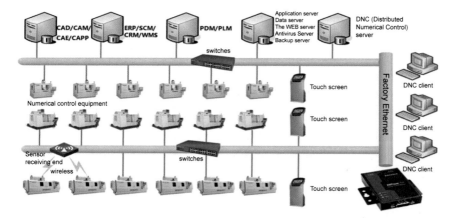

Fig. 4.5 A schematic diagram of interconnection based on industrial Ethernet

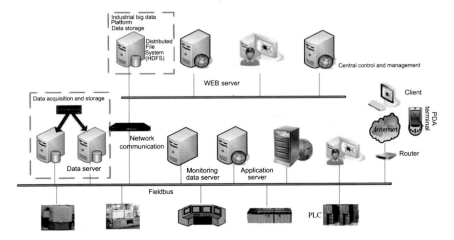

Fig. 4.6 Data acquisition architecture diagram of discrete manufacturing enterprise

interfaces, different control systems and heterogeneous communication protocols into a network. Establish the mapping relationship between OPC service and PLC/DCS data address of the equipment, collect the parameter information of the equipment running time on the production line, and upload it to the database server for analysis, processing and storage through industrial Ethernet or appropriate type of field bus.

In practice, different types of machine tools have different data acquisition implementation paths, as shown in Fig. 4.6a.

For machine tools with network CARDS, there is generally no need to add other hardware, and almost all information can be collected directly through the network card: machine tools start up and shut down; Real-time status of machine tools, such as running status, idle, alarm, etc. Program information, such as which program is running; Number of numerical control program machining; Current speed, feed speed, machine tool ratio; Coordinate information, including absolute coordinates, machine tool coordinates, relative coordinates, residual movement, etc.; Alarm information, can real-time feedback whether the machine alarm, alarm number is what; Spindle power; And so on.

For the equipment supporting PLC acquisition, through the installation of PLC system, data acquisition through PLC, and then can read the data stored in the PLC.

For other types of common machine tools, corresponding sensor hardware and industrial software should be added for data acquisition. At present, the influential Fieldbus types include Foundation Fieldbus (FF), Profibus, P-NET, SwiftNet, ControlNet, Interbus, WorldFIP, etc.

The combination of wired acquisition and wireless acquisition can be adopted. The function of wireless data transmission is to transmit data collected by the sensor to the sink node through wireless transmission. Wireless sensor nodes have radio frequency function. Each node adopts standards-based wireless communication protocol and realizes wireless data transmission through self-organizing zigBee based wireless sensor network. In the equipment of large size and many kinds and large number of monitoring stations, wide distribution area, such as process industry factory, in order to better save node energy, easy to extend the network and the monitoring points distribution, based on hierarchy clustering network topology can be used to differentiate the wireless sensor network (WSN), for a number of distributed production synchronous data acquisition of Field device state parameter that each cluster network is responsible for the Field devices in the network coverage area of state parameter acquisition, at the same time using FPGA (Field programmable gate array) and DSP (Digital Signal Processing) technologies can improve the computing speed and storage capacity of wireless sensor network nodes and reduce network energy consumption. The Wired data transmission part is to relay the data collected from wireless sensor network to the node to the upper monitoring center upper computer and server, and realize visual display combining with the map. Users can like the real environment, in a 3D virtual manufacturing environment overview model choose any path search, can undertake information query for the communications equipment, and conforms to the operation of the device properties, through the way of 3D display processing instructions and parameters adjustment of real time effect, two-way monitoring and interaction, a kind of typical monitoring scenario is shown in Fig. 4.7b.

3. Data services: system middleware and OPC interface often use the way of combining the historical process data and quality inspection data and the comprehensive test data through the middleware server relational database storage, and then the real-time dynamic data transmitted OPC way in the process of production to the real-time database, module respectively by the integrated system to read data from a relational database and real-time database and perform statistical analysis.

4. Application integration: The functional application module will be integrated into the MES system. Users can access the energy management and equipment management module of MES system, including status monitoring, historical status data query, equipment optimization scheme, and analyze the reasons to achieve the purpose of guiding production.

In the complex environment of IM, heterogeneous network fusion is the main problem of the network environment. Industrial IoT network is often a variety of types in common existence, industrial control has higher requirements for time delay and jitter. Each major manufacturer has its own definition of data link, multi-protocol contributed to a lot of information islands, including Siemens Profinet, rockwell/schneider Ethernet/IP, PowerLink, EtherCAT and other standard systems. Devices using different industrial Ethernet networks can coexist, but are difficult to talk to. To achieve connectivity, different networks must be able to connect with each other. To use multiple Ethernet devices in the same network requires a gateway for protocol conversion.

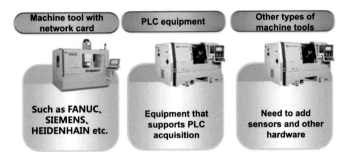

(a) Machine tool type and acquisition approach

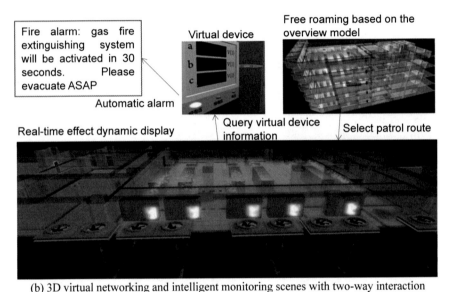

(b) 3D virtual networking and intelligent monitoring scenes with two-way interaction

Fig. 4.7 Intelligent data collection based on industrial Internet. **a** Machine tool type and acquisition approach, **b** 3D virtual networking and intelligent monitoring scenes with two-way interaction

At present, both the idea of designing unified gateways and the idea of Network of Cognitions (NC) are aimed at the fusion of only a few kinds of networks, which cannot meet the needs of dynamic and uncertain environments. In view of the heterogeneity, data volume and large business volume of the IoT network, it is a development trend to develop unified physical and MAC (Media Access Control, physical address) bottom layer and its protocol, and promote the integration of multiple networks.

4.5.2 Network Security

Network security and informatization are two wings of one body and two wheels of driving.

Network security refers to the protection of communication security within and between systems of the intelligent factory through security unit protection and network boundary division technology. In terms of national security, under the background of IM, overall defense against invading, need to maintain safety is higher and higher, intelligent control will become a social norm, widely interconnected production structure and social structure, industrial Internet security become the core of the nation's ability, industrial network control becomes strategic control problems. China should always adhere to the development road of international track, independent control, advanced technology, both offensive and defensive, and strive to create a credible network security environment.

In 2016, General Secretary Xi Jinping delivered a speech at the symposium on network security and informatization, pointing out that: "We should establish a correct concept of network security, accelerate the construction of the security system of key information infrastructure, perceive the network security situation in an all-weather and all-round way, and enhance the network security defense capability and deterrence capability. Not realizing the risk is the greatest risk. Network security has a strong concealment. The perception of network security situation is the most basic and fundamental work. We should comprehensively strengthen network security inspection, find out the bottom of the family, recognize risks, identify vulnerabilities, report results, and supervise and urge rectification. We need to establish a unified and efficient mechanism for reporting, sharing, studying, judging and disposing of cyber security risks, so as to accurately understand the rules, trends of cyber security risks. We need to establish a mechanism for sharing network security information between the government and enterprises and make use of the large amount of network security information in their hands. Leading enterprises should take the lead in participating in this mechanism". General Secretary Xi Jinping has demonstrated through this systematic statement that cyber security issues should be addressed before they arise. It is recorded in The Historical Records that bian Que, a famous doctor, once said, "a disease should be cured before it is discovered", which shows the importance of preventing a disease before it happens. The industrial network system will become more open, cyber security risks are becoming more prominent, and they are increasingly penetrating into political, economic, cultural, social, ecological, national defense and other fields. It is necessary to maintain cyber security in a comprehensive way in both time and space.

[Case: Iranian nuclear power plant]
The man-made computer virus incident at the Iranian nuclear power plant in 2010 has shown that the safety of processing industries such as electricity and chemicals has faced great challenges in the Internet age. Industrial Internet security has become the front line of national confrontation. During the nuclear event in Iran, the

Natanz nuclear power plant was invaded by a virus. Although the computer system of nuclear facilities is physically isolated from the external network, but the "seismic network" worm in the U disk "ferry" way, by Microsoft's Windows operating system into the German Siemens industrial control software, the "seismic network" control system data, change the speed of the centrifuge, eventually, 1000 to 5000 centrifuges are paralyzed or out of control, and send wrong instructions, to make the control system think that all operations were normal. Stuxnet attacks a target, while Flame gathers information. The virus can open the PC microphone, record personnel near computer conversations, capture screen images, to record the conversation of users to use chat software, collect files and remote modify the settings of the computer, even using bluetooth to steal the contents of smartphones, tablets connected to the infected computer.

Effective and reliable security isolation and control should be carried out between the interconnected equipment and systems of industrial enterprises, including: firewall should be deployed between OT system and IT system; the external access to the internal cloud platform of the factory should go through the firewall and provide DDoS defense and other functions. Meanwhile, the network intrusion protection system should be deployed to identify the mainstream application layer protocols and contents, and automatically detect and locate attacks and threats of various business layers. Role authentication and access authorization are required for all devices connected to the internal cloud platform, partner information system and industrial control system of the factory; Through external Network transmission of data using a Virtual Private Network technology (VPN), for example, ordinary mobile users remote access WEB application (Client Site) can easily penetrate the firewall can be used when the SSL VPN, the enterprise all kinds of application of the advanced user Network to Network (Site-Site) connection, adopt more the bottom more powerful IPSec VPN tunnel transport mechanism, prevent data leaks or tampered with, guarantee system and data security.

[Case: GPC's network communication platform planning]
To achieve the strategic objectives of the intelligence platform, Guangzhou pharmaceutical co., LTD. (GPC) decided to network communication platform construction planning, to modify the data center, and the construction of disaster preparedness, disaster backup center, can make the information integration strategy of construction plan, according to the steps to implement. Disaster preparedness center is the guarantee of preventing disasters and reducing losses in the information age.

The Network communication platform is the cornerstone of GPC information integration strategy, and all business systems are running on this platform. The structure of network communication platform must have the characteristics of stability, high speed, flexibility, extensibility and security. The overall requirements of the design are as follows: In the future, Guangzhou will be the core data center, and the ERP system, warehouse management system, financial system and other important business system data of all branches in China will be unified to Guangzhou data center for integrated management. Each branch will adopt the most cost-effective scheme to connect with Guangzhou Data Center; The physical interconnection network adopts the "star" connection mode, while the logical interconnection mode is the "full inter-connection" mode. As the service system requires continuous 24 × 7 operation, it requires unified standards for key network equipment and circuits, and realizes dual-machine and dual-wire standby. The adjusted network reserves the capability of horizontal expansion in the future to avoid the waste caused by repeated investment due to upgrading. Propose specific standards for line selection, equipment purchase, configuration strategy and access control to avoid incompatibility and instability problems caused by different standards. If the branch has the conditions, it can realize the double redundancy directly with the center of the headquarters, while if it doesn't have the conditions, it can connect with the regional center.

The branch interconnection architecture is designed as a three-layer network communication platform, as shown in Fig. 4.8.

The network design scheme is as follows:

1. The core layer of the data center is composed of two high-end switches, the convergence layer is composed of two mid-end switches, and the server equip-ment accesses the core switch through the convergence layer switch. Both the core layer and the sink layer adopt the mode of dual core mutual redundancy.

2. The convergence layer adopts modular switch containing bandwidth of thousand megabytes and 4 layers, access layer switches for correlates with extension module switches, access layer switches via gigabit module and lines connected to the core switch, all servers via network cards and lines of gigabit connected to the gigabit switches, company Intranet client terminal through MB card and lines connected to the secondary switch, through the relevant configuration switches, for example, settings and port optimization of VLAN (Virtual Local Area Network), form a structure of gigabit backbone and MB exchange to the

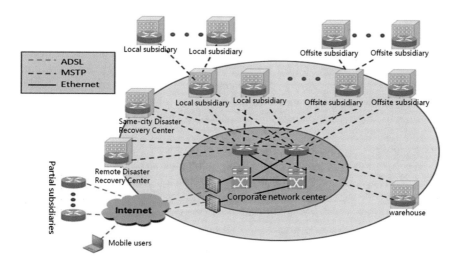

Fig. 4.8 GPC network communication platform with three-layer structure

desktop. In addition, the database server is connected with the storage array through the storage area network to form an efficient storage network.

3. Set up an independent VLAN in the core server area and isolated it from the general server area. The core server is the carrier of the core business application, which must be based on an independent secure and stable network environment. Set up independent VLAN, can effectively block LAN broadcast storm, make the core server area network more stable.

4. Arrange the internal firewall at the entrance of the core server area and open the service port as needed to prevent the core server from being illegally invaded.

5. In the core server gateway layout operations security audit system equipment, continuous and stable operation of the core business application system, depends on the IT operations staff to correct management, optimization, operational audit system to the administrative authority of IT operations staff to each according to his need, prevent illegal users access, and faithfully record the operations staff to the management of core server operation, improve the safety of the server.

6. Set up a high-performance firewall at the Internet access point. Firewall is the barrier of network security, isolating malicious attacks and information stealing from the Internet.

There are three steps for the construction of disaster preparedness center:

1. Establish a city-wide data-level disaster preparedness center. The data level disaster center structure is relatively simple and can adopt an integrated disk-to-disk backup and recovery solution. The scheme by using a mirror image of the source server (system status, application and data), or simply specifying the key data, using real-time replication technology, thus ensuring any source data changes on the server can be real-time replication to the standby server, which is based on the system I/O real-time replication technology, just copy the data changes, so can effectively reduce network resource utilization. When the source server fails, it is easy and quick to create a new source server for the user. When data recovery is needed, snapshot data or data at a certain point in time can be extracted and restored to a different physical server or virtual machine. The linear distance between the production center (headquarters network center) and the disaster preparedness center is within 200 m, which can be connected directly through bare fiber. This can not only ensure network quality, but also reasonable cost savings.

2. Establish a city-wide operational disaster preparedness center. Its structure is similar to the production center, with a complete dedicated line or VPN and Internet access, and a one-to-one server for each application. Physical machines are used as backup for key applications, while secondary critical applications can be disaster recovery applications based on virtualization, with real-time data replication, and can be automatically taken over or manually enabled. The linear distance between the production center and the disaster recovery center is within 200 m, which can be connected directly through bare fiber interconnection. This can not only ensure network quality, but also reasonable cost savings.

3. Establish a remote data-level disaster preparedness center. The construction principle of remote and urban data level disaster recovery center is the same, the only difference lies in the choice of network link. Due to the long distance, the bare fiber interconnection mode cannot be adopted. Considering the factors of cost and line quality, MPLS or MSTP line is the most suitable choice.

4.6 System Integration

This section discusses business integration and application integration across departments and business links, and discusses system security issues.

4.6.1 Application Integration

IM involves the fields of mechanical hardware, software and electronics, forming Complex Product Systems (CoPS). Application integration is to lay out a complex system efficiently. In essence, application integration is the whole process of designing and constructing an integrated open system suitable for specific engineering requirements. One of the shortcomings in the development of China's manufacturing industry is the lack of a sufficient number of high-level system integrators engaged in application integration. The system integrator is a bridge from technology to industry.

In order to achieve application integration, the manufacturing information system must have open interfaces, be seamlessly connected to each subsystem required by the project, and connect multiple manufacturers' equipment together. Application integration is no longer a single professional automation engineering behavior, it is no longer an automated island of application development. In the IM mode, the application of integration technology becomes the key link. Only when the information system is integrated, can the intelligent functions of the whole system be realized, and can the intelligent factory, intelligent production line and IM be realized.

To be open, its integration structure must be standards-compliant. The standard integration structure can be divided into Client/Server (C/S), CORBA or DCOM, Web Services (Browser/Server, B/S), ASP or SaaS, ESB, Multi-agent system (MAS) and other operational architecture patterns.

In the 1990s, it was recognized that too many users or too many functions integrated into a single system would make the system too large and prone to failure. Due to the low hardware cost of Client/Server, Client/Server once became the mainstream method of distributed computing. It is composed of two logical units, the server provides business services, the client requests the server's services, the two form a computing system with obvious responsibilities. However, in a large application environment, its scalability and heterogeneity are facing new challenges.

CORBA (Common Object Request Broker Architecture) is a distributed Object management specification developed by the Object Management Organization (OMG). Software supporting this standard can be used as a programming development tool for distributed systems. It has great openness and flexibility in supporting the interoperability and portability of objects under a heterogeneous platform. The main characteristic of this standard is to realize the soft bus structure. The application module is made into a soft plug-in according to the bus specification, and the integrated operation can be realized by plugging into the bus. DCOM (Distributed Component Object Model) is a series of concepts and program interfaces proposed by Microsoft. By using these interfaces, a client program Object can request a server program Object from another computer in the network. Both CORBA and DCOM, the early service-oriented architectures, suffered from a number of problems. First, they were tightly coupled, which meant that if the code of an object being accessed changed, the code accessing that object had to change accordingly. Secondly, they are divided into the camp with IBM and Oracle as the core and the camp dominated by Microsoft, which are subject to the constraints of the manufacturers.

Web Services are a web-based implementation of Service-Oriented Architecture (SOA), the result of an effort to improve on the shortcomings of CORBA and DCOM. SOA is a software engineering approach that follows open interoperability protocols and implements software systems by combining reusable software resources. The service-oriented architecture of Web Services differs from those of the past in that they are based on standards such as XML (eXtensible Markup Language) and loosely coupled, widely accepted standards such as XML to provide interactivity between solutions of different vendor groups.

ASP (Application Service Provision) and SaaS (Software as a Service) have long been familiar in the IT industry, but there are few applications in the manufacturing industry. China's SaaS projects are mostly focused on CRM, OA, HR and other applications with low entry requirements, but few set foot in ERP. SaaS is an application developed based on a cloud computing infrastructure. With the development of cloud computing and big data analysis technology, the SaaS model has been continuously enhanced. The SaaS model leverages data analysis and data mining to enable more intelligent and networked devices and to optimize integration between devices, software, and services. SAP, IBM, Microsoft and other international giants have successively released information solutions for small and medium-sized manufacturing enterprises in China, and provided various management software functions to customers through their cloud computing service centers through Web services. Enterprises can rent SaaS layer services to solve the problem of enterprise informatization, without considering the management and maintenance of servers, which greatly reduces the cost of informatization. For the average user, SaaS tier services migrate desktop applications to the Internet, enabling ubiquitous access to applications.

The way to implement a centralized SOA is through an ESB (Enterprise Service Bus). ESB is a combination of traditional middleware technology and XML, Web Services (a network based, distributed modular component) and other technologies. The concept of an ESB evolved from SOA. An ESB is a way to provide reliable, guaranteed messaging technology that provides the most basic connection hub in a network. The ESB Middleware product leverages Web Services standards and an interface to a recognized MOM (message-oriented Middleware) protocol, such as IBM'S WebSphere MQ, Tibco's Rendezvous, and WebMethods. Common features of ESB products include the ability to connect heterogeneous Moms, encapsulate the MOM Protocol using the Web Services Description Language interface, and transmit Simple Object Access Protocol (SOAP) transport flows over the MOM transport layer. Most ESB products support direct peer-to-peer communication between distributed applications through an intermediate layer such as an integration broker.

The fundamental appeal of a decentralized SOA is extensibility, which is implemented in the form of micro services. Micro services are more suitable for future applications that have a certain degree of scalability complexity and are expected to increase the number of users, which are generally emerging Internet companies. At the beginning of the startup, it is impossible to buy a large number of machines or very expensive machines, but it must consider to deal with the huge number of users after success, micro service architecture becomes the best choice.

The concept of Agent originated from the academic circle. Researchers introduced the concept of artificial intelligence Agent into the simulation and scheduling control of manufacturing system. MAS is a collection of multiple agents, which has the characteristics of distributed systems and artificial intelligence. With the increase of product categories and the complexity of the manufacturing system, the uncertainty factors in the production process increase significantly, the negative impact of disturbance events such as equipment failure, personalized orders and emergency orders on the operation of the system increases, and the disadvantages of static scheduling control mechanism gradually appear. Different from static scheduling, dynamic scheduling requires real-time monitoring of various disturbance events that may occur during the manufacturing system operation. Once the disturbance event is detected, a corresponding processing mechanism is adopted to change the scheduling scheme according to the current state of the system to ensure the smooth operation of the system. Therefore, dynamic scheduling is the basic requirement of customized and flexible production under IM mode. MAS can implement the real-time dynamic scheduling mechanism of the manufacturing system through the interaction and cooperation of agents to complete complex production tasks according to simple rules. It can also make real-time decisions according to the running state of the system to improve the flexibility of the manufacturing system. Shop floor scheduling is the basic link of manufacturing system control, the quality of scheduling results directly affect the overall production efficiency of the manufacturing system. MAS can also be compatible with the heterogeneous hardware and software in the existing manufacturing system, which facilitates the dynamic reconstruction of the system and improves the reconfigurability.

4.6.2   System Security

The characteristics of the IM system, such as high integration, information fusion and heterogeneous network interconnection, have brought great challenges to system security:

1. The development of China's IM industries is relatively backward, and the comprehensive competitiveness is not strong. Operating systems, major industrial software and chips are monopolized by foreign countries, and the sales and leasing of systems and network services have become control means, making it difficult to achieve safe and controllable. Microsoft, for example, stopped serving Windows XP after Windows 8 was introduced. China's industrial control system security is a big problem, domestic enterprises industrial control system is generally fragile, easy to access through Internet camera equipment, ERP and other systems. According to a survey conducted by the Ministry of Industry and Information Technology in 2014, more than 90% of industrial control systems are foreign brands, and uncontrollable software source code and chips with "back doors" are at risk.

2. The security risks faced by IM systems are different from those faced by traditional information systems. IM system has a long process of business scenarios, and diversity of heterogeneous network protocol, the difference of the equipment, standardized protocols and technology makes public the security vulnerability, all kinds of information systems to the existing manufacturing application expansion and support for new application dynamic join, the security risks of IM system are more prominent and complex.

    The security risks of intelligent systems in the manufacturing industry can be divided from bottom to top:

1) Equipment layer: There are security risks such as channel obstruction, sensory data destruction, Sybil attack, clock synchronization attack security risks and vulnerabilities of software and hardware equipment, such as fire, explosion and other major accidents, short-circuit, overload, gas leak, irregularities, such as direct cause accidents, deviating from the normal and high temperature, electric arc and other parameters.

2) Control layer: There are security risks such as control command forgery attack, control network DoS attack, resonance attack, broadcast storm, and security vulnerabilities in the control system hardware and software.

3) Network layer: There are authentication attacks, cross-network attacks, routing attacks and other security risks and network security vulnerabilities.

4) Enterprise management layer: There are security risks such as user privacy and data leakage, unauthorized access, eavesdropping and sniffing, malicious code, virus/vulnerability attack, and security vulnerabilities in software and hardware of enterprise software platform and management system.

5) Collaboration layer: there are risks such as interruption of data links and website access, interception of trade secrets, tampering with operational data,

data forgery, protocol denial, and security vulnerabilities in the software and hardware of the supply chain management system.

At all levels, there are security risk points such as illegal access, content removal, logic errors, and code flaws. In the network layer, control layer, enterprise management layer, there are often bus abnormalities, eavesdropping and sniffing, password theft, host vulnerabilities, virus attacks, Trojan horse latent, backdoor threats and other information security risk points. In the two levels of equipment layer and control layer, there are always safety risk points such as random failure and diagnosis error.

The security of each level in the system is not independent of each other, but interdependent. The security protection strategy designed for each level is not comprehensive, and it cannot be fully protected only by relying on local and static security strategy. In addition, different application scenarios have different emphasis on security requirements. Security scheme design, therefore, need to keep up with the trend of the development of the security industry, at the same time, make sure to follow the systemic, consistency, gradation, comprehensive, dynamic security planning principle, reasonable use of network isolation technology, access control technology, encryption technology, identification technology, the digital signature technology, invasive monitoring, information auditing technology, safety assessment technology, virus prevention and control technology, backup and recovery technology, the design of intrusion detection and defense system which has more extensively applicable scope, design a more effective access strategy, make effective mobile device cross-domain authentication method will be the future research focus in the system safety.

[Case: analysis to the safe operation of equipment layer of large data center equipment]
In the center of the large data, complicated equipment running status, monitoring objects and the parameter variety, including: CPU, memory, disk, network and ports, power transformation and distribution, UPS, air conditioning and refrigeration, temperature and humidity, chilled water units, water leak, new fan, fire protection, building control, access control, video and other subsystems, and every subsystem has a variety of parameters and millisecond gathering a large number of real-time data. Traditional monitoring generally uses a "passive" mode of operation: monitoring the parameters of various devices and judging whether to alarm based on various thresholds, such as early warning when CPU load continuously reaches 90% or when disk use reaches 90%. It can only play a single alarm function. Obviously, "Solve problems after they happen" is difficult to adapt to the high reliability requirement of big data centers.

The way to solve such problems is to introduce intelligent means to "actively" detect equipment problems, which requires changing the management mode of monitoring passive detection problems. Big data center equipment operation analysis system usually includes intelligent analysis, report output, information acquisition, system management, equipment management and other functions.

The intelligent analysis module can be used to analyze whether the collected and input data have system operation risk, which requires professionals to carry out manual analysis on the data of each professional subsystem in the traditional situation. When the number of big data subsystem devices is large and the analysis time is fast (ranging from a few seconds to a few minutes), manual analysis cannot meet the management needs of big data. This problem is to be solved by the intelligent analysis system of the computer. Its functions include:

1. Matching and reasoning functions of rules

    Through a mechanism, the analysis logic of each professional subsystem can be set up and loaded, and the data collected (including real-time data and historical data) can be calculated (such as the calculation of temperature rise rate), compared and analyzed. For the data conforming to the set rules (conditions), the system extracts relevant data and generates reports and alarm information, or displays the name of the control action to be carried out according to the set logic, or further carries out the automatic control of related subsystem equipment.

2. The building function of the rule library

    According to the analysis logic of each professional subsystem, the system provides a setting interface, which can be set by professionals according to the characteristics of each subsystem. The set of conditions includes a variety of logical conditions, historical records, and the combination of various conditions to judge.

The window of input condition configuration policy is shown in Fig. 4.9: (1) the window to the left is the name of the analysis policy group, which can be understood as the corresponding relation group of a series of input conditions and output conditions; (2) The input conditions on the right can be set separately for one or more input variables, and each input variable can be defined across multiple subsystems; (3) A variety of calculation formulas can be used for the judgment of each input condition, as well as the recorded value in the historical database; (4) The output control object can be set as the adjustment control action of the production of some operation and maintenance events or some sub-equipment; (5) The output control object has successively, delayed execution and other functions; (6) When the intelligent analysis system is put into use, each analysis strategy group can dynamically produce multiple dynamic libraries and quickly execute the analysis logic; (7) According to the above conditions, automatically conduct rule-based reasoning, comprehensively judge what consequences will occur after meeting the conditions, and solve the problem by users' input of emergency plan in advance; (8) Through the interface, the analysis software constantly displays the results of analysis and judgment on the page, and writes the relevant data table based on the results meeting the conditions.

3. Replication and management functions of rule library

Analytical logical knowledge base can accumulate knowledge continuously. If the analysis logic proved to be effective in one big data center A, its rule base can be selectively copied according to the specific situation and applied to another big data center B. In this way, the accumulation of logical knowledge base of analysis and

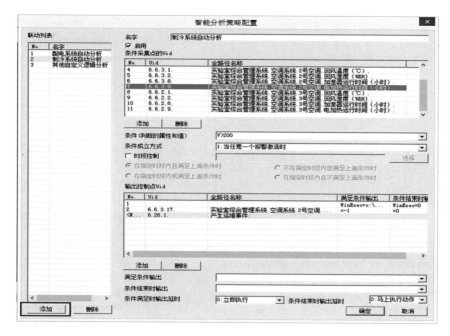

Fig. 4.9 Configuration interface for intelligent analysis policy

the "intelligence" degree of big data analysis system can be continuously improved, and finally a valuable intelligent analysis system of big data can be formed.

4.   Application Examples

More than 30 precision air conditioners of various brands and models are distributed in a big data center. The system monitors in real time and finds out through calculation that there are two variables: the state of no. 1 air conditioning compressor (the collected value is 1 or 0); The air outlet temperature value of the air conditioner. When the air conditioning compressor is in operation, the air outlet temperature of the air conditioning rises continuously and by $1°$ within 2 min (the judgment condition of specific rate can be set by the user), and the historical record keeps rising, then the system outputs the warning information of possible fluorine leakage of the air conditioning, requiring the on-duty personnel to check it on the spot.

4.7   Information Fusion

Originally known as Data Fusion, Information Fusion originates from sonar signal-processing systems developed in 1973 with funding from the Department of Defense of the U.S., and the concept appeared in some literature in the 1970s. In the 1990s, with the extensive development of information technology, a more generalized concept of "information fusion" was proposed. Information fusion technology is more and more popular in the manufacturing industry, especially in the equipment fault detection field.

4.7.1 Data Fusion

In the 1980s, Multi-Sensor Data Fusion (MSDF) technology came into being to meet the needs of military operations. In 1988, the U.S. listed data fusion technology in the C3I (Command, Control, Communication and Intelligence) as one of 20 key technologies that the Defense Department would focus on developing.

Not only is the number or type of sensor important, but data fusion is just as important. Each sensor has different strengths and weaknesses that cannot be overcome simply by using the same type of sensor multiple times. The allusion of "the Blind man touch the Elephant" tells us that to identify an object correctly, information from different kinds of sensors must be fused together.

Generally, the data fusion process includes six steps, including: (1) multi-source sensor system construction and calibration; (2) Data collection; (3) Digital signal conversion and transmission; (4) Data preprocessing and feature extraction; (5) Use fusion algorithm for calculation and analysis; (6) Synthesize and output reliable, accurate, sufficient and consistent effective information.

Multi-sensor data fusion is a common function in human or other logical systems. Like human beings, IM systems naturally combine the information (scenery, sound, smell and touch) from various sensors (eyes, ears, nose and limbs) of the system (human body) with the ability of multi-sensor data fusion, and use prior knowledge to estimate and understand the surrounding environment and ongoing events. The comprehensive and complete information about objects and environments obtained by using multiple sensors is mainly reflected in the fusion algorithm. Therefore, the core problem of multi-sensor system is to choose the appropriate fusion algorithm. For multi-sensor systems, information has diversity and complexity. Therefore, the basic requirements for information fusion methods are robustness and parallel processing capability. In addition, the method is required to have high computing speed and precision, interface performance with the pre-processing system and subsequent information identification system, coordination ability with different technologies and methods, and requirements for information samples. In general, multi-sensor data fusion algorithms should be fault-tolerant, adaptive, associative and parallel.

Multi-sensor data fusion algorithms can be summarized into two categories: random and artificial intelligence. The random class algorithm includes weighted average method, Bayesian network, Kalman filter, clustering analysis and so on. Artificial intelligence has fuzzy logic, neural network, expert system and so on.

Bayesian estimation is the use of bayes' theorem in combination with new evidence and the prior probability of before, according to "a posteriori probability = prior probability × adjustment factor", using new acquired data adjustment factor to get a new posterior probability, without the need for the overall data calculation, suitable for the use of real-time data efficiently modified probability forecast.

The Kalman Filtering algorithm calculates a weighted average of observations and estimates based on experience or models, adjusts the weights based on previous measurements' stability and accuracy, and recursively recurses covariances to estimate the optimal values. In addition, because only the data of the last moment is retained, it is very convenient and memory saving. Therefore, Bayesian estimation and Kalman filtering have great application potential in dynamic and real-time big data environment.

[Case: Realizing data fusion of autonomous driving]
Autonomous vehicle functions such as automatic parking, cruise control and automatic braking are largely realized by sensors. Multi-sensor fusion is the key to the technology, which includes two keys. One is sensor synchronization technology, which can realize high-precision synchronization of micro-second error in time and centimeter error beyond one-hundred-meter distance in space. The second is an algorithm based on fusion data development. The data dimension includes 8 dimensions, including 3Dspace data, RGB color data, laser reflectivity data and Doppler velocity data. At present, ADAS (advanced driver assistance systems) and sensors in most vehicles on the road work independently of each other, which means there is little exchange of information between each other, and they all have separate uses for rear-view cameras, surround-view systems, radar, forward-view cameras, GPS navigation, large central control screens and mobile phone positioning. CMOS chips for cameras in the visible spectrum can run into trouble when exposed to fog, rain, harsh sunlight and low light. Radar lacks the high resolution that imaging sensors currently have. Developers are beginning to realize that the key to autonomous driving is to fuse information from multiple sensors into easily identifiable, higher-dimensional data. Fused data is like a super sensor. Suppose there is an obstacle in front, and each sensor can only perceive the information of local contour. Since the local information is susceptible to noise masking and pulse crosstalk from other vehicles, this obstacle may be filtered out as a whole and ultimately not be recognized. In March 2018, for example, a self-driving test car in the United States hit a pedestrian crossing the road because the pedestrian was wearing a red skirt that was fluttering in the wind and the car sensor misidentified it as a floating red plastic bag, which the car is allowed to drive through under self-driving rules.

If the original data is fused before the sensor makes perception and recognition, the complete contour and behavior of the obstacle can be seen, and the obstacle and its type can be easily identified. The best practice in the industry is to add ultrasonic ranging to the rearview camera and multi-mode front radar to the front camera in the collocation of sensors. By transmitting the video to the central display screen or mobile phone, it provides a 360-degree field of view Angle for the driver and realizes automatic storage by man–machine collaboration. In terms of algorithm combination, probability calculation mainly uses Bayesian estimation, such as the detection rate under the condition that the traffic speed limit sign is blocked. In the visual perception of the module vehicle identification online recognition, traffic signs, lane, mainly used to handle a large number of environmental data of deep learning algorithms, in millimeter wave radar fusion target location with the visual perception module, the effects of using kalman filter to remove noise and get a good estimate on target location.

4.7.2 Data Application

The process of data application includes obtaining data from endpoints, extracting information from data, using various decision models for analysis and calculation, and output of system results. The exponential growth in real-time data convergence from customers, devices, and workflows will provide businesses with new insights. Collecting and analyzing data is important, but this is only the beginning. The key is to use the results of the analysis to discover important patterns and help the organization make the right decisions. Data application values include:

1. Strictly monitor the production process to achieve scientific control: For example, in the past, accurate information of construction site schedule and cost was difficult to obtain; Construction firms can now use 3D laser scanners and drones to collect photographic images, compare and collate the data from the original engineering drawings, and fuse contractor reports so that even a centimetre of difference on a site can be detected. Agricultural drones can also use weather sensing, 3D maps of planting areas, and soil color analysis technology to help adjust the planting, fertilizing, and harvesting processes. In the military aerospace sector, aircraft manufacturers can use the data from the aircraft to create empirical simulation software based on which pilot training can be developed to reduce the wear and tear on the fuselage and human costs associated with real test flights.

2. Accurately grasp user needs and promote product innovation: for example, use big data to mine user needs and market trends and find opportunity products; Based on demand data and 3D model, product collaborative design, design simulation and process optimization are realized. Through the direct feedback and interaction between the production system of the producer and the ordering and demand planning system of the customer, the time to delivery is reduced and the capacity utilization planning is improved. New product development will inevitably lead to changes in the demand for raw materials, and sharing new product development information can enable suppliers to actively participate in the preparation before the product goes on the market, and reduce and resolve investment risks in the supply chain.

3. Real-time monitoring of uncertain factors to avoid the occurrence of risks: For example, in view of the impact of random fluctuations in market demand, sharing sales data can directly reduce the impact of order amplification brought by the bullwhip effect step by step. One of the reasons for the bullwhip effect and inefficiency is that the demand signal of supply chain is predicted independently

by various parties among its members. Sharing sales forecast information can make all enterprises in the supply chain unite together to forecast future sales, so as to improve the competitiveness of the whole supply chain.

4. Enhance user stickiness and improve marketing accuracy: For example, make user portraits according to the browsing data of users' websites, formulate accurate advertising push plans, and realize personalized marketing; In conventional e-commerce platforms such as Taobao, Jingdong, system provides goods and services to end users, cover multiple functions and multiple independent enterprises, but end users deal only with retailers when shopping, they don't know who are retailers outside of the members of the ecosystem, also don't know which order is in production and manufacturing. Ecosystem members can solve this problem by sharing their order status information. With RFID technology, such information sharing is possible, but is rarely implemented.

5. Promote cross-border integration of enterprises to establish a win–win ecosystem.

For example, through the information integration of the whole industrial chain, the whole production system can achieve collaborative optimization and make the production system more dynamic and flexible; In practice, there are different ways to realize Inventory sharing among enterprises, such as Vendor Managed Inventory (VMI) and Collaborative Planning and Replenishment (CPFR) Forecasting. However, there are still various concerns for the manufacturer about the sharing of inventory information, which could lead to information leakage if the supplier also supplies its competitors.

The integration and application of data in the ecosystem still faces many obstacles. First of all, it is about technology and standards. In different nodal enterprises, the level of automation is also different. The level of automation varies. Second, there is the issue of trust. Sharing data is the fuel of the fourth Industrial Revolution. Access to digital information will soon be the key to future success. In some cases, cooperative games provide a solution to this problem, but in reality there are many factors and special considerations that make this problem very complicated.

4.7.3 Data Security

Network is the foundation, data is the core, security is the guarantee. The goal of data security research is to protect the hardware and software resources of the information system and the data in the system from illegal leakage, change and destruction. With the expansion of the system and the complexity of the structure, data security research has focused on the data from the early storage security, data encryption, extensions to the system security, network security, many aspects, such as its connotation from the initial development of confidentiality to the integrity, availability, non-repudiation and controllability of basic theory and implementation technology. IM system is a complex system formed by heterogeneous, heterogeneous and distributed application system and manufacturing resources connection. The security of big data formed by the IM system is a more important issue.

Data security threats include:

1. Interruption: Some components of the system are damaged or rendered inoperable. This is an attack on usability. The main precaution is firewalls. A common cause of interruption is a program on the remote control computer were illegally installed, form a BotNet (botnets), the program is a brief instruction to activate the remote control way, and disguised as legitimate requests to the target computer to send, request a seemingly normal carrier, but brings together the request, a great difficult to resist attacks, the loss of the target system resources, not on the networked manufacturing environment, the real user response, slow performance for the system delay. For example, in IM mode, remote control commands need to be executed quickly, and in 4G network, the instruction transmission time is 100 ms, while in 5G network, the time is reduced to 0.5 ms. If there is a delay due to an interruption after a remote production order is issued, even a one-second delay may cause a significant time anomaly, resulting in a production outage or a quality incident. Slowing down Internet service also drives customers to rival sites, and they never return.

2. Interception (intervention): unauthorized access to system resources, such as the use of software on the Internet to steal confidential data transmitted over the network. This is an attack on confidentiality. The main preventive measure is data encryption technology. If there is no use encryption or encryption intensity is not enough, the attacker may through the Internet, public telephone network, wiretap, within the scope of electromagnetic radiation (between nodes) install interception devices (Windump, etc.) or in packets through gateways and routers (nodes) on intercept data such as way to obtain information, then the data layers of unpacking, get useful information.

The general data encryption model is shown in Fig. 4.10. Symmetric encryption is called if the encryption and decryption keys are the same. In the symmetric encryption method, a key is only used for a pair of data sender and receiver, so the biggest problem in the symmetric encryption system is that the distribution and management of the key is very complex and expensive. For example, for a network with N users, N (N − 1)/2 keys are needed. If the encryption and decryption keys are not the same, it is

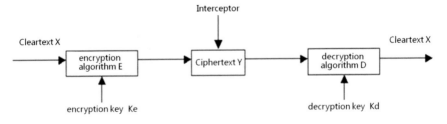

Fig. 4.10 General data encryption model diagram

called asymmetric encryption or public key encryption. In asymmetric encryption, the sender encrypts with the receiver's public key, and the receiver decrypts with a private saved key, and each person only needs to save a pair of keys. Only the private key needs to be kept secret, and the public key can be published. This ease of storage makes this approach adaptable to the demands of the Internet environment.

Today, the security of Internet communications depends in part on asymmetric encryption techniques, such as RSA.RSA was developed by Rivest, Shanir and Adleman to mathematically generate a pair of different keys, which cannot be obtained by any mathematical operation between them, but must be generated simultaneously. Its theoretical basis is: "when only the product of two large prime numbers of mutual elements is known, it is an extremely difficult problem to find the two prime numbers". Specifically, when $n = P \times q$, p and Q are two large prime numbers (also known as prime Numbers, irreducible Numbers). If you only know n, and you want to compute p and Q, this is a very difficult mathematical problem in the world. The foundation of RSA is the decomposition of large prime numbers.

[Case: RSA operation instance]
The following is the whole process of generating public and private keys and encrypting, transmitting and decrypting data:

(1) Select two prime Numbers, where $p = 5$ and $q = 13$. (Two prime numbers are kept secret. In order to calculate with the general tool functions such as EXCEL and MATLAB, a small but not safe prime number is selected).
(2) Calculate $n = P \times q = 65$ (public n. The length of n is the key length);
(3) The Euler function $\varphi(n) = (p - 1) \times (q - 1) = 48$ (treat it with confidentiality);
(4) Select the small prime e as the public key index, as long as $1 < e < \varphi(n)$; e and $\varphi(n)$ are relatively prime numbers. In this case, $e = 29$, because 29 and 48 are relatively prime numbers. (e is disclosed as a public key);
(5) Find the private key d, the relationship between the private key d and the public key e is: e multiply d and divided by $\varphi(n)$ to get a remainder of 1, according to $e = 29$, $\varphi(n) = 48$, set $d = 5$ (d as the private key, is kept confidential);
(6) Destroy p and q.Publish or send n with public key (29, in this case), keep private key (5, in this case);
(7) Data encryption. If the confidential data to be transmitted is clear text $M = 2$, then:

The transmitter is encrypted with the public key: $C = 2^{29} \bmod 65 = 32$; (function: = MOD(POWER(2,29), 65));
The receiver decrypts with the private key: $M = 32^5 \bmod 65 = 2$. (function: = MOD(POWER(32,5), 65)).

You can see, through the public key encryption, can only be decrypted by the private key, the role of confidentiality; If, on the other hand, encryption by a private key can be decrypted by a public key, the encryption is not confidential but acts as a non-repudiation digital signature. To determine the private key by using the public key, I need to determine $\varphi(n)$, and $\varphi(n) = (p - 1)*(q - 1)$, so it need to figure out p and q first, and because $n = p \times q$, it need to decompose the known n into p and q, a process called prime factorization. The prime factorization of n of 1024 bit binary number cannot be done by ordinary computers at present, but only quantum computers in the future will be able to do it in a short time. Based on this mathematical theory, RSA security depends on factorization of large numbers, and RSA can be considered safe when the key length is larger than 1024Bit (128Byte). Since the RSA algorithm usually computes large numbers, the slow speed of encryption and decryption is RSA's biggest drawback, and is only suitable for encrypting small amounts of data, or keys used to encrypt symmetric encryption algorithms, rather than encrypting content directly.

3. Tampering: Tampering with System Resources and Network Transmission Data is an Attack on Integrity.

The main preventive measures are message digest algorithm (hashing algorithm) and digital signature technology. For example, hackers take advantage of a security hole in DNS to replace a manufacturer's IP address with their own, which leads visitors to that manufacturer's e-commerce site to a fake website. This allows the hacker to change the amount of order in the order and change the shipping address. The order was sent to the manufacturer's e-commerce site, which did not know that the integrity breach had occurred and proceeded to fulfill the order after simply verifying the customer's credit card.

4. Forgery (counterfeiting): Unauthorized placing of falsified data into the system is an attack on authenticity (authentication). The main precaution is to rely on certificates issued by trusted third-party certification authorities such as CAS. For example, an online banking espionage Trojan horse induced or redirected, users in its fake website above the input account, password, user's bank account, password and other messages will be sent to the hacker designated mailbox, thereby bringing economic losses.

5. Denial (repudiation): Denying information you have sent is an attack on non-repudiation.

   The main preventive measures are digital signature technology and digital certificates. Due to the ever-changing business situation, once a deal is concluded, it cannot be denied, or it will inevitably damage the interests of one party. For example, when ordering raw materials, the price of raw materials is lower, but after receiving the order, the price rises. If the receiving party denies the actual time of receiving the order or even the fact of receiving the order, the ordering party will suffer losses. A Digital security certificate issued by a trusted third party is a common means to ensure the non-repudiation of every link in the electronic transaction communication process, so that the interests of both parties are not harmed. Non - repudiation is usually the result of authentication, non-forgery and non-tampering.

[Case: Blockchain and not copy, not tamper]
The blockchain technology can be applied in many fields, such as the authentication of notes. In reality, a file or note is unique. How to digitize and verify its uniqueness in the operation of mortgage and return is a big problem. Blockchain can be used to

achieve this purpose, so that a document cannot be copied or tampered with, nor can IT be transferred into two copies. In the past, one piece of document could be copied into two pieces under traditional IT. Blockchain can be used to truly simulate the uniqueness of such document bills in real society. At the 2017 Tencent global partners conference, Tencent Blockchain as a Service (TBaaS) was officially launched. Built on the basis of financial cloud, it provides users with Blockchain services under the scenarios of intelligent contract, supply chain finance and supply chain management, cross-border payment/settlement/audit and so on.

4.8 Emerging Industries

4.8.1 Personalized Customization

The future production mode will evolve from mass production to mass customization and then to personalized customization. Today, the production organization of manufacturing enterprises still generally has the characteristics of Ford assembly line, which cannot meet the needs of mass customization, let alone the needs of personalized customization.

As shown in Fig. 4.11, the method of mass production line is uniform and can be repeated continuously, but with obvious defects:

1. The arrangement of each process on the production line is fixed, and the products move from one process to another. It is difficult to adjust the production line if a high degree of customization requires a change in the process, resulting in the need for additional special processes.

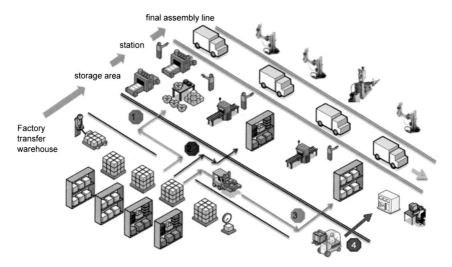

Fig. 4.11 Assembly line of mass production

2. Before the production of the order, it is necessary to consider the rhythm balance of the production line, carry out sequencing optimization, and then produce in the sequence, and the parts reflecting the personalized needs of customers will also be assembled in the sequence of the order. Due to the rigidity of the sorting plan, the complexity of sequence adjustment is beyond the scope of manual work, which requires the support of industrial software, which is often rigid and difficult to adjust and customize.

3. Sorting plan requires communication with suppliers in advance. Once the actual sequence of production orders is abnormal and changed, such as product defects, order changes and design changes, equipment failures, etc., complex adjustments and interventions are required, as well as a certain buffer time and inventory adjustment.

4. On the assembly line, production tasks are usually decomposed into undecomposable action elements and then distributed, resulting in monotonous work content and lack of interest and creativity.

In a word, the traditional way brings about the investment in hardware and software, limited personalized product configuration, low tolerance for anomalies and errors, and small space for improving flexibility and reducing costs. The production line needs to be transformed and upgraded from "rigid production flow line" to "adaptive flexible production mesh".

With the support of CPS, the autonomous interaction between products and machines, machines and machines, and between people and machines can realize a higher degree of flexible customization. This mode of production has the following characteristics:

1. The production order can be determined by the products and equipment themselves. In other words, the products or equipment themselves decide the next production.

2. The production route of each order can be adjusted flexibly.

3. The supply of corresponding raw materials and parts can adapt to the changes of production in real time.

4. Logistics no longer follows the fixed route, but can be adjusted flexibly and has certain intelligence.

In order to meet these characteristics, the following technologies are required:

1. Realize the identification of semi-finished products online through Near Field Communication (NFC) or visual recognition, and select the appropriate process according to the identification results.

2. Data extraction of production orders and reconfigurable adaptive process planning can maintain balanced production through rapid mold change during frequent switching.

3. Dynamic production sequence and line combination, and dynamic production logistics.

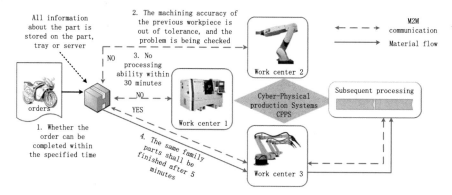

Fig. 4.12 Adaptive mesh flexible production line

4. Cooperate with the production and transportation rhythms and cycles of suppliers to realize the dynamic production of parts supply.
5. Online product quality testing to eliminate downtime testing.

Generally speaking, in the "adaptive flexible production mesh", the optimal processing route is dynamically formed through the interaction between product and equipment, and between equipment and equipment to handle abnormal events and solve uncertain problems such as personalized orders and supply disruptions, as shown in Fig. 4.12.

In the production mode of personalized customization, mass production is still indispensable, as is multi-variety and small-batch production. Therefore, there will be three production modes, among which the most complex is personalized customization. If the problem of individual piece customization can be solved, all kinds of problems of multi-variety small-batch production and single-variety mass production will be readily solved. The core of IM is to solve the problem of single-piece customization.

4.8.2 Remote Operation and Maintenance

China now has more than 600 million Internet users and 700 million smart terminals, and the booming development of the mobile Internet has accelerated the transition from manufacturing to services. The "industrial Internet" advocated by the United States links people, data and machines to form an open and global industrial network. Its connotation has gone beyond the manufacturing process and the manufacturing industry itself, spanning the entire value chain of product life cycle and covering more industrial fields such as aviation, energy, transportation and medical treatment. In the industrial Internet environment, manufacturing enterprises can monitor, analyze and improve the design and manufacturing of equipment through network data, and improve product reliability and efficiency through remote operation and maintenance.

Under the mode of "Internet+", traditional enterprises need to innovate their business models constantly and find a suitable service mode to impress customers. Remote operation and maintenance is a major way to make servitization, transformation and upgrading. For example, one of the problems encountered when China's high-speed rail exports to the United States is that in China, the railway authorities adopt the arrangement of running high-speed trains in the daytime and organizing maintenance at night. However, in the United States, there is a shortage of workers willing to perform maintenance at night, so the high-speed rail must solve the shortage of fault monitoring and operation and maintenance through long-distance operation and maintenance.

Remote operation and maintenance of production equipment is the main function of various industrial Internet under the cloud manufacturing mode. The current industrial Internet platform provided by large industrial enterprises and leading software enterprises is shown in Fig. 4.13. GE first defined the concept of industrial big data and developed the Predix platform. How to build China's local industrial Internet platform in the window period of the rapid development of industrial Internet platform is an important strategic issue that should be considered by all circles.

[Case: remote service solutions]
Orbotech's products include PCBs manufacturing process control systems and AOI (Automatic Optical Inspection) systems. In general, if AOI machines have problems, customers will call Orbotech's service department and ask service experts to use their domain knowledge to diagnose and solve the problem, and service experts will send field engineers to solve the problem. If the customer is far away, as in remote parts of China, and the time spent on the road and repairing it can take days.

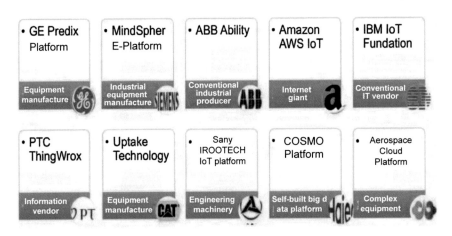

Fig. 4.13 Industrial Internet platforms of large industrial enterprises and leading software enterprises

In order to facilitate the implementation of remote service solutions, companies also need to ensure that end customers can control the remote behavior of the instrument. "The device has a lot of moving parts, so there's a chance that someone will get hurt while using it," says the company's chief information officer. "If someone orders the device remotely, the user will get hurt."

Orbotech decided to use PTC Axeda software after learning about proposals from major remote service software vendors. It has advantages in IoT connectivity and device management, such as flexible and lightweight software, and it can connect to customers through a firewall while ensuring security. Using the Safe mode of The Orbotech instrument allows service personnel and developers to remotely control the instrument in the machine's observation mode. In this safe mode, the engine of the instrument will not operate.

Orbotech also put their own integrated service management system to the PTC Axeda software, move the service of the database information management system in the remote software, so if the customer instrument problems can give service professionals complete information, set up a two-way flow of information, improve the diagnosis ability of each sending, reduce the wrong parts. The platform allows Orbotech engineers in China to work with the Israeli R&D department to solve problems together. This collaboration can also increase knowledge sharing with customers through instrument control and effective training.

4.8.3 Network Collaborative Manufacturing and Horizontal Integration

The model system of manufacturing pyramid is the skeleton system of automatic manufacturing, as shown in Fig. 4.14. International standards in most manufacturing industries are highly mature, and the architecture of IT/OT integration tends towards standard specifications, as shown in Fig. 4.14. The pyramidal centralized

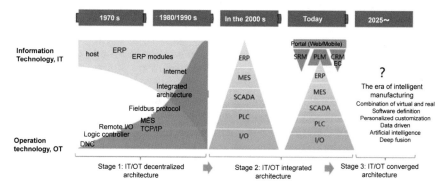

Fig. 4.14 Evolution of the IT/OT integration architecture

control architecture was finalized around 2000, however, and must be revised to accommodate fundamental changes in future manufacturing patterns.

In the customization model and dynamic market environment uncertainty, to complete implementation of IM system function, the traditional manufacturing system architecture that control paradigm based on pyramid hierarchical model will be made by a kind of new manufacturing services based on distributed paradigm that the new paradigm in Germany "industrial 4.0" also known as the cyber physical production system (CPPS). Due to the introduction of various smart devices, devices can be connected to each other to provide a network service. At each level, there is more embedded intelligence that can use virtual-controlled cloud computing. With these capabilities, it is possible to use new methods to control more extensive automation across hierarchies.

Specifically, as shown in Fig. 4.15, the traditional pyramid hierarchy of centralized production automation will be decomposed, and a CPS-based, distributed, decentralized, non-hierarchical, cloud virtualization service system will be the future direction. The architecture will shift from the traditional centralized automated pyramid hierarchical architecture to distributed cloud network architecture, from a set of fixed static systems to networked dynamic systems, and the new service-oriented paradigm will eventually transform the manufacturing system into a fully connected and integrated system. All manufacturing functions within these three dimensions and within the manufacturing pyramid can be virtualized and hosted as services, with the exception of some manufacturing functions that are very demanding in timing and security at the workshop level. Its information flow has the characteristics of network, wide information area, large data volume and diversified structure, and its workflow has the characteristics of virtual team and virtual enterprise, open innovation, process matching and collaborative ecosystem. MAS and blockchain will be important supporting technologies to facilitate this transformation.

Under the mode of virtual enterprise, the leader enterprise responds to a market opportunity, according to user requirements, determine product functions and development general task, and then to evaluate the task and preliminary decomposition, determine the role of need, through tendering/bidding select partners to form a virtual enterprise, complete the task allocation and implementation. The key tasks of design scheme, manufacturing task decomposition and partner selection must be mutually adaptive, which is a process of constant communication and multiple adjustments to obtain the optimal design and manufacturing scheme and the best enterprise portfolio. The above description is the essential characteristic of agile manufacturing model.

Rapid, severe, and uncertain change is the most unsettling market reality that companies and people must cope with today. New products, even the whole markets, appear, mutate, and disappear within shorter and shorter periods of time. The pace of innovation continues to quicken, and the direction of innovation is often unpredictable. Product variety has proliferated to a bewildering degree. Agility is a comprehensive response to the challenges posed by a business environment dominated by change and uncertainty.

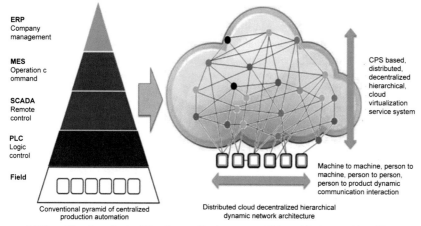

(a) Transition from the traditional centralized, automated hierarchical pyramid architecture to a distributed cloud network architecture

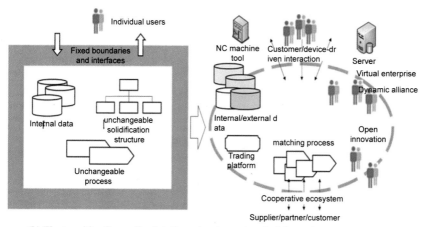

(b) The transition from a fixed static system to a networked dynamic system

Fig. 4.15 Evolution of the production system architecture. **a** Transition from the traditional centralized, automated hierarchical pyramid architecture to a distributed cloud network architecture, **b** the transition from a fixed static system to a networked dynamic system

For a company, to be agile is to be capable of operating profitably in a competitive environment of continuous, unpredictable, changing customer opportunities.

Successful agile companies, therefore, know a great deal about individual customers and interact with them routinely and intensively. Neither knowledge of individual customers nor interaction on this level was relevant to mass-production-era competitors. As suppliers of standardized, uniform goods and services, mass-production-era competitors relied on market surveys that created an abstraction: the "average" or "typical" customer. However, individuality could not be accommodated in a mass-production competitive environment.

By contrast, offering individualized products—not a bewildering list of options and models but a choice of ordering a product configured by the vendor to the particular requirements of individual customers—is the feature of agile competition. Success lies in formulating customer value-based business strategies for competing in the highest-value-added market, that is, in the most profitable, and the most competitive markets of today.

The global manufacturing network presents China with innovation opportunities. Multinationals, meanwhile, will produce the intentions of the separation of the technology is more obvious, its means of control of the manufacturing industry chain are from the indirect control is given priority to with joint venture and cooperation, to the "assets control" and "technical control" as one of the "direct control", "overall control" transition, deepened China related industry each link of MNC's technology dependence, for example, represented by Guangzhou Honda sino-japanese joint venture, the Chinese government has believed that as long as the product produced in China, China can control technology in the future, but the reality is far from being the case, Localization rate and technology digestion have always been the two major interest game points of China's auto industry and multinational companies. China's local manufacturers have formed a typical business model of low-end export-oriented and high-end import dependence. For example, 89% of the manufacturing equipment for integrated circuits and chips are imported, leading to marginalization in the division of labor in the global manufacturing network.

In order to promote China to climb up the global value chain with IM in the global manufacturing network, we should first strengthen the support for the development of the domestic industrial software industry, and promote the development and application of industrial software, especially the IM operating system platform. Secondly, innovate production mode and business mode, and build IM industry ecosystem with industrial alliance mode. Finally, build the intelligent value chain, enhance the enterprise intelligent management ability, and improve the added value of data.

Network collaborative manufacturing mode is the embodiment of horizontal integration of industrial system. According to the German "Industry 4.0" strategic plan implementation proposal, horizontal integration refers to "the integration of various IT systems used in different manufacturing stages and business plans, including the allocation of materials, energy and information within a company, as well as the allocation (value network) between different companies". That is to take the product supply chain as the main line and realize social collaborative production through the automatic and orderly flow of data between enterprises.

The horizontal integration of the industrial system is to connect the company with the upstream and downstream factories and customers through the industrial Internet. First of all, horizontal integration enables customers to have a closer relationship with the factory and to customize and adjust the goods purchased by customers in more detail, so as to meet the highly personalized needs of customers. Secondly, it makes factories more closely connected with each other, and can respond to the demand more quickly. Market supply and demand information is more unobstructed, and the price is more reasonable. Third, horizontal integration also integrates logistics into the network, making logistics and information flow more closely integrated, greatly improving logistics efficiency and reducing costs. Finally, horizontal integration also links the industrial system with data enterprises, smart grid and other enterprises, enabling factories to receive faster and automated support services.

4.9 Integration of Three Dimensions

Through the IM implementation described in Chaps. 2, 3 and 4, the integration of the three dimensions is realized respectively. The realization of automatic and orderly flow of data, and integration in three dimensions, is the main work content of IM implementation. The three dimensions of integration include horizontal integration of industrial system, end-to-end (or point-to-point) integration of operation system, and vertical integration of manufacturing system. As shown in Table 4.1 and Fig. 4.16.

Professor Thomas Bauernhansl, one of the architects of Germany's "Industry 4.0" plan and director of the Institute of The Fraunhofer Institute (the world's first industrial institute), estimates that the three integrations of intelligent manufacturing will increase industry's overall value-creating performance by 30–50%. Among them, the cost of inventory will be reduced by 30–40% due to the reduction of spare

Table 4.1 Fusion of three dimensions of intelligent manufacturing

Mode	Fusion content	Main object
Vertical integration	It is the integration within the enterprise and the basis of the other two integrations. The ERP at the top layer, MES as the middle bridge and PLC at the bottom layer have changed from fixed production line to dynamic production line	ERP, MES, industrial control, underlying equipment
End to end integration	The integration among enterprises around the product life cycle is the integration of product value chain, including raw material supply, R & D design, production and manufacturing, sales service. For example, the cooperation of automobile and mobile phone manufacturers	PLM、ERP、SCM、CRM
Horizontal integration	It is based on the value network and is the concept of ecosystem. Various IT systems applied in different manufacturing stages and business plans are integrated to realize social and ecological collaborative production. For example, the production and matching of aircraft carriers	Factory, users, partners, logistics

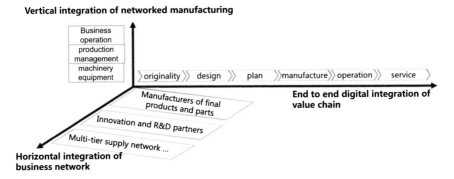

Fig. 4.16 The dimensions that drive business growth

inventory and the elimination of demand distortion. Manufacturing costs will be reduced by 10–20% due to increased manufacturing efficiency and employee flexibility; Transportation costs will be reduced by 10–20% due to high automation; The cost of complexity control will be reduced by 60-70% due to wider control span and reduced fault detection. The cost of quality control will be reduced by 10% to 20% due to real-time control; Maintenance costs will be reduced by 20–30% due to spare parts optimization, state-oriented maintenance, dynamic prioritization, and so on. In the long run, IM will bring down the cost of industry.

4.10   Practical Cases: Haier's "Integration of People and Orders" and the Internet Factory Ecosystem

By 2018, Haier has completed a total of five industrial lines, 28 factories and more than 800 processes of intelligent transformation, and built seven intelligent connected factories, including Shenyang refrigerator, Zhengzhou air conditioner, Foshan Washing Machine, Qingdao water heater and Jiaozhou air conditioner. Haier's implementation of the Internet factory has achieved initial results. The overall efficiency of the Internet factory has been greatly improved, the product development cycle has been shortened by more than 20%, the delivery cycle has been shortened from 21 days to 7–15 days, and the energy utilization rate has been increased by 5%. Haier's IM transformation mainly includes the following three aspects:

1.   User-centered and "integration of people and orders"

Haier has created a new management model in the Internet era: "integration of people and orders". A person is an employee. An order is not a narrow order, but a user's demand. "The integration of people and order" is to link the needs of users and employees together. Haier has been exploring for 13 years since 2005. Currently, COSMOPlat is the world's only industrial Internet platform for users to participate in interaction, and the core of its operation is "users". At present, users are mainly channel dealers, who can participate in the whole process of product design and development, production and manufacturing, logistics distribution, iterative upgrading and other links. Collaborative design customization mode is to integrate the fragmented needs of users, from inventory production to user production, and users can participate in the whole process of design and manufacturing, from a simple consumer to a "producer and consumer".

Haier's Internet factory provides not only industrial products, but also service solutions. The focus is more on user sensors than product sensors. What the Internet brings is zero distance between users. The platform is decentralized, disintermediated and distributed. Haier has also set a "non-entry rate", in which all products are delivered directly to users' homes instead of entering warehouses after being offline. The rate of non-entry for Haier products now reaches 69%.

2.   Build an interconnected factory ecosystem with parallel inner and outer rings

The relevant person in charge of Haier group explains the transformation of interconnected factories: subversion of traditional manufacturing system, made by large-scale transformation for mass customization, the user personalized requirements gathering, connected the plant through the big data to realize mass customization and personalized production mix, through the IoT under the support of people, machine, connectivity, meet the demand of the user personalization. Haier's proposed connected factory is not only a factory concept, but also an ecosystem, which includes enterprises and partners. Haier's process has changed from the series mode under the traditional enterprise ERP control to an open enterprise ecosystem in parallel.

The interconnected factory ecosystem has the following three capabilities:

One is to connect vendor resources and solutions. By integrating R&D resources and rapidly allocating resources, Haier can turn the relationship among employees, users and suppliers into a win–win business ecosystem. It only needs to put its R&D needs on the platform, find scientific research resources through two-way matching and even wait for solutions.

The second is to achieve zero distance between the user and the factory and realize the real-time interconnection of the whole process of the user. Users' personalized orders can be directly sent to Haier's global supply chain factories, so as to reduce the intermediate link between production and order processing and transfer the value of the middle part to users. Take a customized refrigerator as an example. After users log on to Haier's customized platform through PAD and put forward customized demands, the demand information will immediately arrive at the factory and generate orders. The factory's "Haier Personalized Customization Execution System" automatically matches the door panels, boxes and other components required by the order, automatically arranges production, and transmits instructions and information to each production line to finally produce customized refrigerators. Users all over the world can customize the personalized products they need through their mobile terminals at any time and anywhere, and participate in the design and manufacturing in the whole process.

Thirdly, the whole process is transparent and visible. The production and distribution of orders can be pushed to users in real time, and users can also query them in real time. Through product identification and tracking, users can realize real-time visualization of any link from their customized orders to the production of the factory, and then to the logistics. A "3 No" conference held in haier, namely, no one, no scene, no products, through the interconnected factory installed multiple cameras will broadcast live to the world, regardless of where the user lives, can all watch haier interconnected via cell phone or PC real-time factory production situation, haier passed in this way the concept of "transparent factory".

For example, in Haier's Jiazhou Interactive Air Conditioning plant, it takes the user as the center to build the capability of end-to-end information fusion, and through layout simulation, line balance simulation and automatic simulation, it creates the optimal layout with high flexibility and modularization. Through value stream simulation, logistics simulation, Human Factors Engineering simulation and process simulation, efficient self-optimized product manufacturing is constructed. In the equipment layer, through manipulator, integration line, image recognition, automatic detection and other flexible equipment to support unmanned customization production; In the control level and workshop layer, use ten thousand sensors, produce million level data every day, such as temperature, pressure, position of self awareness, self learning, self diagnosis, information-driven AGV, product type suspension conveyor, automatic sorting equipment, 3D storage equipment, such as structure of "a flow" shattered, zero wait, zero inventory of the intelligent logistics system; In the factory layer, iMES can automatically arrange production according to the user's order, with iMES as the core, PLM/ERP/WMS/MES/SCADA this five big system realize the interconnection from the device, sensor, network to top-level information system, each customized product can answer "which I should be sent to", finished product automatic parking, looking for park or car, realize the optimization of distribution path, automatic response to a user personalized orders. The entire factory becomes an IM execution system, and the role of employees is transformed into managers of machines. The purpose of interconnected plants is to achieve high quality products, shorter time-to-market and greater flexibility, which are contradictory to the traditional factory model based on mass manufacturing.

3.   Transform from a traditional enterprise into a platform enterprise

The precondition of an Internet factory is the transformation of enterprise organization, from traditional enterprise organization to a platform-based enterprise. Haier has transformed from a closed triangle organization, leading command and control to a staffs facing users directly and creating value for users. Internally, the enterprise's capabilities and resources are provided in the form of a platform, and employees can call them independently, thus self-organizing and self-driven. The organization becomes flatter and becomes a platform for employees to start their own businesses. Haier COSMOPlat platform has carried out social promotion of seven capacity modules, including interaction, design, procurement, logistics and service. At present, the platform has been connected to ten thousand enterprises. In

2018, the transaction volume of COSMOPlat platform will reach nearly 400 billion yuan.

To be specific, the whole supply chain, including production, manufacturing, logistics, procurement and other links, has been transformed from a traditional chain of departments to a common user oriented department micro. These department micro and user micro are parallel. If the user micro does not create user value, then these purchasing micro or manufacturing micro is not valuable. For example, Haier's "Car micro" platform is built on the basis of logistics capability. For example, an employee of Haier bought a car and joined Haier's logistics distribution system. He or she found partners through the information platform and formed a pair with them. They grab orders, deliver goods, install and maintain goods every day, and share the profits in proportion, forming a community of interests, sharing risks and super profits. They have become a "micro enterprise" and a "rapid reaction force". The platform is also open to the public, allowing idle vehicles in the community to help them with deliveries. In 2014, there were 90,000 vehicles on the platform, with more than 30,000 service outlets, and logistics promised to arrive within 48 h. By the end of 2017, Haier platform had more than 200 start-up micro, 3800 micro nodes and 1.22 million micro stores. To these micro businesses, the CEO, Zhang ruimin, is a shareholder rather than a leader. After 13 years of change, Haier eliminated more than 10,000 middle-level managers and turned itself into an entrepreneurial platform.

In June 2016, Haier acquired GEA of the United States. In the decade before the merger, GEA's sales were down 11%, and in 2017, their profits grew 22%, several times faster than their revenue. In the past, GEA, like most businesses, was in tandem. R&D, manufacturing, sales and other departments are not open to each other. When all the departments are connected as a whole, a series becomes a parallel. When it becomes "parallel," everyone is trying to figure out how to make the product create user value. After such a big international merger and acquisition, no Haier employee entered GEA. Instead, the original management of GEA was allowed to make decisions by themselves. On December 6, 2017, the Institute of Electrical and Electronics Engineers (IEEE), one of the four international standards organizations, approved a proposal for mass customization of international standards led by Haier.

Thinking Exercise

1. What has Haier done in setting international standards? What does this mean for Haier?
2. What has Haier done in terms of business model reform? What innovations?
3. This paper tries to analyze the characteristics and innovation of "people and orders in one" from the perspectives of user demand, enterprise architecture and system engineering. Try to analyze the intelligent characteristic level of Haier's Internet factory ecosystem.
4. What work has Haier done in vertical integration and horizontal integration?
5. What should be the goal of enterprise IM transformation? What are some common mistakes?

Homework of this Chapter

1. In the Central Research Institute model, what are the differences in the division of labor and cooperation between Central Research Institute and Divisional R&D Institution? Why is that?
2. Please briefly describe the impact of IM (Intelligent Manufacturing) on employees.
3. Should Midea and Haier, or Alibaba and Tencent, or Huawei and ZTE, take the lead in establishing the industrial Internet? Which corporate platform is more promising?
4. Why does it sometimes need to take millisecond-level data acquisition to achieve the target in the monitoring of Feng Aluminum company?
5. Why did Feng Aluminum company take six identical machines to be monitored at the same time to find an effective energy saving measure?
6. The neural network for Google energy management include external factors such as wet bulb temperature, outdoor humidity, wind speed and direction. Why?
7. What is the significance of software autonomic control?
8. Which level in the IM technology maturity model begins to have requirements for system integration?
9. Which level in the IM technology maturity model begins to require data security?
10. Try to analyze industrial Internet security from the perspective of national security strategy.
11. What can the key generated by the RSA algorithm do to the data? What do these two different keys do?
12. Where was blockchain first used? How does it prevent copying and tampering?
13. What are the benefits of remote operations and maintenance? What are the difficulties in achieving a two-way flow of information and instruction?
14. what is the role of personalization in emerging formats?

Reference

1. Sing. The interaction between intelligent manufacturing and energy transformation in new industrial revolution. Sci Manage Res. 2017;5: 45–8 (in Chinese).

Chapter 5
Comprehensive Implementation Path of Intelligent Manufacturing

5.1 Inspiration Case: Toyota's Lean Manufacturing

1.   Lean R&D

On January 8, 2018, Akio Toyoda announced at the consumer electronics show that Toyota would transform itself from a car manufacturing company into a mobile travel company, and its competitors would no longer be car companies, but technology giants like Google, Apple and Facebook. "I am determined to create new ways to transport and connect with our customers," he said. "Technology is changing rapidly in our industry and the R&D race is on." In order to build a better mobile ecosystem, Toyota will also work with Amazon, Didi, Uber, Mazda and Pizza Hut. The e-Palette, which can provide a variety of services, is expected to be used in the 2020 Tokyo Olympics, a disruptive platform for its products. The e-Palette by low chassis, box of barrier-free design, spacious interior space, and through the combination of different chassis and car body, carrying all kinds of equipment, and can realize different functions, play a shared travel, mobile selling cars, freight cars, and other roles, for sharing rides, hotels, retail service partners such as the needs of different uses. In addition, the e-Palette will tap into Toyota's Guardian, an advanced autopilot assistance system, and carry smart Connectivity technology for broader commercial uses.

In order to meet the challenge of new product platform development, Toyota attaches great importance to standardization of R&D process, establishing balanced R&D process, reducing variation with strict standardization, and establishing highly flexible and predictable output. One approach is: they put a lot of work for outsourcing, the adoption of a standardized project list and knowledge base, is also set up a set of design engineers design and assembly line manufacturing standards, make any other company's engineers also take part in the project successfully, and quickly into the state of R&D. Toyota integrated suppliers into its R&D system. For all engineers to construct a "spire" type knowledge structure, the database provides standardized components and ensures simultaneous access to design data, as well as production process formulation and procurement during product design. American

C. Lai, *Intelligent Manufacturing*,
https://doi.org/10.1007/978-981-19-0167-6_5

counterparts often outsource, but for unknown reasons, often suffer poor results and high transaction costs, which they blame on their suppliers.

The company will provide a large room for each R&D project, and the R&D staff will live together. Establish an appropriate organizational structure to find a balance between technical expertise within functional departments and cross-functional integration. The chief engineer can also post key information about the entire project (such as technology, finance, and schedule) in this room, and he manages the team with a simple, visual communication approach. The chief engineer of Toyota is a leader and technical integrator, who has the final decision-making power on major issues related to product projects. He can not only represent the voice of customers, but also bear the ultimate responsibility for the success or failure of the product. Unlike the general project manager of the company, the latter only controls the personnel and duration of the project. Many companies imitate Toyota's lean design model, but most don't understand why Toyota requires such a chief engineer.

Thoroughly analyze the options at the beginning of the development process, when there is the most room for design change. Many auto companies, including Toyota, didn't make a hybrid years ago when they were preparing to develop the Prius. At the time, Toyota's President told the chief engineer that he would have a total of four months to complete the overall design of such a hybrid so that the Prius concept could be on display at the show. The chief engineer who received the task found about 80 hybrid engines by visiting customers and other ways, and then selected them from 80 to 10, and then from 10 to 4. Finally, after computer simulation of these 4 engines, one engine was finally selected. It was from these concepts, then fairly abstract, that Toyota engineers made their decisions.

2.　Lean Production

Two lean principles advocated in Toyota are still the mainstream of the core management idea: Jidoka; JIT.

(1)　The principle of automation (Jidoka, 自働化)

Toyota believes that "Jidoka = human + automation". Toyota uses this word " 働" having a herringbone to express a tight combination of human and automatic machines. The Jidoka refers to when there is something wrong with the production, equipment or production lines with automatic stop or active ability will make it stop. People call it the art of stopping. The main purpose is to prevent bad products flowing into the next process.

Switching from one operation to another in just-in-time (JIT) production mode can be a huge quality hazard, as operational errors often occur during frequent switching. A JIT production line often switches hundreds of times a day to balance the delivery of customized orders, which is a dangerous business. Whether punctuality can be implemented depends on whether there is an automation platform that can guarantee the quality. Jidoka equipment operation mode is to solve this problem.

The features of Jidoka rely on personnel with judgment ability or machines with human intelligence. Jidoka equipment can automatically stop when processing a bad product, find the cause of the abnormal in time, avoid causing more accidents. For example, a minor error in the program may result in a collision or a safety accident. The shortage of water and electricity may cause short circuit of equipment, which will lead to excessive busbar current in the whole plant, thus causing tripping or fire hazard. If the lubrication is insufficient, the equipment may be sintered. If fixture wear is not found, then the artifacts size will be out of tolerance, the product in use may occur mechanical seal failure, leakage accident. All this major breakdown is generated due to failure to find glitches. Therefore, one of the purposes of Jidoka equipment is predicting and preventing failure accidents in time.

Jidoka principle requires workers and equipment to be able to judge whether process is qualified, if not qualified, equipment or workers will judge, and to stop it, will not allow defective artifacts to flow to the next working procedure. To deal with defects, the implementation of "three not" policy "do not accept, do not make, do not pass". The main methods are as follows: adding detection devices in automatic equipment to make the machine possess some intelligence of humans; to detach people from jobs that can be replaced by machines and to avoid human error. Including the following technical methods:

1) One-human multi-machine and U-shaped production line

Smooth production can make the employees clearly feel the necessity of Jidoka, which is the premise of Jidoka. One-human multi-machine needs three conditions: one is the appropriate equipment layout design, U-shaped production line is the common layout design; Second, multifunctional work; The third is the continuous reevaluation and revision of the standard operation.

In a U-line, because the finished one unit comes out at the exit, the raw material of another unit comes in at the entrance. If both export and import operations are performed by the same person using standard operations, the quantity of products in the production line can be effectively controlled instead of being increased or decreased at will, thus effectively controlling the quantity of products flowing online. The station distance and the distance between the lines of the production line should be shortened as far as possible. In a U-line, Workers only need to turn around to operate the opposite machine, so as to reduce the waste caused by workers' walking and material movement, as shown in Fig. 5.1a.

2) Error-proof and fool-proof technology

Designed to prevent wrong operation, so that there is a mistake or defective goods can not be installed on the fixture, or the machine will not be processed. If the previous process is found unqualified in the later process, the operation of the previous process shall be stopped; If there is any omission in the operation, the post-process will stop; Design tools and facilities based on ten principles of error prevention. For example, based on the principle of insurance, in order to prevent punch operators from accidentally getting their hands hurt, design two operation buttons that must be pressed at the same time. Based on the principle of automation, the elevator will not start to work until it senses gravity; Based on the principle of isolation, defective products and good products are stored separately in different areas, and small parts of different models are placed separately in different boxes.

(a) Applicable U-line spacing   (b) limited trolley dimensions and part heights

Fig. 5.1   Compact production line and trolley with a size limit

Visual management is the application and extension of error—and—fool—proof technology. Visual management keeps workers from being tied down by equipment. Specific measures include: designated parts in designated locations, not misplaced parts; Set the maximum and minimum limit indicator, and realize first-in, first-out control through orderly arrangement; Production management board showing production status; A flag distinguished by different colors or symbols shows the current operating state of the device. To realize the current state, for anyone to judge whether normal or not, and to clarify abnormal disposal management methods are the three levels of incremental goals.

3)   Andon technology

Collect information about equipment and quality on the production line, so that each equipment or workstation is equipped with call lights. If problems are found in the production process, the operator can pull the lamp rope (Andon cord) and turn on the lights, or let the lights turn on automatically, so as to attract the attention of others and solve the problems in time. When the production cycle time (Takt Time) can't keep up with the requirements or there is a non-standard situation, they can call for help with the light; Using laser technology and camera technology to judge the installation position of each component, and then record it into the quality system, the production line will decide whether to stop or continue to flow down according to this judgment.

Focus on the issues that the lights are pulled the most often rather than the lights stay on for the longest time. Frequent pulling of the light means that the root of the problem has not been found. Moreover, frequent pulling of the light will make the team leader spend a lot of time responding, and the production line will frequently stop, causing a variety of wastes. This also applies to manual production lines when a problem occurs and the equipment is shut down. If there is a malfunction or a defective product, the operator himself stays in place and notifies his superior.

The closer to the source of the product life cycle, the more valuable Jidoka is. If we can follow the Jidoka principle to improve in product conceptual design phase, the greatest improvement in quality will be achieved at minimum cost. Therefore, the Jidoka principle has the connotation of concurrent engineering.

Strategically, based on the Jidoka principle of "stop if there is a problem, correct the error and then develop", Toyota firmly believes in the concept of "tree-ring operation", believes that all plans to expand capacity and increase sales must be organic and sustainable, and only when market demand exceeds capacity can new production lines be built. But that solid growth philosophy has repeatedly led Toyota to come a step or two behind its rivals in the market. When Volkswagen entered China for nearly 20 years and almost monopolized the passenger car market, Toyota only launched its first car on the mainland in 2002: the Toyota Vios, which sells for as much as 195,000 yuan.

(2)   The principle of Just In Time

Just In Time (JIT) refers to "the production of the required product in the required quantity only when necessary". Specific implementation approaches include:

1)   Minimum moving line and minimum area

The goal of layout design is to minimize the moving line and the workshop area. In order to control the workshop area to the minimum, the width of the road should be controlled as far as possible when designing the road. The one-way channel is generally controlled at about 1.8 m, and the two-way channel is controlled at 2.5 m. To two-way channel two cars passing by, therefore, to control the width of a car at 1.2 m, height is also limited. As shown in Fig. 5.1b, the main purpose is to speed up the car. The parts do not fall off, the point from the ground to the highest parts is in control of 1.5 m high.

The engine company and its supply chain structure adopt the "1 + N" mode: centering on the engine, a circle with a radius of 15-min drive is constructed, and the main suppliers are evenly distributed throughout the entire circle. The engine company, as one of the suppliers of vehicle factory, is also within a 15-min drive radius of the vehicle factory. The size of the inventory depends on the proximity of parts suppliers and the frequency of delivery.

2)   "One piece flow"

"One piece flow" is the main line of layout design. "One piece flow" production refers to the process and movement of parts one by one through various process equipment in accordance with a certain working order, with only one work-in-progress (WIP) or finished product in each process at most. Its characteristics are as follows: after finishing the processing in the previous process, it can immediately "flow" to the next process for further processing, instead of batch processing and batch moving. There is no need to stop and wait, as shown in Fig. 5.2. The whole factory is like using an "invisible conveyor belt" to connect various processes and production lines, forming an integrated "one piece flow" production of the whole factory. Around the fundamental principle of minimum waste, the flow minimizes WIP inventory.

Fig. 5.2 "One piece flow" production line

3)   Kanban-based pull production

Toyota's production execution system is a production indication system, with no reverse control. In the system design stage, kanban recycling frequency, recycling mode and kanban calculation method should be considered. Logistics goes from parts purchased by suppliers to machining and assembly lines, 80% of which are on the assembly line, so when designing distribution logistics kanban, the focus is on kanban from logistics to assembly line. As shown in Fig. 5.3a, a traditional logistics kanban is a piece of paper with information inside a plastic bag on the side of a parts box, which is sent to the assembly line along with the parts. When the assembly line operator begins to retrieve the parts, kanban is placed in the Kanban recycling box and waits for delivery person to retrieve it next time.

The important indicators of Kanban calculation include: receiving capacity of shelves, transportation capacity of logistics trolley and sorting time. As shown in Fig. 5.3b, the receiving capacity of the shelf at the side of the production line is determined by its length, depth and height, while the length is limited by the spacing between two stations of the production line, the height is slightly lower than the height

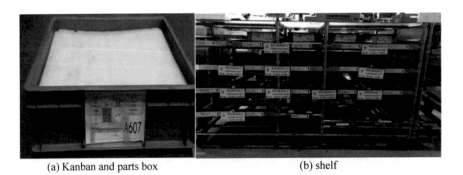

    (a) Kanban and parts box   (b) shelf

Fig. 5.3 Shelves, Kanban and parts boxes **a** Kanban and parts box, **b** shelf

of human eyes, and the depth is determined by the size of parts and the convenience of access. The size of the shelf should be designed according to the above constraints to find the optimal value.

Reducing shelves will be able to compress storage area and space, on the one hand to reduce inventory management costs related to space size; Secondly, it can improve the convenience of handling and management of spare parts, so as to reduce various problems caused by storage loss such as poor storage, overdue storage and "dead goods". Finally, compressing the shelf area next to the production line will provide preconditions for shortening the length of the production line and reducing the station of the production line, which can significantly reduce the production cost. The parts in the parts warehouse change dynamically with the variety switching and flexible production, so it is necessary to design and optimize the storage scheme at any time.

In Toyota's post-pull production mode, the order system will be reflected in the production management system, and the information of parts required by the production management system will be sent to various suppliers and manufacturers, as well as to the manufacturing system of various terminals. The system is connected together. Toyota's marketing model is to sell to order, that is, to produce only when there is an order. Therefore, Toyota will never pressure dealers with sales tasks, but will patiently collect demand and help dealers tide over difficulties. Toyota's average inventory is only 0.7 months, which its rivals can't even imagine.

The operation optimization of parts sorting area is a big problem of logistics distribution. The common practice of factories is to adopt the way of "who sorts who distributes". Under this practice, quality assurance is relatively difficult, because the source of the problem is the outflow source of the problem. Secondly, a prominent problem of this method is that the operation process and content are relatively complex, and the personnel training is more difficult. It takes about 2–3 weeks for a new employee to get on the job, which is much longer than the 3–5 days for a position on the production line. In order to solve this problem effectively, the management idea of separating the sorting operation from the distribution operation is put forward, and the sorting operation is built into a production line, which can change with the change of production volume. Meanwhile, the problem of too long personnel training cycle is also solved.

After comprehensively considering the shelf capacity around the production line, the transportation capacity of the logistics trolley and the sorting time of the parts sorting area, the kanban number of certain parts can be effectively calculated.

(X)—(Y)—(Z) is the basic way of kanban expression, where X: How many days are there for every delivery; Y: How many deliveries per day? Z: How many times has the kanban been delayed before it is sent back? Such as:

1-10-1: Daily delivery; 10 times a day; The last kanban was sent back after delay once.

2-1-0: Delivery every 2 Days; Only one delivery on the delivery day; Kanban sent last time is returned this time without delay.

The calculation method of the number of Kanban is as follows:

Number of Kanban = {number of daily use × X × [(Z + 1) /Y] + safety stock}/number of receptions per box

Usually at the end of each month, that is, around the 22nd, publicly release next month's production plan. Logistics personnel need to calculate the number of Kanban, increase or decrease. Before or after 28 days (except February), specify the number of kanban required for each component in the next month, and submit it to the kanban management room after confirmation by the logistics section chief. Kanban management room staff will add or decrease the number of kanban in the kanban management cabinet. Monthly changes, 12 times a year, are carried out in this way to constantly adjust the number of Kanban, so as to minimize the inventory around the production line.

4)   Flexibility

Flexibility = general + special. The overall area of SPS (Set Parts System) is in accordance with the "general + special" way of the overall layout. The same components of each type of machine are placed in the same area for distribution. The parts of the selection are not based on signal indication, can be directly selected. This is the flexible use of JIT. For different parts of the machine, it is important to use signal light or raster control, because no operator can be guaranteed to make no mistakes and maintain a high level of concentration for eight hours a day. To be a manager is to reduce the risk of mistakes, to make the operator's work simpler and clearer, so that the chance of mistakes will be reduced to zero. The load of workers needs to be balanced. For example, in the engine assembly station, the engine is heavy or light, and the weight should be matched at this time, which is also a manifestation of humanized management.

5)   Standardization

The "TNGA" (Toyota New Global Architecture) imported by the company has the concept of nest ecosystem. Its core idea is chassis generalization. Through the chassis to promote modularization, standardization and generalization to reduce costs, the parts above chassis are open to each region to design for local and regional demand. In the future, the rate of parts co-ownership will reach 60–70%, and multiple cars can share parts to reduce costs. Before the introduction of TNGA, Toyota engines had to respond to local regulations and local environment. There are more than 800 types of engines, each of which has to be developed and tested in different regions. After the introduction of TNGA, the engine is reduced to more than 200 kinds, and the cost is greatly reduced.

6)   Collaborative value co-creation by customers

Customer collaborative value co-creation is to create the value of products and services by cooperating with customers, which is the new connotation of the principle of JIT. Toyota believes that the current most crucial competition is to compete for customer connection interface, and the customer's link to determine the future. If the automobile manufacturer does not transform, cannot connect to the customer, the enterprise will become the OEM factory, even does not need the brand. Big Internet companies will be able to acquire traditional car manufacturers, so Toyota plans to connect factories and sales, including tracking the entire product life cycle and mobile travel platforms based on big data platforms.

The link with the customer needs to rely on the application of the Internet, need to develop a network of information mobile terminal platform to support the car owner and users travel. The platform that can most impress customers is to cater to the changes of customers' lifestyle and build it around the life of car owners. As people in different countries have different lifestyles, the platform should be re-developed locally. For example, WeChat APP can be used in China to solve the problem that people often refuse to answer marketing calls. Toyota does not have the power to change the way customers live, but WeChat can. Only by connecting with customers can Toyota have real big data. Manufacturing big data is mainly equipment data, which is relatively static. But people change rapidly, so marketing big data is dynamic.

To support the lean production model, Toyota has more than 50 information systems operating around the world. For example, GPPS (Global production planning system), quality control system, equipment maintenance and other auxiliary management systems. Most of the systems are not yet connected. They are individual entities, and it is very difficult to integrate these data. The next step is to focus on data fusion and big data processing. The data collected from multiple systems will be fused to assist the comprehensive decision, making the decision more convenient and efficient, and avoiding the big impact of system change on the enterprise.

Thinking exercise

1. What are the principles and benefits of Lean R&D in Toyota?
2. What are the principles and benefits of Lean production in Toyota?
3. What is the relationship between lean production and IM? What are the similarities and differences?
4. Please explain the ten principles of error prevention and give examples of application.

5.2 The Introduction

The IM implementation of a single dimension is applicable to small businesses. For large enterprises, IM implementation should be cross-dimensional, with each dimension advancing simultaneously in each step of the path. Integration for a single domain or a single dimension can be very complex, and in fact many people work on a single domain for a long time. However, it can be observed that in this IM ecosystem, industry organizations engaged in a single dimensional integration are expanding, so the integration scope is bound to involve more dimensions. Manufacturing paradigms, including Continuous Process Improvement (CPI), Flexible

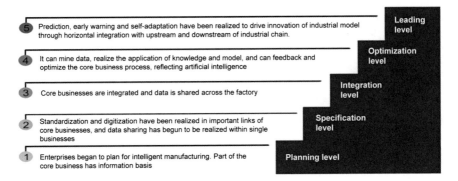

Fig. 5.4 Intelligent manufacturing capability maturity level [1]

Manufacturing systems (FMS), Design For Manufacturing and Assembly (DFMA), and Design For Supply Chain Management (DFSCM), depend on the exchange of information between different dimensions. Refer to the three dimensions of the IM ecosystem in Fig. 5.19.

As shown in Fig. 5.4, the Capability Maturity Model (CMM) for IM is divided into five levels, which define the five stages of IM and describe the path for an organization to gradually move towards the final vision of IM. It represents the current implementation degree of IM and is also the result of IM assessment activities.

In order to gradually improve the capability maturity of IM, this chapter proposes a five-stage integrated implementation path of IM, as shown in Fig. 5.5.

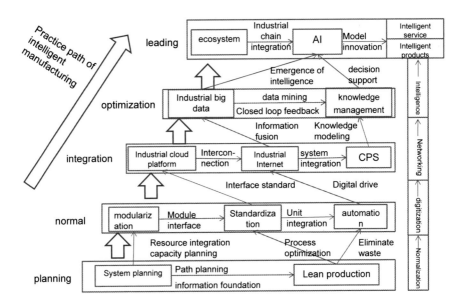

Fig. 5.5 Integrated implementation path of intelligent manufacturing

5.3 Planning Level (Level 1): System Planning and Production Improvement

At this initial level, the enterprise has the idea to implement IM and starts to plan and invest. Part of the core manufacturing links of those enterprises have realized process information and started manufacturing activities based on IT, but this only laid a solid foundation for IM, and has not really entered the category of IM.

5.3.1 System Planning

More than a decade ago, enterprises successively introduced the ERP system originated from the West. Nowadays, under the encouragement of the slogan of "replacing people with machines", manufacturing enterprises have successively implemented various "IM" or "intelligent logistics" systems in factories. However, through relevant tracking and investigation, it is found that most projects have a low success rate due to blind motivation and lack of planning, just like the ERP system in those years.

IM project is a system engineering, which needs a scientific decision-making and implementation process. The demand for the IM system must come from the development of the enterprise itself, and any of the following decisions are incorrect:

1) Blindly follow others: follow the IM system implemented by others;
2) Government policy-oriented. Enterprises have no needs, but with the financial support of the government, the system can be built with less money or without spending money. However, it should be considered that a large amount of money should be invested in the future upgrade, transformation and daily maintenance of the system.
3) Whitewash. Although there is no demand, in order to enhance the corporate image, some enterprises have implemented a system beyond what they can bear.

The overall planning and design of the IM system should consider both the vertical dimension of space and the horizontal dimension of time. The spatial dimension is considered because the IM system is a part of the whole management system, and it must be associated with many systems to realize "intelligence or automation". For example, the intelligent logistics system in the workshop must exchange data with the inventory management system, material supply system and production planning system in order to realize the functions of automatic feeding and batching. Secondly, the time dimension is also an important factor. Enterprises are constantly developing and changing, including changes in the market, technological iteration, business model changes, etc. To predict such changes, the system can better serve the present and future enterprises. If changes are not foreseen and uncertainties are not taken into account in the planning, design and implementation of the system, large-scale secondary reconstruction will be necessary in the future.

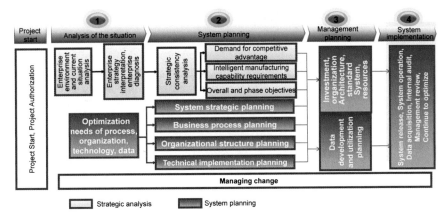

Fig. 5.6 A typical process of intelligent manufacturing planning and implementation

To sum up, system planning is an important step in the construction of the IM system, which needs to be completed by a team mastering mechanical engineering science, information software, electronic control and modern management science and engineering theory and practice skills. The team is an interdisciplinary team of experts. In addition, the IM system is embedded in the entire management system and closely related to the enterprise's development strategy. It is necessary to maintain strategic consistency and cover the needs of the enterprise's future development with foresight.

5.3.1.1 System Planning Method

Intelligent Manufacturing is a system engineering, and the "four-step" of IM planning is: status analysis, system planning, governance planning, and system implementation, as shown in Fig. 5.6.

5.3.1.2 Status Investigation and Evaluation

After the launch of the project, We found out the real needs through the status quo investigation and determined the upgrade path of IM. A large number of enterprises have the conditions of manufacturing informatization, and also know the urgency of promoting IM, but some of them are not clear about their own needs, let alone the foothold and path of IM. They are in the fuzzy state of "want to use but don't know how to start". Therefore, the most urgent task for enterprises to implement IM is to clarify their needs.

Whether the real needs are identified makes all the difference. For example, for driverless cars, the real demand of car owners is to drive without worry. For unmanned supermarkets, the real demand of customers is to avoid queues; For unmanned factories, customers' real demand is low cost, high quality and high efficiency. The status quo investigation and analysis mainly includes the following steps:

Positioning: Firstly, select and determine the maturity assessment model (as a whole or as a single item) suitable for the enterprise to be evaluated. According to the characteristics of the industry where the enterprise is located, cut out 27 evaluation domains (see the matrix diagram shown in Fig. 1.13 for details) and determine the evaluation domains. Then collect internal and external environmental information and existing strategies of the enterprise, analyze, classify and confirm the information.

Sorting: considering the needs put forward by the enterprise in the positioning stage, sort out and set different questions and indicators for each evaluation domain, state, analyze and deduce the questions and indicators, demonstrate their necessity and comprehensiveness, and set scoring principles.

Diagnosis: according to the scoring principles set above, the maturity requirements of 27 evaluation domains are scored and diagnosed from the perspectives of process, organization, technology and data, and then calculated according to the domain weights set by different industry models to form the evaluation scores and total scores of domains and classes level by level, and finally determine the maturity grade according to the score segment. Through the maturity evaluation, enterprises can recognize the existing problems and gap between the current situation of the enterprise and the maturity model and benchmark enterprises. According to the gradual improvement method of the maturity model, and according to the current situation and objectives, enterprises can formulate the scheme and promotion path of intelligent manufacturing, carry out pilot in a certain field, and gradually expand the promotion after success.

5.3.1.3 System Planning and Design

System planning includes system strategic planning, business process planning, organizational structure planning and technology application planning. Draw lessons from the planning process of industry best practices and standards, the steps of system strategic planning and system architecture planning are shown in Fig. 5.7. Strategic planning system is mainly based on the enterprise management strategy, and formulate the demand of sustainable competitive advantage, determine IM capacity requirements, and then determine business process optimization, structure, technical implementation scheme, and data management requirements. Strategic coherence should be maintained between the preceding and subsequent steps of the planning process.

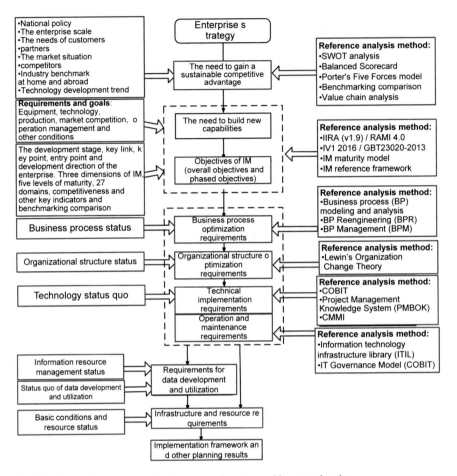

Fig. 5.7 Steps of system strategic planning and system architecture planning

5.3.1.4 System Governance Planning

IM is the deep integration of information network technology and extended big data, artificial intelligence and manufacturing. Therefore, the implementation of IM needs the guarantee and promotion of organizational structure and standard system.

System governance is a kind of structure and institutional arrangement for realizing IM and maximizing the interests of enterprises. The role of system governance in improving enterprise operation and management efficiency and creating enterprise value is mainly reflected in the following aspects: (1) make more reasonable use of enterprise information resources to promote the overall enterprise income and value maximization; (2) Accelerate the improvement of corporate governance structure

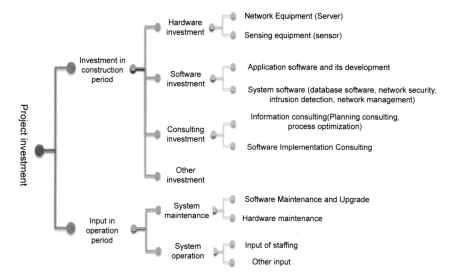

Fig. 5.8 Investment cost composition of intelligent manufacturing project

and steadily promote the construction of modern enterprise system; (3) Avoid enterprise risks to the maximum extent to ensure the smooth realization of the enterprise's overall strategic goals and IM goals.

The decision-making scope of system governance planning generally includes the following five aspects:

1. The Investment

The object of investment and the amount of investment are important decision objectives.

As shown in Fig. 5.8, the general structure of the project budget is that software costs account roughly for 1/3 of the total cost, accordingly, hardware sensors, servers and network construction costs account for 1/3, and finally internal training, business investigation, data sorting, project summary and other expenses account for remaining 1/3. Software projects also require the IT department (or outsourcing) and the business department to invest manpower or man-hours.

[Case: IM benefit analysis of a power supply manufacturing enterprise in Huizhou]
A power supply manufacturing enterprise in Huizhou approved the construction of the IM system in 2015. After three years of development and implementation, the total project cost is RMB 28 million ¥, as shown in Table 5.1.

Key equipment of the project includes automatic production line and software system, the cost of which is shown in Table 5.2.

Table 5.1 Total budget unit of project investment (10 K RMB$)

Subject	Investment (10 K ¥ RMB)
Labor cost	440
Information software cost	565
Input of information equipment	20
Automation equipment input	1675
Other expenses	100
Total	2800

Table 5.2 List of key equipment of the project

No.	Name	Supplier	Price (10 K ¥ RMB)
1	ERP K3 (including upgrading and customer service)	Kingdee K3	135
2	MES (including intelligent material cabinet, PDA, computer and other hardware and installation training services)	TCL	105
3	Intelligent warehouse light picking system (WMS)	TCL	50
4	Lean material distribution (WMS upgrade/APS)	TCL	60
5	Industrial Internet of things and big data platform	Universities	50
6	Equipment management system	Universities	30
7	Energy management system	Universities	30
8	Lean manufacturing knowledge base and industrial engineering system	Universities	60
9	CAD	PTC	45
10	Automatic production line	Multi-suppliers	400
11	Network server	Multi-suppliers	20

Project results:

1) Increased production efficiency by 24%: realized lean manufacturing and quality traceability, reduced 20% material personnel and warehouse keeper through flexible automatic production scheduling, and reduced production and line change time to 0.5 h; In SMT and DIP workshops, industrial IoT and big data platform at the workshop level were established to realize automatic production lines. The number of operators was reduced from 30 to 8, which was better than the average level of 15 in the same industry with the same output value. The production efficiency is calculated by the statistical method of UPPH (units Per Hour Per Person). UPPH = total output/actual man-hour input. The total labor cost of production workers in 2015, 2016 and 2017 was 36 million yuan, so the labor cost was reduced by 8.64 million yuan if the sales volume remained unchanged.

Fig. 5.9 The light corresponding to the position of the material needed

2) The just-in-time delivery rate increased by 12%: Automatic feeding was realized through the industrial big data platform and the intelligent storage system with light picking function, as shown in Fig. 5.9.

The feeding speed was shortened from 55 to 30 s; Just-in-time delivery rate = just-in-time deliveries/total deliveries, calculated on an annual basis. According to the compensation loss of 100,000 yuan caused by each untimely delivery, the average annual delivery of 25 times, three years to reduce a total loss of 900,000 yuan.

3) 10% reduction in energy consumption per unit product: The main energy consumption of the project is electricity consumption. Energy consumption per unit product = total electric charge of the company/total quantity of shipments of the company. The total electricity cost of the production system in 2015, 2016 and 2017 is calculated to be 10.5 million yuan, saving 1.05 million yuan in electricity cost.

4) The defective rate is reduced by 20%: the defective rate is calculated by using the statistical method of PPM, PPM = the number of defective products/the total number of production × 1 million; This index is obtained by implementing the product quality traceability system. Based on the loss caused by defective products, which is about 3 times the value of the product, the cumulative sales in 2015, 2016 and 2017 are 187,1793 million yuan, and the defective rate before transformation is 35PPM. Therefore, the loss was reduced by 39,300 yuan in three years. Therefore, the income = 864 + 105 + 90 + 3.93 = 10,629,300 ¥.

To sum up, combined with the total investment of 28 million yuan, the project is expected to recover the cost within 7.9 years, which fails to meet the expectation of recovering the cost within 5 years. However, the company believes that 7.9 years is still shorter than the depreciation and service life of the software and hardware of the IM system purchased. The project also improves the image and competitiveness of the enterprise. Therefore, the project is considered to be a preliminary success, and it decided to invest in the second-phase project.

2. Organizational Pattern and Structure

The organizational pattern includes three types: centralization, decentralization and hybrid structure. The "centralization" and "decentralization" of organizational structure have both advantages and disadvantages. The key problem is the degree of support of the organization for business. For medium and large enterprises, the preferred organizational model should be centralized governance.

In the process of system governance organizational structure design, the internal and external development environment of the enterprise should be combined to select the organizational structure suitable for the actual development of the enterprise. The mainstream system governance organizational structure types include the following three categories.

(1) IM Department + development, operation and maintenance outsourcing

In this organizational structure mode, the original information management department or the newly established IM department of the head office serve as the "management center" of the enterprise information work, and the positions are divided according to professional and technical expertise, but the system development, operation and maintenance work are completely outsourced to the system manufacturers outside the company.

The advantages of this mode are as follows: the number of department personnel is small, and the organizational structure is refined; The main work is "coordination and management", and the technical application and system R&D are all outsourced, which frees the department staff from the complicated technical work of system development, operation and maintenance, and concentrates on "management". Disadvantages include: no grouping according to the major, so that one person has to do multiple jobs, and the division of work is easy to be unclear; The system development, operation and maintenance work are completely outsourced, which can easily lead to technology development and product application being completely controlled by others. It is not conducive to the cultivation of its own IM technical force, and will also cause a large amount of investment loss to external manufacturers.

(2) Fully IM technology enterprises

In this mode, the company needs to establish a new IM technology enterprise with independent accounting and self-financing, which is fully responsible for the IM project construction, system operation and maintenance of the whole company and even external enterprises. There is no leadership between it and all departments of the headquarters. Instead, the relationship between Party A and Party B is formed through the signed project contracts.

The advantages of this mode include: technical service work is more specialized and standardized; The "technical service quality" can be completely controlled through the "Contract terms". Once the service of the technical enterprise is not up to standard, Party A of the contract can completely refuse to pay and thus improve the service quality. Fundamentally transform the IM construction department from a "cost center" to a "profit center". Disadvantages are as follows: due to the complete realization of internal marketization and the promotion of IM construction entirely by "contract", the overall direction of system implementation is easy to be unclear and unified leadership and supervision are lacking; After the enterprise, the IM department is easy to become "mercenary", easily resulting in false service prices, cutting corners and other phenomena.

(3)   IM Technology Center

In this mode, enterprises upgrade the original information management department to "IM Technology Center" or "IM Business Division", and usually set up several offices under the center, such as planning standard management office, project management office, system security management office, procurement management office and technology office.

The advantages of this mode include: the implementation of professional division of labor within the center, and the improvement of the professional level of various business work; Through continuous project construction, the center will cultivate a skilled IM team. System operation and maintenance, development funds are not easy to outflow, completely feed back to the center. Disadvantages include: the whole organization is huge, and the position setting in the center is complicated; Technicians have a heavy workload.

In order to choose the right organizational structure, you need to consider whether the organization focuses on stability, flexibility, or cost, and the weight of each goal, as well as whether the system is purchased externally, developed internally, or developed internally. This requires a distinction between:

If the company from the strategic attaches great importance to the construction of IM, IM for a company to realize the strategic target has extreme importance, is necessary to set the IT department to higher-level organizations, such as IM technology center, to establish the project team combined with functions of matrix organization operation architecture, and the corresponding set of scientific and reasonable performance evaluation mechanisms.

When the company developed into a sufficiently powerful group company, and is in the leading role in the industrial chain, the company's IM department has enough ability to provide high-quality overall information service solutions for all subordinate units within the group company, and has spare power to export IM technology capabilities to external stakeholders and build a closely coordinated industrial chain ecosystem. At this time, the IM department can be transformed into an independent and self-financing IM technology service company, become a "fully IM technology enterprises", which is not only responsible for the IM construction tasks of the company, but also actively seek business. For example, Midea Group's "Midea cloud smart digital" company is a completely IM technology enterprise. In this way, we can not only avoid the passive situation of being completely outsourced by others, but also focus on developing our own unique core competitiveness of IM.

If the company is a traditional manufacturing enterprise, and lacks adequate resources, good technical support and system arrangement in the field of intelligent, and the board of directors of the company lacks enough ability to handle large IM centers, even independent companies, and there is also no relevant development plan, the mode of "IM Department + development, operation and maintenance outsourcing" can be adopted.

3.   Standard System

A good governance system can strengthen the standardization of architecture, technology and suppliers across the organization, create a solid guarantee for business operations, and at the same time solidify the idea of system governance. In the design process of system governance standard system, the following principles should be generally followed:

1)   Compliance. During the development of the system governance standard system, the relevant national laws and regulations, various management systems of the enterprise, the framework of COBIT (Control OBjectives for Information and Related Technology) and SOX (sarbanes–oxley) Act generally adopted internationally should be strictly followed.

2)   Applicability. In the development of system governance standards system, we should also consider the actual situation of company itself and its industry.

3)   Feasibility. The system Governance Standards System should clearly and easily define the various processes of system governance, especially the operational activities of key processes. Refer to the ITIL (Information Technology Infrastructure Library) standard.

4)   Completeness and Soundness. The construction of system governance standard system should standardize the main activities of enterprise information system operation management, covering the main information service processes such as data management, project management, service management and quality management, and comprehensively manage the people, finance and goods involved in the process.

In short, in the construction of the system management standard system, to highlight the "unified control, unified technical standards" system governance core idea, through the establishment and perfection of standard system, standardize the company headquarters and subordinate units and their IM system construction and operational work behavior, and with high quality and efficiency to ensure the smooth operation of the company's management system.

4.   The Resources

The resources required for IM projects mainly include hardware and software investment, business personnel input, existing systems and equipment, and required management, business, technical capabilities, data utilization capabilities and infrastructure. The resources and capabilities required by the project are the important basis for project classification and priority setting.

Accumulating data is an increasingly important resource. According to the actual needs and technical conditions, effective storage and management of data should be designed and selected. Common methods include: Data encryption; Database and data warehouse storage; Cloud backup services, etc. As for the source of resources, enterprises should be good at utilizing the combination of internal and external resources, and usually reach service agreements with a few preferred service providers.

5.3.2 Lean Production

Lean Production (LP) is a management science derived from Toyota's Production System (TPS). The goal of lean production is to enhance the enterprise's market adaptation and rapid delivery capacity, and at the same time reduce the production cost, so that enterprises can achieve higher economic benefits in production. Lean production is first geared to the needs of many varieties of small batches designed for personalized needs, its two core are JIT and 自働化 (Jidoka). The Lean production system is shown in Fig. 5.10.

JIT's core lies in the pursuit of inventory-free production, the pursuit of rapid response, eliminating waste. 5S activity, 10-min mold change, balanced production and Kanban management are its core management tools. One purpose of Jidoka is through low cost automation equipment to reduce work intensity, and to achieve zero defect; Secondly, it is humanization saving. "human-saving" is achieved as "labor-saving". In the case of saving the caretaker of monitoring equipment operation, the

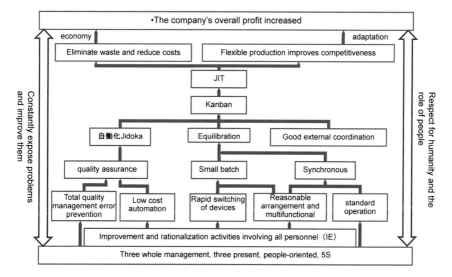

Fig. 5.10 Lean production system

equipment will automatically stop when abnormalities occur. Its idea can be summarized simply as: "an abnormal immediately stop, absolutely no defective production", "people do not do the machine's guard slave". "ANDON" rope pulling device, labor-saving chair, error-proof (fool-proof) design and other measures are concrete measures. For specific implementation, see the inspiration case in this chapter

So far, lean has evolved into a concept and method involving marketing, R&D, supply chain, production, business process and even the whole value chain management, driving the transformation of the global industry, from manufacturing to services, the idea, methods and tools of pursuit of "creating value to eliminate waste" promoted the production resources to optimize configuration, quality, efficiency and speed of response.

Traditional lean factories face new challenges. Modular Supply, Set Parts Supply (SPS) and JIS (Just In Sequence) are new practices for modern enterprises to improve their core competitiveness in the environment of high-speed mixed flow production.

In electronics industry, the average inventory cycling time for Chinese companies is 51 days, compared with eight days in the US. In the textile and apparel industry, the average inventory cycling time of Chinese companies is 120 days, while that of industry benchmark ZARA is 15 days. Even with the same profit margins, Chinese companies will have lower returns on their investments. IM cannot be based on this inefficient mode of production. Lean is the first step that must be taken. Lean hardly requires a large additional investment by the enterprise, and can be rewarded by simply redeploying production resources on top of existing ones.

Lean production and IM complement each other, and their correlation is as follows:

1. They both have the same purpose, which is to improve efficiency and benefits.
2. Lean production focuses on management, and principles of lean thinking fully conform to the characteristics of IM, which can guide the development of IM; Meanwhile IM focuses on network information technology, can provide intelligent tools for lean production.
3. Lean production and IM can also be considered as the relationship between material foundation and superstructure. The essence of lean production is to eliminate non-value-added activities in the production process, while IM makes value-added activities automatic, flexible, intelligent and efficient after eliminating non-value-added activities.

Therefore, the relationship between lean production and IM is complementary to each other. Their deep integration and superposition of effects can greatly improve efficiency and benefits.

5.4 Specification Level (Level 2): Standardization and Digitization

When entering the specification level, the enterprise starts to invest in the equipment and system supporting the core business on the basis of the formed IM plan. Through technological transformation, the main equipment is equipped with the ability of data acquisition and communication, and the automation and digital upgrading covering the important links of the core business is realized. By developing standardized interfaces and data formats, the information system can realize internal integration, and the data and information can be shared within the business, and the enterprise has stepped into the threshold of IM.

5.4.1 Modular

1. Modular Products

Modular production network has gradually become the core of the modern new international division of labor system. More and more industrial products are transitioning to modular systems and promoting specialization of production, which lowers the barriers of entry for developing countries to integrate into the global production network.

Cars and computers are the first industries to realize modularization. Typical automobile companies such as Volkswagen and GAC need about 500 suppliers, while Smart modular cars only need 25 first-level module suppliers. Future auto parts suppliers will provide automotive power, chassis, body, interior and electronics, totally five complete automotive system assembly. Faced with the competition from Apple, IBM built a modular computer structure to solve the contradiction of increasing product diversity and reducing manufacturing costs. The modular computer led to fierce competition and rapid innovation in operating systems, microprocessors, peripherals and so on, while the outsourcing of core components fostered industry giants Microsoft and Intel, as well as direct competitors such as Acer, leading to the "hollowing out" of IBM and the decline of the computer-manufacturing business.

IM enterprises need to integrate internal and external resources, change the mode of cooperation with suppliers, and transform parts supply into modules collaborative production, so as to realize the modularization of product parts. The advantages of modularization of components are as follows: first, the complexity of products can be reduced, product diversity can be increased at a lower cost, and product development cycle can be shortened through collaborative design. The second is conducive to module manufacturers to make fine modules, accelerate product innovation, and also accelerate technical iteration through the replacement or acquisition of module manufacturers. The third is making the parts easy to be replaced and personalized, and also facilitating the user to participate in the customization; Fourth, the product dimension is reduced and the management range on the supply chain is reduced, which is conducive to the standardization and intellectualization of the production process. Therefore, the breakthrough point for enterprises to realize IM is to establish collaborative innovation mechanism and improve product modularization level.

However, compared with integrated architecture, modular architecture often lacks overall coordination and performance optimization, and its weak link is often found in the interface between modules.

[Case: VW's MQB modular platform]

At the Frankfurt Motor Show 2017, Volkswagen (VW) Group President Martin Winterkorn said: "The modular strategy will give us an advantage in production, investment and so on." According to the plan, VW's modular production line capacity will reach 3 million vehicles per year in the future. Starting from 2010, the MQB modular vehicle technology platform will be successively applied to front-drive models of VW brand. This platform has the following characteristics: First, it can effectively reduce the development cost of a single model and shorten the cycle of development of a single car; Secondly, modular production can be compatible with a variety of power combinations and even new energy powertrain. The Passat B8 will be VW's first B-class car to unveil its MQB-B modular production. Due to the advantages of modular production, the B8 has become the most derivative model in Passat history, including the station wagon version, the four-door coupe version, the convertible version and so on.

[Case: BMW's "Strategy NUMBER ONE"]

There are 350 models of BMW cars, 175 kinds of interior, 500 kinds of configuration, and 90 kinds of painted colors, which means that there is a theoretical combination of species of 10^{17} BMW cars. This greatly increased the difficulty of product development and had a significant impact on the entire business process of BMW. Hence BMW's "Strategy No. 1", which includes a full modular integration of models and powertrains, has helped the group cut billions of euros in costs. The core of the project is to establish a unique product structure concept on the central data platform that can support product configuration and product change management, and achieve direct unified management across departments and the entire product life cycle. Departments engaged in product management, product architecture, product configuration, product design, parts procurement, prototyping, and even partners work on a unified data platform. Herbert Diess told the press that BMW is planning 45 new models, including 30 rear-drive cars and 15 front-drive models. "In addition, BMW will achieve internal synergies between the BMW, MINI and Rolls-Royce brands." Rolls-Royce will continue to share some elements with BMW, such as electronic systems and electrical components; The powertrain will be significantly improved on the BMW system. The project allows BMW to reduce the order lead time from 28 to 12 days. Among them, it takes one day to freeze orders, six days to prepare plans, two days to assemble, and finally three days to send cars to dealers.

2. Modular Equipment

From modular design, modular procurement to modular production, fixed-cycle production and assembly lines will be replaced by modular mobile devices. The modular processing line in the IM mode is devices connected by wireless network, which has the following advantages:

1) Modular construction brings higher agility and flexibility of production line;
2) Combine more workstations with minimum input, or add processing capacity as needed conveniently.

For example, in the automobile manufacturing process, each product component corresponds to a standardized and modular device. When the combination of product components changes due to personalized design, the device can flexibly combine with the changes and communicate in real time through WIFI or Internet. The equipment "tells" the car body parts that it is ready for production and realizes self-organized production; Auxiliary systems assist workers in various manual installations, and optical signals indicate which parts must be taken out; The device uses Bluetooth communication to identify workers and adjust operation instructions according to their knowledge level and preferred language. Check the completeness and accuracy of the installation steps by the auxiliary system to reduce errors and workload. In this way, the mode of production will be changed again.

Modularization is a popular trend in industrial facilities engineering, and this is for a good reason. When planned from the early stages of engineering and combined with advanced work packaging, the benefits of optimized modularization include reduced costs, shortened schedules, and safer work environments, relentless optimization. Vista Company has engineered some of the most successful large-scale processing facilities in the Canadian energy sector. A key success factor has been its relentless optimization of modularized designs in facility engineering.

Modular Kitchen Equipment is easy to install & service with modular flexibility. The modular design format of "Silverlink 600" enables fast food restaurant to create a custom-built appearance in the kitchen. Countertop units can be sited on a range of pedestals and stands allowing store managers to mix and match according to their requirements and the space available. Better still, as their operation grows, Silverlink 600 adapts quickly and easily to keep pace with consumer's changing needs.

3. Modularized Production Network

From the perspective of technological innovation, product innovation in modularized production network is no longer concentrated in a single enterprise, and the source of innovation tends to be scattered, which significantly reduces the threshold and risk of innovation, and better solves the incentive mechanism and trust mechanism. On the other hand, modular production will lead to intensified competition and the formation of certain market forces, which will promote the development of the local manufacturing industry in the direction of innovation and upgrading. The institutional and structural characteristics of modular network organizations are conducive to the promotion of knowledge flow and technological innovation. For example, Cisco can acquire its own modules through "A&D" (acquisition and development) by purchasing the most cutting-edge technologies from the outside to achieve rapid innovation. In turn, knowledge flow and technological innovation further enhance the level of modular division of labor, promote the upgrading of modular network organization, effectively reduce the risks brought by the modular process, and enhance the competitive advantage of the organization. Through a large number of practices at home and abroad, it has been proved that modular network organization is not only an important part of the formation of regional innovation system, but also an important engine to drive regional economic development and realize industrial upgrading.

5.4.2 Standardization

"Intelligent manufacturing, standards first". Standardization is the basis of automation and the premise of IM.

1.   International Reference for IM Standardization

Through the efforts of various countries, great progress has been made in the formulation of IM standards. At the core of Germany's "industry 4.0" is the establishment of a Cyber Physical System (CPS) for the interconnection of humans, equipment and other resources, and this interconnection must be based on a standardized system. The Working Group on Industry 4.0 believes that the implementation of "Industry 4.0" requires action in eight key areas. "Standardization and reference architecture" is at the top of the list. To this end, "Industry 4.0" will develop a package of common standards, make cooperation mechanisms possible, and optimize production processes through a series of standards. At the same time, the German government believes that standards are not formulated from the top down by the government, but emphasizes that enterprises take the lead and develop from the bottom up. In December 2013, the German Association for Electrical, Electronics and Information Technology published the "Industry 4.0" standardization roadmap, providing technical standards and specifications as well as routes and plans to achieve the standard.

Japan also recognizes the importance of standards. Advanced companies such as Toyota's experience is, "lean manufacturing" should be after removal of waste and standardization, in order to obtain sustainable results. Its intelligent manufacturing was realized through the path of "modularization—standardization—automation—intelligence".

The "Industrial Internet" of the United States covers the entire ecological chain of industrial manufacturing enterprises with connectivity among devices, big data collection and analysis technology as its core content, and provides norms and guidance for the development of the IM industry through connectivity interface standards.

The standardization of data acquisition interface is the key content of IM standards. The "Industry 4.0" organization of Germany and the Industrial Internet organization of the United States have listed OPC UA (OLE for Process Control Unified Architecture) as the standard specification for realizing semantic interoperability. OPC UA provides a standard for how to encapsulate an information model and open platform communications (the new interpretation of OPC), and OPC UA's object-oriented approach makes these developments easy.

2. Standardization of IM in China

Manufacturing standardization has a long history in China, which can be traced back to the Warring States Period.

[Case: Qin's standardized weapons]
According to an archaeological study, weapons left over from the vassal states of the Warring States Period are of different lengths, different weights, and in various forms. There is no unified standard and no standard weapons. Only the weapons of the State of Qin, no matter where they are found, are astonishingly identical in shape and size. In ancient ruins of Qin Emperor, there are about 40,000 three-edged arrows, all of which are within a millimeter of deviation from each other, just like the weapons produced by modern assembly lines. This makes the parts and arrowheads of the crossbow used in the war interchangeable. In addition, it also makes the soldiers' daily training exactly the same as the weapons used in the war, ensuring the training effect and efficiency. Millions of troops waged war for years, so it can be imagined that their consumption was huge. Qin dynasty provided and maintained the demand for weapons in the war with its own strength. The material support was the advanced and efficient standardized production technology of Qin.

China attaches great importance to the IM standard system. The integration of modernization formulated by China has now become an international standard. In order to solve the problems of missing standards, lagging and overlapping standards, we should give full play to the basic and guiding role of standards in promoting the development of IM, to guide the standardization work of IM in the current and future period of time, according to the reference model and system framework of IM standard system issued by the state, combining with the characteristics and needs of industries and enterprises of a country and the current lack of interdisciplinary, cross-industry system integration standards, and in accordance with the principle of "common first, urgent need first", formulated a number of IM industry standards, including the "foundation", "safety" and "management", "testing evaluation" and "reliability" five types of general standards, and "intelligent equipment", "smart factory", "intelligent services", "industrial software and data", "Internet industry" five types of key technologies standards, give priority to making breakthroughs in machine equipment data transmission interface standards, logistics standards and standard containers, product module design standards, IM implementation process specifications. The key jobs are as follows:

(1) Standardized operation process and operation mode

Automation can only be developed with standardized job processes and methods. For example, automatic welding and automatic assembly, assuming that parts and components vary in variety and operation mode is not fixed, it will be difficult for robots to cope with, and automation will be difficult to achieve, even if the implementation cost is high. Some enterprises deliberately create this kind of disharmony.

One is to prevent the technology from being imitated. The other is to create non-standard automation in order to sell high prices, which makes it difficult for small and medium-sized enterprises to accept, then IM can only be implemented by large enterprises. On the contrary, if standardization is realized, the volume can be increased, the cost and price of automation equipment can be reduced, supporting enterprises gradually develop, a benign ecosystem is formed, the whole industry will naturally rise, the penetration rate of automation will be higher, intelligence will be achieved naturally. Domestic enterprises lament the high degree of automation in the automotive industry, wondering why such a complex product can be automated, but such a simple product as home appliances is difficult to achieve, an important factor is standardization.

[Case: Midea standardized packing boxes]
Midea standardized semi-finished product packing boxes, so that a kind of packing boxes can be transferred and used between different enterprises, making the logistics between partners smooth, eliminating movement waste and environmental pollution caused by unpacking, decomposition, secondary packing and carton, and creating an ecological circle. Midea has also further promoted the use of standard containers such as EU boxes and wheels to improve the consistency of partner quality.

(2)   Standardization of equipment communication and data acquisition

International vendors typically do not provide data at the core of automated equipment, meaning that technical secrecy prevents much of the information from being collected, only by means of paid collection from designated industrial software. What can be provided through a standard interface is often only simple data, such as running data, which is not sufficient. Intelligent diagnosis in the future, including intelligent feedback, intelligent guidance and self-learning, requires some in-depth data, such as not only pointing out that the temperature is high, but also pointing out specific parts, because different parts have different fault causes.

At present, the level and capacity of China's intelligent basic industries cannot meet the needs of enterprise transformation, and it is difficult to accelerate enterprise transformation only by relying on market traction. The industrial network standards and protocols, key equipment, intelligent control equipment, corresponding intelligent core technology is subject to the United States, Germany and Japan and other developed countries, inadequate equipment openness objectively brings high costs to the transformation and upgrading of enterprise application, restricted the transformation of enterprise from the return on investment and other economic interests level.

This constrained and paid-for data collection makes it difficult for Chinese enterprises to conduct big data analysis and cloud computing processing, so it is difficult for them to connect single machines and equipment into industrial operation systems in an all-round and whole process at low cost, so as to realize data-driven and supply–demand matching. Both the government and enterprises should focus on solving basic bottlenecks such as data integration and connectivity encountered in the current promotion of IM, and collaborate in the R&D of remote procedure call

(RPC) and data acquisition protocol, as well as related technologies such as radio frequency identification technology, real-time positioning and wireless sensing, so as to produce Chinese standards and promote their application.

5.4.3 Automation

One of the characteristics of IM is the wide application of automation technology. There is always a degree of intelligence in automation. At present, the advanced unmanned factories in developed countries are the result of the comprehensive upgrade and leap of automation technology. Their basic characteristics are that the main production activities are controlled by computers, and the production lines are equipped with robots without a large number of workers. Logistics automation is also crucial to the realization of IM. Materials transfer between processes can be realized through logistics equipment such as AGV, rack manipulator, hanging conveyor chain, etc., and material supermarkets can be equipped to deliver materials to the line as far as possible.

With the advance of technology and the increase in labor cost, automation is an irreversible trend. Enterprises need to combine their own situation to plan the direction of automation. Start with the part with the largest return on investment and the easiest to achieve. First, consider how to use power assisted equipment to reduce the labor intensity of workers, and then optimize the corresponding production process to advance step by step.

As shown in Fig. 5.11, the classic automation pyramid model shows a clearly defined hierarchical structure. Information is obtained from the field equipment layer on the site up through three levels of PLC control, SCADA remote control and MES operation command, and is obtained by ERP and other top-level management software. ERP and other software make decisions through data analysis and send

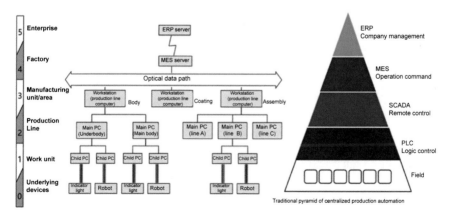

Fig. 5.11 Centralized production automation pyramid

instructions back to the bottom. Although this model is widely accepted, the flow of data is not smooth. To provide the underlying data more directly, people turned to the open OPC UA standard. The standard provides a pattern whereby OPC UA "clients" at the management and enterprise levels can invoke data directly from OPC UA "servers" at the equipment level.

Automation and information are the largest part of the investment needed to realize IM. Enterprises should be careful before doing automation transformation, need to analyze the necessity of automation, return on investment, automation equipment reliability, risk, extended flexibility, etc., reasonable return on investment period should be controlled in less than 5 years, to prevent the cost not falling but rising. After investment, many enterprises find that the equipment is not mature, failure is frequent, or the newly purchased equipment is not easy to configure, so it is not as flexible and convenient as manual operation. As a result, expensive automation equipment is put on hold, and ineffective investment will even bring down the enterprise. Enterprises with few product varieties and large production batches can realize a high degree of automation and even set up black lamp (unmanned) factories. Small-batch enterprises with multiple varieties or customized orders should pay attention to the combination of humanization and man–machine, and make use of human flexibility to deal with the uncertainty of orders and production.

5.4.4 Digitalization and 3D Data Model

1. Digitalization

Enterprise must do a good job on digitalization while IM transformation. Data is the most important asset for enterprises in the future and the driving force for innovation of production mode and business mode. Automatic and orderly flow of data is a necessary condition for completing the transformation to intelligence.

Digital management refers to the process of making clear measurement, scientific analysis and precise qualitative analysis of business work through a sound basic statistical statement system and data analysis system, and recording, inquiring, reporting, publishing and storing in the form of data statements. It is one of the modern enterprise management methods. The goal of data-based management is to provide a real and effective scientific basis for managers to make decisions. Take the manufacturing data as an example, if the manufacturing data is analyzed by Stream of Variation (SoV), and it is found that the quality problem of a product is caused by the upstream material and design process, it can prompt the upstream to detect the link with the problem based on this result, and then change the selection of materials, improve the design process, and finally achieve the optimization of the product, and even realize the upgrade of the entire industrial chain.

Digitalization is expected to play an important role in the supply-side reform. To realize real IM and give play to the value of data will not only change the internal management, production process and design process of enterprises, but also solve the dilemma of effective docking between the supply side and the demand side. Data from the demand side, such as relatively open personalized and customized requirements and individual consumers' hobbies and behaviors, will change the supply side with relatively closed data. Data analysis can even create more consumer demands, and then connect these data and feed them back to the production and R&D process on the supply side. In this model, the impetus for supply-side reform is provided by demand-side data.

At present, the bottleneck problems of data are mainly: (1) dumb terminals. There are a large number of dumb terminals in the "man machine material method environment (4M1E)", making it more difficult to obtain customer data; (2) Multiple protocols. Diversified equipment types, software types and interface protocols, resulting in the closure of production process data and process data; (3) The supply side and the demand side are not effectively connected, so the big data on the demand side is difficult to collect completely.

2. 3D Data Model

The 3D product model in virtual manufacturing must contain geometric data, material data, process data, production management data and equipment data, as well as the production process data, detection data and experimental data. Therefore, the 3D data model is very complex. 3D data models can carry people's design knowledge, process knowledge, manufacturing maintenance and guarantee knowledge, as well as various mathematical and physical parameters. Therefore, 3D digital technology is the basic technology of IM.

In the process of product design, process analysis and manufacturing, people first establish 3D digital virtual bodies. With 3D data model, physical production and process manufacturing become simpler. This model is available from the design process to the maintenance guarantee, and only a single model is needed for a single product. Only when a 3D data model is established, can the digital factory simulation software be used for the simulation of equipment and production line layout, plant logistics and human–machine engineering to ensure the reasonable structure of the factory. In order to establish the digital map of the factory, it is convenient to gain insight into and predict the status of the production site, and assist managers at all levels to make correct decisions. To conduct virtual experiments, such as virtual wind tunnel experiments, to reduce the number of physical wind tunnels and reduce costs.

Because the previous 3D models could not express the data of process manufacturing and testing, 2d drawings must be generated after the 3D product design to guide the production. In fact, the conversion of 3D models to 2D drawings will generate a lot of extra work and is prone to errors. Therefore, in 1996, Boeing joined 16 manufacturing companies in the world to promote the establishment of a working group of The American Society of Mechanical Engineers (ASME). It took seven years to establish the MBD standard, which is based on the definition of model and realized the representation of process manufacturing materials and test data on the

3D model. This means that there is no need for the 2D drawing process, and no need for 3D to 2D conversion. Therefore, the development of Boeing 787 has truly realized paperless design, with significant improvement in design and manufacturing quality and efficiency.

With digital virtual body, the design process, manufacture and experiment of products can be completed in cyberspace. In this process, design problems, process problems, manufacturing problems and test problems are constantly found. After problems are found, virtual models are directly adjusted to solve them. After all problems are solved, they are mapped to the production process and test process.

[Case: China's combataircrafts and C919]

As China's first digital prototype, with a fully digital shape, the Flying Leopard can reduce the frequency of wind blowing by about 60%, and finally make its modified aircraft fly in two and a half years. Since 2010, all new aircraft developed by the Chinese aviation industry have adopted this technology. The C919, which made its maiden flight on May 5, 2017, originally had a 500-aileron design and would have required 15,000 wind tunnel tests, which would not have been possible without CFD. Based on the virtual wind tunnel test, it was optimized from 500 ailerons to 8 sets, and then started to do physical model, and then went to the physical wind tunnel test all over the world, and the physical wind tunnel test spread all over the United States, Britain, France, Germany, Netherlands, Ukraine, Russia and other countries, and then optimized to 4 sets and then to 1 set. A large modern commercial aircraft usually consists of more than four million parts and dozens of complex systems supplied by suppliers from all over the world. It is far more difficult and complex than ordinary engineering projects to fit these systems and parts together perfectly.

5.5   Integration Level (Level 3): Interconnection, Integration and Networking

After entering the third stage, enterprise investment mainly shifts from the infrastructure, production equipment and single information system, to Internet integration. The important manufacturing business, production equipment and production unit have completed the digital and network transformation, which can realize information system integration among the core businesses such as design, production, sales, logistics, service, etc., and begin to focus on the sharing of data within the factory, and finally complete the preparation work for intelligent upgrading.

The focus of this phase is the integration of information islands and the realization of data sharing throughout the factory. The integration of data collected by machines and human intelligence will promote the optimization of the whole plant and the realization of enterprise management objectives, and the economic benefit, employee safety and environmental sustainability will be greatly improved.

5.5.1 Industrial Cloud Platform

In the past several years, cloud computing, along with big data and artificial intelligence, became a hot word of national concern, especially the term "enterprises go to cloud "." go to cloud" means that enterprises can access computing, storage, software, data and other computing services provided by cloud service providers anytime and anywhere through high-speed Internet. For a long time in the past, enterprises preferred to adopt self-installed servers, self-built data centers and other ways to carry out information construction. After the cloud, the combined cost drops to about one-third to one-half of what it was before.

Cloud manufacturing is an IM mode based on the concept of cloud computing, all product-related planning, production, warehousing, maintenance and other information are stored in the cloud for continuous access by other products or systems, and remote work or remote services can also be realized. In recent years, driven by the national strategic policy of IM, cloud manufacturing mode has been greatly developed. Cloud manufacturing service platform is the realization carrier of cloud manufacturing mode. The Cloud manufacturing service platform has received great attention and investment from all sectors of the industry, universities and research institutes. Among them, Sany Heavy Industry's Root Interconnection, Aerospace Cloud Network, Foxconn Cloud for small and medium-sized manufacturing enterprises, and Midea Cloud for upstream and downstream partners of the ecological chain are typical representatives. Gartner predicts that the global public cloud service market will grow by 12.6% and reach US \$331.2 billion in 2022. Small and medium-sized enterprises in China generally have problems such as lack of capital, small scale and low level of information. Cloud manufacturing service platform can effectively solve these problems, and its difficulties mainly include the business model and market mechanism of cloud manufacturing.

Cloud manufacturing can provide users with on-demand full life cycle manufacturing services with low cost and dynamic scalability. Cloud manufacturing has the following characteristics: (1) Cloud manufacturing takes cloud computing technology as the core and is a new service-oriented manufacturing mode; (2) Cloud manufacturing is user-centered and knowledge-supported. With the help of virtualization and servitization technologies, a unified manufacturing cloud service pool can be formed to conduct unified and centralized intelligent management and operation of manufacturing cloud services, and allocate manufacturing resources/capabilities according to needs; (3) Cloud manufacturing provides a new platform for collaborative manufacturing, management and innovation in the full life cycle of product R&D, design, production and service, which leads to the transformation of manufacturing mode and further changes the industrial development mode.

Common ways of industrial cloud Service include industrial SaaS (Software as a Service), IaaS (Infrastructure as a Service), and PaaS (Platform as a Service). PaaS solution is the development trend of industrial software in the future. Developers only need to focus on their own business logic, not the underlying technology. Cloud computing centers provide PaaS services in a transparent manner.

By establishing heterogeneous resource pools of IT hardware and software, flexible allocation and rapid delivery capacity, industrial cloud provides cloud infrastructure, various tools on cloud, cloud-oriented transformation of business system, and provides information services for small enterprises to purchase or lease, indirectly promoting the application of industrial software in small enterprises. Enterprises pay according to the actual use of resources, which can significantly reduce the cost of enterprise information construction and promote the integration and sharing of enterprise data resources. For example, GE carries out remote operation and maintenance of equipment assets in the power, medical, aviation and other industries based on the Predix platform, which significantly reduces or avoids non-accident shutdown of equipment, improves operation efficiency, saves a lot of maintenance costs for customers, and speeds up the transformation process from providing products to providing services.

[Case: Ariba Business Cloud]
Ariba, founded in 1996, is an old provider of cloud electronic purchasing software and service in the U.S. It launched the world's first procurement automation solution in 1997 and the world's first electronic sourcing solution in 1998. Ariba's network platform can directly connect with suppliers on Alibaba's network or other networks, as shown in Fig. 5.12, as well as carry out data collection, price comparison, demand forecasting and supplier insight. On the Ariba platform, 2.9 million enterprise operations each day generate massive amounts of data. On average, a 1% cut in procurement costs means a 10% increase in overall profits. Enterprises can achieve cost savings ranging from 1% to 8% by using Ariba business cloud data analysis, or by competitive bidding or finding new suppliers, suppliers' sources, and purchasing departments can increase their work efficiency by 20%. SAP acquired Ariba in 2012 for a total of about $4.3 billion.

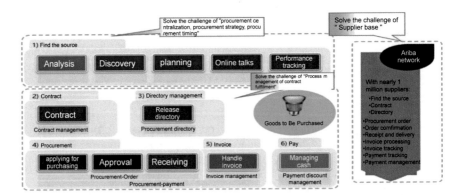

Fig. 5.12 Ariba business cloud. *Data source* SAP official website

5.5.2 *Industrial Internet*

Industrial Internet is an extension of industrial cloud platforms. Its essence is to build a more accurate, real-time and efficient data acquisition system by superimposing emerging technologies such as IoT, big data and artificial intelligence on the basis of traditional cloud platform. Including storage, integration, access, analysis, management and other functional platforms, to realize the modeling and reuse of industrial technology, experience and knowledge, and finally form a manufacturing ecology with resource enrichment and collaborative participation.

The Internet is all about connecting: connecting people, connecting people to services, connecting people to devices. The connection of resource elements has historically been a major driving force of the industrial revolution. With the development of the Internet today, a new network, IoT, has evolved based on information carriers such as traditional telecommunication networks. Compared with the Internet, the IoT emphasizes the connection between things, while the industrial Internet aims to realize the comprehensive interconnection between people, machines and things. Manufacturing intelligence and the industrial IoT complement each other. Without data, there is no manufacturing intelligence. The value of the industrial Internet, on the other hand, lies in providing connectivity and massive amounts of data. The interconnection constructed by the industrial Internet includes the Internet of Everything (IOE) between people, production equipment, equipment and products, and virtual and real life. The data generated includes products (size, BOM), operations data (business, quality, production, procurement, market, inventory), value chain data (customers, suppliers, partners), etc. IM mainly consists of four links: perception, calculation, judgment and response, each of which is inseparable from connection and data.

Development and the industrial Internet technology is one of the most important aspects of the implementation of IM, China's "twelfth five-year" science and technology of Manufacturing informatization engineering plan clearly put forward to develop the Manufacturing technology, in embedded systems, near field communication (NFC) and sensor networks such as constructing the modern Internet of Manufacturing Things (IoMT), to enhance the management of the product and service information, real-time acquisition, dynamic data related to the production site and perception and intelligent processing and optimization control, improve the production process control. In addition, through situational awareness and information fusion, rapid launch of new products, dynamic response to market opportunities and real-time optimization of production supply chain can also be achieved, and new business and manufacturing models can be generated to achieve multiple benefits such as economy, efficiency and competitiveness.

[Case: Alibaba builds the R&D platform to the workshop]
With the continuous development of the industrial Internet, more and more Internet engineers are appearing in the workshop. Internet companies "take the R&D platform to the workshop". In Guangdong, the case of using "ET industrial brain" in cooperation with Ali Cloud to realize intelligent transformation is emerging rapidly. Jingxin,

an antenna manufacturer, has increased the debugging efficiency of its products by up to 50%. Disen Thermal Company has started to predict the boiler's health, with the goal of giving 6–12 h' warning in advance. Trina solar company has increased the rate of its batteries by 7%. China Strategic Rubber Company increased the qualified rate of mixed rubber by 5%. GCL Photovoltaic increased its premium rate by 1%. They all attribute their success to the fact that "what used to be judged only by experience can now be described and judged from a data point of view whether a part is operating in a normal state."

5.5.3   Cyber-Physical System (CPS)

Cyber-physical Systems (CPS) is the integration of computing processes and physical processes, as well as an intelligent system integrating computing, communication and control. As shown in Fig. 5.13, CPS gave rise to the big data, and thus promoted the second outbreak of AI. "Made in China 2025" proposed, "IM based on CPS is leading the innovation of production and manufacturing mode, and China's manufacturing industry faces great opportunities in transformation, upgrading and innovation and development."

Although the new generation of information technologies has different focuses, they are still interrelated. Figure 5.13 shows their logical relationship. The primary function of the Internet of Things (IoT) is to be responsible for the automatic collection of all kinds of data. CPS maps the interactive operation and internal evolution of complex objects in virtual space to capture the complete life cycle of the entity system. The continuous, detailed perception of objects in the physical world leads to the explosion of data volume and the expansion of physical space. Large amounts of structured and unstructured data are generated, thus forming the big data. The increasing data amount and complexity of the structure require more memory and storage on the cloud. In turn, the parallel computing capability of cloud computing

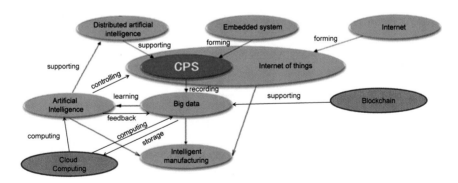

Fig. 5.13 The relationship between CPS and big data and artificial intelligence

also promotes the efficient and intelligent processing of big data. Artificial intelligence based on big data deep learning aims to obtain the law of value, cognitive experience, and knowledge wisdom. The training of the AI model also needs the support of large-scale cloud computing resources. The constructed intelligent model can also react to the IoT to control various front-end devices of the IoT more optimally and intelligently. Blockchain solves the security problem of information leakage and tampering, provides essential support for the IoT, big data, cloud computing, and reshapes trust mechanisms.

Cyber comes from a book by Norbert Wiener. As he wrote "Cybernetics: Control and Communication in Animals and Machines" in 1948, he couldn't find a word for control. He thought of the ancient Greek verb "kybereo"—"steering, guiding, controlling," and Norbert used cybernetics to express the concept of control. "Cyber" was coined by the U.S. Department of Defense in 1991 during the War in Somalia to reduce casualties by using remote control. The idea behind Cyber is that humans can connect to machines, creating systems that provide an alternative environment for interaction. Cyberspace is a time-related collection, consisting of interconnected information systems and human users who interact with these systems. The common point of all definitions is that the core of Cyberspace consists of four interconnected levels of "hardware and software infrastructure—data—people—activities" on a global scale. The choice of which of the four levels to define reflects the state's intention in formulating a strategy for cyberspace. Some countries, such as New Zealand, Turkey and Saudi Arabia, defined Cyberspace as an information and communications infrastructure. Therefore, the protection focus was only on infrastructure. Some countries, such as China, India and Russia, define it as a complete set of facilities, data, people and activities. In addition to protecting infrastructure, data and users, they also focus on protecting and managing related operational activities. Academician Wang Chengwei suggested translating it into a "control domain".

In July 2007, the report titled "leading under the challenge—information technology R&D in the competitive world", listed CPS as the first of the eight key information technology, the others are software, data, data storage and data flow, network, high-end computing, network and information security, man–machine interface, NIT and social sciences, raising it as a national strategy. The report defines the core of CPS technology as "Control, information Communication, and Computing with common implications," or 3C. Academician He Jifeng (2010) pointed out its five functions: computing, information communication, precise control, remote coordination and autonomy. As you can see, information content is only a controlled object, not a complete representation of Cyber.

CPS opens up the interaction between the physical world and the virtual world, as shown in Fig. 5.14. In many modern enterprise, regardless of the Intranet or extranet, is just an independent system or computer network virtual space, how they are in harmony with the physical entity enterprise and the coordinated and efficient work accurately, how to enhance the adaptability of such systems, autonomy, functionality, reliability, security, availability and efficiency, is a new system engineering.

As shown in Fig. 5.15, with the advent of IM and big data, a new CPS-based automation architecture is emerging. In an intelligent network, each device or service

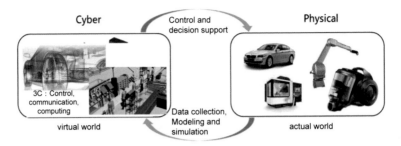

Fig. 5.14 Relationship between physical world and virtual world in CPS

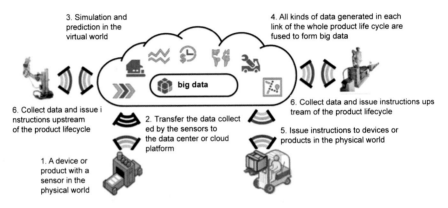

Fig. 5.15 An automation architecture based on CPS

can automatically start communication with other services. Various services (such as production scheduling) automatically subscribe to the required real-time data, and sensor data is directly sent to the cloud.

CPS consists of 5 levels, namely, Smart Connection Level, Data to Information Conversion Level, Cyber Level, Cognitive Level and Configuration Level, as shown in Fig. 5.16.

When the digital virtual body is modeled according to the physical entity, the CPS is formed. In this case, the digital virtual body has two functions:

First, when the digital virtual body is embedded in the physical entity, the physical entity gains intelligence. A desk, for example, used to be non-intelligent and "dumb," if sensors are installed, connected to an embedded system with CPU and software and wireless communications equipment. Then if the table is turned over, it can tell you through phone APP, that's going to be a smart table. A microphone, for another example, can automatically process sound levels and even record audio and convert it to text for display on a projection. The basic logic of CPS is to make people's tacit knowledge (thought, algorithm and reasoning) explicit, and then to embed knowledge into software, software into chips, chips into hardware and hardware into objects (physical devices), which form intelligent products and intelligent devices. The core

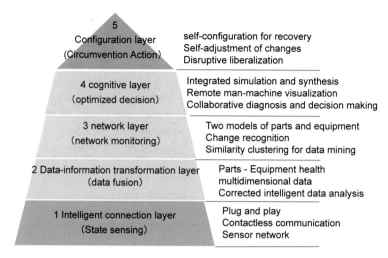

Fig. 5.16 Five levels of CPS

of enterprise transformation and upgrading is intelligent products and equipment. CPS includes embedded software-intensive systems found in almost all high-tech products today, such as equipment, automobiles, airplanes, buildings, production systems, and so on.

Secondly, in the process of product R&D and factory construction, the entire production line, flow, transportation and assembly line transmission process can be modeled, simulated and optimized on the computer, so as to guide product development and plant construction. CPS includes global networks, such as the Internet, and the data and services provided on the Internet.

5.6 Optimization Level (Level 4): Big Data and Knowledge Management

Enters the optimization level, enterprise prepares optimizing production processes and business processes by big data, knowledge base and expert system.

Optimizing level, the enterprise production system, management system and other supporting systems have completed comprehensive integration, realized the digital modeling of plant level, and started to personnel, equipment, products, network data collected and produced in the process of production of data fusion, prepared by big data, knowledge base and expert system, optimizing production processes and business processes. From level 3 to level 4 is a process of quantitative change to qualitative change. In this stage, through data integration, big data analysis, knowledge management, and the formation of the nervous system, the IM ability of enterprise is driven and improved rapidly.

5.6.1   Industrial Big Data

1.   Development, Definition and Characteristics of Big Data

In 2015, humans created a total of 4.4 ZBS (1ZB = 1.18×10^{21} B), and that number is doubling roughly every two years. In the 1980s, the concept of "big data" was first put forward by American Alvin Toffler in his book The Third Wave, which was used to describe the large amount of data. According to the definition of National Institute of Standards and Technology (NIST), big data refers to data with large data size, fast acquisition speed or diverse forms, which is difficult to be analyzed effectively by traditional relational data analysis methods, or requires large-scale horizontal expansion to be processed efficiently. The author of The Age of Big Data believes that big data refers to the analysis of all data rather than random analysis (sampling survey). In simple terms, big data is often defined as data sets that "exceed the capabilities of common software tools to capture, manage, and process." Big data has several features, of which the most famous are Volume, Velocity, and Variety. Apart from Veracity, Valence, and Value, for short, the "6V" feature.

Although "big data" has become a hot spot, less than 10% of the data is analyzed each year. In the future, automated artificial intelligence software will be able to identify and extract relevant information from scattered data. And the power of this data analysis will spread from commercial applications to the hands of ordinary people. Big data-driven has become the mainstream model of current artificial intelligence. China's huge number of Internet users and network applications, rich data resources in various industries, huge market demand and open market environment provide the most suitable soil for the development of artificial intelligence, and big data has become the biggest resource advantage of China's manufacturing industry.

Big data refers to the industrial field, around a typical IM model, from customer requirements to the sales, orders, planning, R&D, design, technology, manufacturing, procurement, supply, inventory, shipping and delivery, after-sales service, operations, scrap or recycled and manufacture and so on, the whole life cycle of products produced by each link all kinds of data, and related the floorboard of the technology and application. Industrial big data is the core of industrial Internet and the foundation and key to realize intelligent applications such as intelligent production, personalized customization, networking collaboration and servitization extension. Industrial big data is an important strategic resource for the transformation and upgrading of China's manufacturing industry.

2.   Hierarchy and Process of Big Data Analysis

Big data applications that currently improve manufacturing processes are broadly divided into three tiers.

The first level is descriptive analysis. It refers to paying attention to what is happening now. the "excavator index" reflecting the national economic activity level for the reference of national leaders in decision-making.

It refers to paying attention to what is happening now, analyzing the correlation or causation between factors, and displaying it with data visualization technology, so that people can grasp the basic situation of things. For example, the Poisson distribution of human joint sizes in different countries, genders and age groups can be measured through the census, so as to provide human factors engineering data for product design. Find out the individual preferences of each type of user through the public opinion monitoring system; Collect the feedback and experience of end users into the big data platform to build knowledge rule base for future product quality, design and other aspects of work; Monitor the manufacturing process of products in the production line, display real-time data or processing parameters many years ago, assist to find the source of product or equipment failure; Digital wind tunnel, virtual simulation analysis of product performance; Obtain the detailed data of national construction machinery operation rate and start time in real time, and fuse to form the "excavator index" reflecting the national economic activity level for the reference of national leaders in decision-making, etc.

The second level is predictive analysis. It's about predicting what's going to happen next, based on the description. For example, multiple sensors are placed on each machine to monitor, diagnose and predict equipment failure. Through the prediction results of big data, the number of potential orders can be obtained, and then directly into the product design and manufacturing and subsequent links; Accurately understand the market development trend, user demand, industry trend and other data, so as to provide a basis for the development of enterprise strategic planning. No matter in which industry big data technology is applied, its most fundamental advantage is the ability to predict.

The third level is guidance analysis. For example, find the best design solution by matching user portraits, designer portraits and product portraits. It is based on the current situation, based on big data prediction, combined with mathematical optimization model, to provide decision support or optimal solution. This is one of the superlatives. For example, find the best design solution by matching user portraits, designer portraits and product portraits. Understand the processing time of each product on each machine tool through data analysis, and then make the optimal scheduling plan of machine tool task assignment; Based on the optimal model under various algorithms, the design resources and module business resources on the platform are integrated to generate the most appropriate time and the shortest cycle production plan. Optimize the distribution path, deliver to users in real time, and so on.

The process of big data analysis can be represented by "intelligent device, intelligent system and intelligent decision". To put it simply, first the intelligent equipment is responsible for collecting data, then the software analysis tools in the intelligent system carry out data mining and analysis, and finally the intelligent decision visualization presentation can guide the production and optimize the manufacturing process. When the Cyber world composed of intelligent devices, intelligent systems

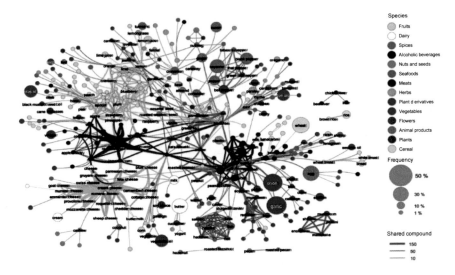

Fig. 5.17 Flavor network based on big data

and intelligent decisions and the physical world composed of resource elements are deeply integrated to form CPS, the foundation of big data analysis has been formed.

[Case: Application of big data in food flavor]

In the past decade, some chefs and food scientists have noticed a pattern: food pair-ups with shared flavor compounds are more popular than that without shared flavor compounds. For example, the famous pairing of blue cheese with chocolate has a whopping 73 flavor compounds shared between the two ingredients. To confirm the above rules, Ahn et al. (2011, 2013) conducted relevant studies on this. They constructed a flavor network containing 381 kinds of food ingredients and 1021 kinds of flavor compounds, as shown in Fig. 5.17.

Note: This figure is the schematic diagram of the flavor network presented by Ahn et al. In the figure, each dot represents an ingredient, the color of the dot represents the type of the ingredient, and the volume of the dot represents the frequency of the ingredient appearing in the recipe. The larger the volume of the dot is, the more times the ingredient appears in the recipe compared with other ingredients. Moreover, the lines between the dots indicate that they share compounds. The thickness of the lines represents the number of shared compounds.

Flavor network is the derivative of big data. By establishing the connection between two kinds of materials, we can judge whether these materials should appear in the recipes of a specific region according to the strength and weaknesses of the connection. It provides people with a more intelligent and scientific way of thinking for choosing recipes and developing new foods.

3.   Key Points of Big Data Analysis

(1)   Combination of "small data" and "big data"

The size of big data often means that its value density is usually very low. Some data are small but have a high value density. Some scholars estimated, according to this law, the enterprise PLM/ERP/CRM/SCM system and relational database accounted for 20% data volume and 80% data value, such as product drawings, test analysis, the processing technology of each drawings need to be repeated reading, and big data accounts for 80% data volume and 20% data value, due to the data volume is beyond the limit of the human ability of reading, big data be available only through machine learning and data mining, such as the working condition of equipment data, audio, data, text data, and so on are the data that all humans can't read. However, these "small data" and "big data" are interdependent. Only by combining small data in a relational database can the value of industrial big data be mined. It can be said that "without the outline of small data, we can't expand the scope of big data".

[Case: Hitachi Elevator's Big data system]
"Internet + Elevator" brings new imagination to the elevator industry during the sluggish period of growth. Hitachi Elevator (China) Co., LTD., headquartered in Guangzhou, achieved nationwide coverage of elevator wireless remote monitoring system in 2013, and took the lead in establishing a global elevator big data system in China in 2015. The main functions of the system are as follows:

1)   Data storage

The basic data of the system is stored in Oracle relational database, and there are three tables used: elevator basic information table, fault information table and user information table. Elevator basic information data mainly includes elevator equipment serial number, elevator brand, elevator installation location, installation community, maintenance company and so on. Fault information data mainly includes elevator fault equipment, fault occurrence time, fault end time and fault type, etc. The user information mainly includes the user information of elevator remote supervision system. There are millions of pieces of data in a single table of fault information in a single database, and more than 50 gigabytes of data in all relational databases.

All the real-time monitoring big data uploaded under the remote supervision of the elevator is stored in HDFS (Hadoop Distributed File System).

2)   Data import and export

The data import and export module is mainly responsible for importing the basic elevator information, user information and other fixed information into HDFS in a full amount at one time and cleaning the data, while the fault information data in the database will be updated and increased every day, so incremental import is needed every day.

The data import and export module also exports and returns the data mining results of the data service layer to the relational database.

3) Data preprocessing

The raw data contains a lot of "dirty data". For example, the basic information of the elevator is entered manually, and incomplete or incorrect data may occur during the data entry process. The elevator fault data is uploaded to the server database by GPRS. If the network condition is not good, the collection of fault information will be incomplete. An elevator failure will be reported several times in just a few minutes, and this should be removed and treated as a failure. Therefore, it is necessary to preprocess the data.

4) Data mining

The data mining module mainly uses two methods, one is cluster analysis, the other is association analysis.

By k-means clustering mining, the community and fault types can be clustered based on the elevator repair time, and the maintenance situation of the community and the repair time of various kinds of faults can be analyzed. Through Apriori association rule mining, the association between fault types and cells can be found, common faults of each cell can be analyzed, and decision support can be provided for maintenance personnel of the fault cells. Based on the above analysis, the common types of elevator failures, the fault repair efficiency data of each maintenance company, the failure rate of each community, and the frequent failures of a particular community are obtained.

Through the elevator remote monitoring system and the application of big data technology, Hitachi elevators can make proper and timely maintenance measures for the parts that have been warned in advance, so as to eliminate the safety risks before the accident, and promote the shift of Hitachi elevators' business focus from "selling elevator products" to "providing customized services".

Through the elevator remote monitoring system and the application of big data technology, Hitachi elevators can make proper and timely maintenance measures for the parts that have been warned in advance, so as to eliminate the safety risks before the accident, and promote the shift of Hitachi elevators' business focus from "selling elevator products" to "providing customized services".

(2) Complete restoration of scenario

Big data not only depends on the size of the data, but also depends on whether the "big information" contained in it can be restored to obtain a complete "big scene".

The value of big data lies not only in itself, but also in the way it is processed. If you can't use the data to restore the complete situation, it will only become a bunch of useless data. In "Sitting alone in autumn night" written by Wang Wei in the Tang Dynasty, "Fruit falls in mountain rain, grass insects chirp under lamp", a relatively complete scene can be reappeared: there is a house in the deep mountain, and someone is sitting in this house. It rains at night, and he hears a dozen ripe fruit from the trees outside the window, and it falls with a bang. In autumn, many insects in the grass are crying under the rain. The author is in the house under the light in the rain, through the audio-visual feeling of life in the mountain fruit and grass insect's body, desolate in the quiet rain at night. Such an environment, there is scene, sound and light, it is living and moving, there is big information. These ten words are small but informative. Therefore, big data not only depends on the size of the data, but also depends on whether the "big information" contained in it can be restored to obtain a complete "big scene", and the "big wisdom" can be obtained through mining and processing the laws contained therein.

[Case: Jingdong's complete ecosystem]
In Jingdong's complete ecosystem, each link has an exceptionally rich application scenario, which provides a rich "imagination space" for big data innovation, which is reflected in Jingdong's current unbounded retail strategy. In jingdong recent unmanned convenience stores, 360 degrees using the data from jingdong online accumulated large portraits and networks or relationship chain, restore the user details and behavior rule of life, including the user's professional, address, age, family members, and even with free WIFI access for commuting traffic track, forming a complete life scenes, this can make targeted marketing strategy, for the user to accurately select and recommend commodities, and synchronous online dynamic pricing. The intelligent camera in the store can analyze the in-store rate, gender ratio and other data of shoppers through the traffic funnel, and integrate with the online data to provide a data basis for merchants to manage their stores and choose to purchase goods.

(2) Memory calculation

Due to the increase in data volume, the traditional business suite based architecture has become difficult to adapt to the new requirements, including:

1) Performance bottlenecks exist in databases based on multiple traditional hard disk operations;
2) An architecture that separates information technology (IT) from operational technology (OT);
3) Serious redundancy exists in the internal data structures of each system;
4) Lack of support for big data processing;
5) Inflexible deployment mode.

In recent years, multi-core and multi-CPU servers have been able to achieve high-speed communication in the kernel through memory or shared caching. Memory is no longer a bottleneck resource. Since 2012, there have been servers with more than 2 TB of memory. In 2010, SAP launched the high-performance analysis tool HANA, which provides the in-memory database with large memory and adopts column storage in the in-memory database to facilitate data compression, so that more data can be put into memory for memory calculation and the calculation speed can be improved.

[Case: memory computing for Peugeot Citroen (PSA)]
PSA faced the following business challenges:

1) When assembly plant supplies, the final customer order of the assembly plant's assembly line can only be obtained 4 h before the delivery deadline, so the time is very urgent.

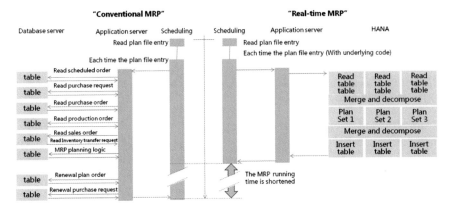

Fig. 5.18 Memory computing technology accelerates the MRP running speed

2) Due to the variety and complexity of product configuration, the traditional MRP operation takes 11–21 h, which greatly exceeds the requirement of 4 h, and is not conducive to reducing inventory. And it takes too long to make the report, which causes waste.

3) Due to the low profit level of the auto parts industry, the profit level of the group is only 5%, so even 0.5% profit change is crucial for the group.

As shown in Fig. 5.18, based on memory computing technology, MRP operation is reduced to 1 h on SAP HANA. This reduces inventory, saving millions of euros, while also allowing production lines to be balanced once an hour, increasing productivity.

5.6.2   Knowledge Management

In the era of IM, knowledge is the wisdom of R&D, but also the cornerstone of innovation. Knowledge Management (KM) has become the source of the core competitiveness of modern enterprises. Knowledge management refers to the transfer of appropriate knowledge to appropriate knowledge users through the acquisition and sharing of knowledge through the construction of enterprise knowledge management system.

In McKinsey's <2025: The 12 Disruptive Technologies that Will Define the Future of the economy>, Knowledge Worker Automation (KA) comes in second. KA realigns the division of labor between man and machine by transforming engineering knowledge system into "engineering intelligence" and driving industrial software and industrial infrastructure, helping to liberate knowledge technicians from repetitive labor. McKinsey argues that the technologies that will have the most economic impact over the next decade or so will be those that have made good progress, such as the automation of knowledge workers, including computerised query and response systems, to handle the bulk of client inquiries and diagnostic requests for maintenance. According to McKinsey estimates that by 2025, each of these subversive technology contribution to the value of the global economy more than $1 trillion, ranking top five as follows: mobile Internet (3.7–10.8), KA (5.2–6.7), IoT (2.7–6.2), cloud computing (1.7–6.2), advanced robot (1.7–4.5), the value of the Numbers in brackets are the technical contribution forecast, the unit is trillions of dollars.

Design is the soul of manufacturing enterprises, and the core of manufacturing digitization is the digitization of design and manufacturing. Modern design is knowledge-based design, which itself is a highly knowledge-intensive work. It can be said that there is no design without knowledge. The knowledge accumulated in the process of design and manufacturing must be collected, extracted, sorted and stored according to the requirements of digital design through knowledge management, and then applied to the digital design of products.

The entire development process of Boeing 787 uses industrial software, including CAD/CAE/CAE system such as CATIA, which has been accumulated by Boeing for decades, including the key knowledge and experience of aircraft design, optimization and process, etc., and the technical system is integrated into these software, which constitutes Boeing's core competitiveness. In order to implement knowledge management in manufacturing enterprises, besides relying on enterprise knowledge management system, some measures and strategies must also be taken from organizational and institutional aspects.

The implementation steps of knowledge management are as follows:

1. Carry out enterprise research: investigate the current situation of enterprise operation, including: enterprise knowledge (implicit and explicit); employee knowledge (implicit and explicit); Current situation of enterprise knowledge management.

2. Knowledge collection: (1) Classification of knowledge: enterprise knowledge (key/core position identification), explicit knowledge and implicit knowledge; (2) Knowledge collection and check: Design knowledge record mode and relevant forms, and implement knowledge collection. Collect and sort out relevant knowledge of enterprises/individuals before implementing knowledge management; Knowledge management in daily work; establish the job manual; (3) Knowledge archive: select appropriate knowledge expression and organization methods for knowledge in various fields, such as rules, cases and models, such as tradeoff curves; Explicit tacit knowledge; (4) Management mode: Design knowledge management mode; Establishing knowledge management system; To establish written and electronic knowledge files; Implement dynamic management.

3. Knowledge refinement, sharing and Systematization: The refinement of core knowledge; The confidential management of knowledge; To establish channels for knowledge dissemination and sharing; Systematization of explicit knowledge.

4. Application of knowledge: knowledge classification; Knowledge applica-
    tion domain division; Knowledge owner identified; Knowledge application
    guidance; Knowledge scene.
5. Knowledge innovation: to establish the field and direction of knowledge inno-
    vation; Design knowledge innovation evaluation system; Establish incentive
    mechanism of knowledge innovation.

After the implementation of knowledge management, it will be realized: 1. Establishment of knowledge management system; 2. Knowledge structure establishment;
3. Tacit knowledge mining and explicit implementation; 4. Improve the efficiency
of knowledge workers, improve the level of enterprise management and technology,
and build the core competence of the enterprise.

The trade-off curve is a relatively simple tool for systematically preserving and
passing on knowledge. A sensitivity analysis performed on a daily basis is a trade-
off, such as: what would happen to Net Present Value (NPV) if development costs
were reduced by 15%? If the development time is increased by 25%, what will be
the impact on NPV and how much will be equivalent to the cost increase? If adding
a function point in the product design scheme will delay the product development
for 2 months, how much should the sales volume increase to exceed the profit and
loss balance point? And so on. Many companies use ERP systems to develop Dupont
financial analysis models to support this type of sensitivity analysis. All you need to
do is input these changes into the financial analysis model to calculate NPV. Many
companies simplify the results of tradeoff analysis into a linear relationship, such as
increasing development cost by 1% and reducing NPV by a few percent, and then
make tables to assist designers in making decisions. Toyota Motor Corporation, on
the other hand, is adept at using tradeoffs in product development.

[Case: Toyota's Tradeoff Curve]
Toyota engineers use tradeoff curves to analyze the relationship between different
design features. In a trade-off curve, one performance or cost of a product or
subsystem is represented on the Y-axis, while another performance is represented on
the X-axis. For example, when designing an engine, the trade-off curves for speed
and fuel economy can be used for analysis and parameter selection. In a Toyota car
development project, a discharge system supplier for Toyota provides more than 40
different kinds of prototype samples, used for change of different parameters, and
then tested to trade-off curve drawing, so that you can understand the tail exhaust
muffler of the relationship between pressure and engine noise, and then make the
best choice. "The main difference between Toyota and American companies," says
an American engineer at Toyota, "is that Toyota has so much accumulated knowledge, with this notebook (pointing to the trade-offs), I can design a pretty good car
body." "We didn't understand the difference between success and failure in previous
products," says another engineer at an American company. "Then we hired a former
Toyota employee who knows thousands of tradeoffs. It's unbelievable how much he
knows!".

Siemens has long accumulated experience in the establishment of knowledge
transmission and sharing channels and systematization of explicit knowledge.

[Case: Siemens' Global knowledge Management System]
Siemens vice president has an ambitious plan: Use the Internet to disseminate the knowledge of 460,000 colleagues around the world, so employees can borrow the knowledge and technology of others. At its core is a WEB platform called shareNet that includes chat rooms, databases, rule libraries, case libraries, and search engines. Employees can store information they think is useful, search or browse, and get more information by contacting authors. Simple knowledge can be refined, explicit, and stored in the form of "if–then" production rules, while comprehensive knowledge across application domains can be explicit and stored in the form of "large granularity" cases. A successful case was when a Malaysian team was competing for a broadband network project and found in shareNet's case base that a Danish team had carried out the same project and won the contract with the technical guidance of the knowledge owner's Danish team. The Swedish team raised the alarm via shareNet, got technical data from colleagues in the Netherlands and won a project to build a telecommunications network for two hospitals. For tacit knowledge that is difficult to make explicit, for example, experienced pre-sales technical support personnel can choose appropriate technical solutions and implementation paths based on their intuitive judgment of environmental changes and client supervisors' intentions. Such knowledge is suitable to be taught through a "mentoring system" and teaching by words and deeds. In Germany, two-thirds of high school graduates take part in apprenticeship programmes. In order to motivate employees to upload knowledge to the system and motivate teachers to impart experience to apprentices, effective knowledge innovation incentive mechanism is needed.

Comac is worthy of reference in knowledge collection and refinement, knowledge scenization and intelligent push.

[Case: Dual-screen innovation by Comac]
Comac's "Second screen" vividly depicts the company's employees adding a new computer screen in addition to their daily work computer screen for information reference, data support and knowledge reference. This is a management innovation that Comac is actively promoting, namely "dual-screen innovation". The first step is to establish an electronic library of structured knowledge, to classify, scientifically sort and store knowledge in an orderly manner, including the compilation of manuals of various jobs, tasks and procedures, the compilation of knowledge process maps and the sorting out of tacit knowledge. At the same time, the corresponding knowledge management system should be established, such as assessment, evaluation and incentive system. The second step is problem-oriented, creating a scenario-based knowledge application platform. Based on the asset-based database, the work platform is formed to match, connect and combine the asset-based knowledge and work flow, so as to realize the standardization of knowledge, modularize knowledge for different scenes, and directly improve the efficiency and quality of the fragmented assets for the work scenes. The third step is to realize the intelligent service of knowledge with intelligent push function as the goal. There will be a personalized reference prompt for the optimal scheme during the task execution and an automatic

error correction function at the end of the task. Based on the standardization, modularization and scenization of knowledge, knowledge push can achieve thousands of aspects and improve the accuracy of push. The Shanghai Aircraft Subsidiary alone contributes more than 37,000 knowledge points every year, with each person contributing 12 knowledge points. The tooling design task cycle of Shanghai Aircraft Manufacturing Center has been shortened from an average of 22 working days to 14, and the design efficiency has been increased by 36% on average.

5.7 Leading Level (Level 5): Artificial Intelligence and Collaborative Innovation

Leading level is the highest degree of IM capacity building, at this level, the analysis of the data used throughout all aspects of business, all kinds of production resources to optimize the use of equipment to achieve autonomy between feedback and optimization, the enterprise has become an important role in upstream and downstream industry chain, then personalization, network coordination, remote operations have become the main mode of enterprise business, enterprise become IM benchmark in the industry.

The fifth stage of the IM will stimulate the process and product innovation, change passive consumers accept the current situation of mass production of products, such as the customer can "tell" the factory what function cars should be produced in the personal computer configuration, or how to adjust tailored clothes. These are the collaborative innovations that were achieved by introducing artificial intelligence.

5.7.1 Ecosystems

Software defines the ecology of the entire manufacturing industry. The essence of the new enterprise capability is the construction capability of the ecosystem. Like Wintel system 30 years ago, iOS system and Android system 10 years ago, intelligent manufacturing as the future form of manufacturing industry is also forming its own ecosystem. Siemens, Intel and other companies are building their own data collection, equipment interconnection, industrial software and cloud computing systems which build the IM industry ecosystem.

As shown in Fig. 5.19, the IM ecosystem model proposed by the US National Institute of Standards and Technology (NIST) covers a wide range of manufacturing systems and gives three dimensions shown in the IM system. Each dimension (such as product, production system, and business) represents an independent full life cycle. The manufacturing pyramid is still at its core, where the three life cycles converge and interact. IM covers product ecosystem, production ecosystem and business ecosystem.

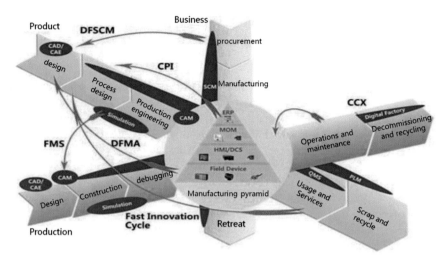

Fig. 5.19 Three dimensions of intelligent manufacturing ecosystem [2]

1. Product dimension: involving information flow and control, Product life cycle management for smart manufacturing system (SMS) includes six stages: design, process design, production engineering, manufacturing, use and service, scrap and recycling.
2. Production dimension: Focus on the design, construction, commissioning, operation and maintenance, decommissioning and recycling of the entire production facility and its system. The term "production system" here refers to the organization of resources and creation of products and services from a variety of collections of machines, equipment, and ancillary systems.
3. Business dimension: Focusing on the interaction between suppliers and customers, since e-commerce is crucial today, so that any type of business or commerce transaction will involve the exchange of information between stakeholders. Standards for interactions among manufacturers, suppliers, customers, partners, and even competitors, including common business modeling standards, manufacturing specific modeling standards, and corresponding messaging protocols, are key to improving supply chain efficiency and manufacturing agility.

The Manufacturing Pyramid: The core of the IM ecosystem, where the product life cycle, production cycle and business cycle all converge and interact. Information from each dimension must be able to flow up and down the pyramid to function for the vertical integration of the manufacturing pyramid from machine to factory and from factory to enterprise. Along each dimension, the integration of manufacturing applications helps improve control at the shop floor level and optimize factory and business decisions. These dimensions and the software systems that support them ultimately constitute the ecosystem of the manufacturing software system.

[Case: Midea building ecosystem]

Midea company already built a marketing platform based on big data and the Internet and after-sales service system. Through the two big platforms, Midea clearly grasps the customer information, product details, and tells the customer when products need maintenance station information, build the **products ecosystems** covering client, terminal, platform and third-party applications; Through IT subsidiary and cooperation of "Midea Cloud wisdom data" company, to promote manufacturing standards of Midea, such as logistics container standard even intelligent factory system solution, the IT and OT based on many years of experience in Midea were output to partner, close collaboration with small and medium-sized enterprises was realized based on a consistent standard implementation, the **production ecosystem** was constructed on the basis; Further, with large data analysis, the information to aid in the production of products, cultivating industrial robots, improving robot arm intelligent equipment suppliers' ability to fast response, through the interaction of upstream and downstream links to build **business ecosystem** covering customers, manufacturing, suppliers, realized the automation of production process of Midea.

The IM industry ecosystem is the commanding height of competition. With the rise of the "We Economy", leading digital companies are finding it far more effective to operate as an ecosystem than to go it alone. The competition among enterprises is gradually changing from competition among individual enterprises to competition among supply chains and even ecosystems. Kevin Kelly writes in Out of Control: "The wave of corporate alliances, especially in the information and web industries, is yet another aspect of the growing co-evolution of the world economy, where instead of eating or competing with rivals, you form alliances and live in symbiosis. The future of control is partnership, collaborative control, human–computer hybrid control, where humans share control with our creations." As different industries are reshaped by platforms into interconnected ecosystems, the boundaries of future industries will be blurred.

The formation of the ecosystem is accompanied by the organizational structure transformation of the core enterprise. It has four modes:

1. Flattening and self-organization of traditional enterprises. For example, after the reform, Haier is composed of more than 2000 independent operators, platform-based enterprises, maker members and personalized customers.

[Case: Huawei's "iron Triangle" management mode]

In a network analysis meeting convened by customers in 2009, there were 8 employees in Huawei, and each employee explained problems in their respective fields to the customer. The customer complained on the spot that they did not know what to do and finally lost the order. This has driven a new organizational change throughout Huawei. To establish a business core management team with account manager, delivery manager and product manager as the core, they put forward the "iron triangle" management mode. The construction system of Huawei's "iron triangle" mode includes two levels, one is the project "iron triangle" team, the other is the "iron triangle" organization of the System department. The project "Iron Triangle" team is the most basic organization on behalf of Huawei to directly face customers and the front-line operation and combat unit, which is the core component of Huawei's "Iron Triangle" mode. The three organizations included AR (Account Responsibility), SR (Solution Responsibility), and FR (Fulfill Responsibility). After the reform of Huawei, "decision is made by those who hear the sound of guns", changing from "division-level combat" to "class-level combat".

[Case: Business model and organizational structure of Handu Clothing Company]
Supply chain type and product type need to match. Innovative products refer to products with uncertain demand, short product cycle and high marginal profit. The market reaction speed is the focus of attention, so the responsive supply chain appropriately matches it. Functional products refer to products with relatively fixed demand, long product cycle and emphasis on material functions. Minimizing material cost is a very important goal, so the physically efficient supply chain matches with it. In these two supply chain models, speed and cost are a pair of contradictions, difficult to do both. Through the connection of Internet technology and the superposition of digital lean thinking, Handu Clothing has solved the contradiction and the problem of mismatch between supply and demand.

In the summer of 2017, 95% of Handu's clothing products were sold out, compared with 60–70% for traditional clothing companies. Nearly 100 new products are launched every day, exceeding the global benchmark -Zara, and nearly 30,000 new products are launched annually. The production mode is small-batch, multi-style and high-frequency production. 40% of the orders are chasing orders and there is basically no inventory. In terms of organizational structure, a flat organizational system with "product team" as the core has been established. In 2017, there are nearly 300 product teams, and 3–5 people in the product team are responsible for non-standardized links: the team leader is responsible for product design and development, one is responsible for production inventory management, and one is responsible for commercial marketing. Standardization links, such as customer service, marketing, logistics, etc. shall be uniformly taken charge by the enterprise. Independent accounting and self-financing shall be realized in operation, and "responsibility and rights" shall be unified in the minimum business unit.

In the face of uncertain demand, Handu Clothing co., Ltd. deals with the real demand through small batch and high frequency production, and takes the dynamic "burst, flourishing, flat, stagnation" algorithm as the core of operation. In particular, according to inventory turnover, sales volume, the number of customers into the store, web page retention time, such as data types, and dynamically adjust the algorithm weights. Two weeks after product launch, the team uses the algorithm to determine which of these four categories of "burst, flourishing, flat, stagnation" does this product belong to, and then give additional purchase of burst style products, clearance sale for flourishing style products. Decisions such as which product advertisement should be placed in the highest priority position to achieve greater profits, and which location of hot style product should be placed in the warehouse to make the shortest picking path, etc., are also made through the "burst, flourishing, flat, stagnation" algorithm and big data analysis support to realize rapid response to fragmented single product demand.

2.   Big Business Alliances

In 1984, WINTEL formed the computer ecosystem, which was composed of "Intel CPU + Microsoft operating system + Microsoft development tools + Application software". Application software including Chinese input method, mail system, antivirus software, video playback, hundreds of thousands of applications, tens of thousands of kinds of printers, cameras, scanners drive.

3.   Large Enterprises Build Platforms

Typical examples of this form of ecosystem include the Apple APP ecosystem, the FACEBOOK ecosystem, and the Taobao (contractual, service, and rule relationships) ecosystem, all of which have achieved commercial success due to their successful construction of the ecosystem.

In 2007, Apple Inc. built an ecosystem composed of CPU (ARM) + operating system (Apple) + communication module (Broadcom) + APP Store + APP for smart phones, making smart phones quickly replace functional phones. After the three products of watch, TV and car came into the vision of Apple Inc. 's senior management, the scope of the ecosystem was greatly expanded, and the production system was a crucial influencing factor for the decision-making of each product. Please refer to the inspiration case in Chap. 6.

In 2013, ROS (Robot Operating System) will create an ecosystem by combining Robot + ROS + chip solutions + APP applications. The development of the Robot industry will follow the development path of the PC, which is of great significance for promoting the development and popularization of the Robot industry. Based on The Turing OS 1.5 version, the robot application service has been developed, covering many fields such as remote operation, games, education, tools, social networking, and so on.

Xiaomi formed the MIUI operating system based on the secondary development of Android and built an open ecology on it. So far, it has incubated more than 90 companies, gathered more than 120,000 developers and more than 300 million users. In 2017, the income of Xiaomi ecological chain exceeded 100 billion yuan, and the ecology of its industrial platform began to take shape.

Huawei has been planning its own operating system, called Hongmeng, since 2012. Ren Zhengfei, Huawei's founder, said that the technical difficulty in developing an operating system is not so much the building of ecology as the building of ecology. The advantage of Android and iOS lies in their good software ecology, which enables developers to create high-quality applications and make users willing to pay for and use them, thus forming a virtuous circle in which developers are willing to continue developing.

4. More Complete Chains Including "technology + patent + standard + tools + business Model"

Technology leaders, patent and standard makers, data and tool owners, and ecological builders will have the right to establish industrial competition rules, dominate the structure of the industrial value chain, and determine the direction of industrial development. The business model includes leasing and supply chain finance. Through the establishment of leasing law, financial leasing can reduce the one-time capital investment of enterprises and enable them to use IM equipment such as industrial robots at a higher price and discount in advance. This is the ecosystem model of the future.

5.7.2 Artificial Intelligence

Artificial Intelligence is now in the knowledge processing, pattern recognition, natural language processing, game, automatic theorem proving, automatic programming, knowledge base and expert system, intelligent robot, and other fields have achieved practical results, also has been widely used in manufacturing, such as product of intelligent design and manufacturing, intelligent collaborative robot and intelligent process planning, intelligent scheduling, smart metering and intelligent management, intelligent decision-making, and so on. With the rapid development of sensor, radio frequency identification (RFID) and other sensing technologies, network communication technologies and big data, various artificial intelligence technologies that have been incubated for decades have begun to give full play to their advantages in manufacturing practices. Only by perfecting the theoretical research of artificial intelligence technology and applying the mature artificial intelligence theory to every link of complex manufacturing system can IM be realized truly.

The advent of vast amounts of available data and a surge in computing power, as well as breakthroughs in "deep learning", have ushered in a new phase of Artificial Intelligence. Artificial Intelligence is the most important tool for China's manufacturing transformation and upgrading. After the completion of digitization and networking, and the accumulation of big data, the application of Artificial Intelligence in the manufacturing industry has a real foundation, and starts to become the core of the entire era of intelligence.

The combination of big data and artificial intelligence has had a significant impact on the industry, with e-commerce and the automobile industry being the two most significant industries. In popular e-commerce platforms, intelligent digital assistants interact with customers online and provide personalized marketing suggestions and recommendations. In the auto industry, Tesla uses artificial intelligence and big data to power autopilot.

Artificial intelligence is generally regarded as the generation of computing systems with human-like intelligence by simulating, extending and extending human intelligence. Its main goal is to make machines capable of performing complex tasks that previously required human intelligence, or were impossible for humans to perform. For example, banks can use artificial intelligence technology to mine the data of financing customers, complete the work of customer credit rating, risk monitoring and reasonable capital allocation. Transport systems can use them to predict congestion ahead of time and to locate solutions where appropriate; In the field of manufacturing services, which the system can be intelligent analysis equipment fault, fault position and the reason that will happen in machine vision can be used to help production line to identify abnormal products, measuring target object size, auxiliary multi-joint industrial robot to find the best quality products, is the shortest, using the least movement trajectory scheme;In the field of purchasing, supplying and selling of enterprises, thousands of enterprises can find their matching enterprises in the open market to distinguish who is the reliable supplier and who is the purchaser of products. Through semantic recognition, the machine can dig out the market's product demand and change, product iteration cycle, cutting-edge technology direction, the emergence of challengers and the elimination of competitors from the mass market opinion data. In terms of human resources, the intelligent system can analyze whether a qualified job seeker corresponds to a resume, and make decisions for salary, performance management, training management, employee relations, etc. In terms of finance, accounting, investment and financing, etc., the application fields include handwritten reimbursement vouchers, automatic account checking, financial risk identification and capital planning. Through continuous learning of artificial handwriting, the error rate is even lower than that of manual record.

In general, there are three main representative schools of artificial intelligence. The history is shown in Fig. 5.20.

1. Symbol school. The logical method is the basic tool, which can also be called logicism. Semantic network, generative system and Agent are its typical representatives. The symbol school believes that the cognitive process of human beings is the process of the operation of various symbols. So computers should also operate on symbols. So cognition is computation. Knowledge representation, knowledge

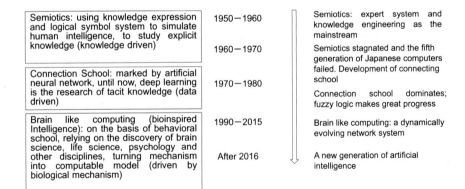

Fig. 5.20 Development history of artificial intelligence

reasoning and knowledge application are the core of artificial intelligence. Knowledge can be represented by symbols, cognition is the process of symbol processing, and reasoning is the process of solving problems by using heuristic knowledge and heuristic search. The knowledge system is good at reasoning and suitable for explicit knowledge management.

At present, generative knowledge representation has become one of the most widely used knowledge representation methods in artificial intelligence. Many successful expert systems adopt generative knowledge representation methods. The basic form is the production form P → Q or IF P THEN Q, P is the premise of the production form. It gives the prerequisite for the use of the production form, composed by the logical combination of facts; Q is a set of conclusions or operations that indicate the conclusion to be deduced or the action to be performed when the premise P is satisfied. The meaning of the production is that if the premise P is satisfied, the conclusion Q can be deduced or the operation specified by Q can be performed.

Most relatively simple expert systems represent knowledge in terms of production, and the corresponding systems are called production systems. The production system consists of a knowledge base and a reasoning machine. Knowledge base is separated from a reasoning machine. This structure facilitates the modification of knowledge without having to modify the program and makes it easy to explain the reasoning path of the system. The knowledge base consists of a rule base and a database. A rule base is a collection of production rules, and a database is a collection of facts.

[Case: expert system for chemical equipment breakdown diagnosis]
A chemical plant turned breakdown maintenance into preventive maintenance in order to improve the safety of production, therefore established the chemical equipment fault diagnosis expert system, its purpose is: according to equipment sensing parameters and the symptom, rapid execution of logic analysis, to predict and judge the fault level, and related suggestions are given. The production rule base is organized according to the equipment classification. Only a few fault diagnosis rules are summarized here to illustrate the characteristics of the production rule. Specific rules are as follows:

- IF the mixer pressure >160 kPa AND temperature >60 °C, THEN 1 h after the material deterioration
- IF the reactor vibrates >60 db OR vibrates db rising speed >0.5db/min, THEN the bearing is at high temperature
- IF the bearing high temperature, THEN 1 day after the power failure
- IF the grounding resistance of the distribution cabinet is >10 Ω, THEN the grounding system fails
- IF the grounding system fails, THEN the equipment carries static electricity
- IF the device has static electricity, THEN the risk of electric shock and explosion is high
- IF the risk of electric shock and explosion is high, THEN belongs to the first type of failure

- IF the power fails, THEN belongs to the first type of failure
- IF the material deteriorates, THEN belongs to the first category
- IF the pressure area of the reactor is relatively large in excess deformation, THEN belongs to the first type of fault
- IF the tensile zone of the reactor is relatively moderate in the over-limit deformation, THEN belongs to the second type of failure
- IF the tensile zone of the reactor is relatively small in excess deformation, THEN belongs to the third type of fault
- IF it belongs to the first type of fault, THEN the structure is seriously damaged, work should be prohibited and repaired immediately
- IF it is the second type of fault, THEN the structure damage is serious, but it can be used under a certain load and repaired later
- IF the fault belongs to the third type, THEN the structure is the local damage, does not affect safety, does not need to be repaired temporarily.

The chemical plant installs sensors for common faults on each key device, forming a sensor network to obtain data related to faults in real time. When the vibration of the reactor was detected to be 60 dB, the reasoning machine obtained the following prediction results through chained reasoning: bearing high temperature; Power failure after 1 day (type I failure); It should be repaired immediately. For this reason, the factory made production adjustments in advance and arranged the equipment to be shut down for maintenance at the end of the day before the failure, which improved the safety and reduced the safety cost.

Q&A: Please indicate what results and recommendations will occur when the vibration of the reactor was detected to be 61 dB according to chained reasoning.

2. Connection school. It is an artificial neural network as the basic tool, which can also be called connectionism, parallel distribution processing. Intelligence emerges through simple network connections. It comes from bionics, especially the study of human brain models. Deep learning, Extreme Learning Machine (ELM) and self-organizing feature mapping (SOFM or SOM) are the new generation of technology representatives. Such models are often discovered on a data basis and can be called data-driven models. Data-driven models are good at prediction and recognition, but the process is difficult to understand. They are suitable for the management of tacit knowledge, which is hidden in a large amount of data.

In Fig. 5.21, As shown in Fig. 5.21a, a neuron usually has multiple dendrites, which are mainly used to receive information from other neurons. There is only one axon, and there are many axon endings at the end of the axon that can send the same message to multiple other neurons. Axon terminals connect to dendrites of other neurons, transmitting signals. When the cell receives the signal input and accumulates, resulting in the difference between the potential inside and outside the cell membrane exceeding the threshold potential (-55 mV), then the cell becomes the active cell, and produces a 100 mV electrical impulse outward, also known as the nerve impulse.

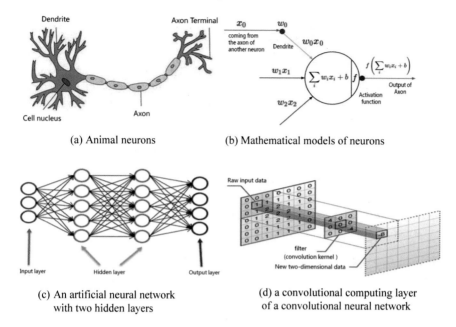

(a) Animal neurons

(b) Mathematical models of neurons

(c) An artificial neural network
with two hidden layers

(d) a convolutional computing layer
of a convolutional neural network

Fig. 5.21 Artificial neural network model **a** Animal neurons, **b** mathematical models of neurons, **c** An artificial neural network with two hidden layers, **d** a convolutional computing layer of a convolutional neural network

Many single neural networks are the so-called "neurons" linked together, so that the output of a "neuron" can be the input of another "neuron", each neuron can accept multiple input stimulus. When the weighted input summation received by a neuron exceeds a certain threshold, the neuron will "activate" an output, as shown in Fig. 5.21b. Commonly used activation functions are Logistic (Sigmoid), Tanh, ReLU (Rectified Linear Unit) and Softmax. The activation function is usually a nonlinear function. If the activation function is linear, then the neural network is equivalent to a single-layer network no matter how deep it is because the combination of linear functions is linear. ReLU is the most commonly used and simplest activation function, which is defined as: $f(z) = \max(0, z)$, in which z is the weighted sum of the input stimulus.

Neural network is to make its collective behavior have various complex information processing functions through the interconnected structure of these neuron components and the weight coefficient reflecting the correlation strength, so that the intelligence of single simple neurons emerges through the connection. Information is stored in a non—symbolic distributed manner in the neural network. Figure 5.21c is an example of a neural network with two hidden layers.

The first layer is called the input layer; The hidden layer (there are two hidden layers in the figure) takes the output of the previous layer as the input of the next layer; And the output of the next level will be the input of the next level. The hidden layer adjusts the weights of those inputs until the neural network error is minimized.

These weights assigned to the input can be automatically obtained by training. By adjusting the internal structure of the network, we can learn the relationship between input and output.

In the traditional sense, the multi-layer neural network has only an input layer, hidden layers and an output layer. Scholars have realized that the depth of the neural network is very important to its performance. However, the deeper the neural network, the more difficult it is to train it. Moreover, it also faces problems such as high requirement for computing power, network overfitting and accuracy degradation.

The network structure of deep learning is a kind of multi-layer neural network. In a deep neural network, the hidden layer extracts salient features from the input data that can be used to predict the output. With a large number of hidden layers, deep neural network has the ability to extract deeper features from data. While the most famous Convolutional Neural Networks (CNN) in deep learning is a feature learning part which is added on the basis of the original multilayer Neural network, and the part imitates the classification of the human brain in signal processing. Specific operation is in the original full connection layer joined the front part of the connection of convolution and dimension reduction layer, layer in convolution filter is utilized to extract the characteristics of the object, as shown in Fig. 5.21d, and then in the pooling layer characteristics, and take out the most obvious characteristics of local, finally through all connection neural network output the final conclusion. With the translation and sliding of the data window, the filter carries out convolution calculation for different local data. The specific calculation process of the current position in the figure is as follows: $4 \times 0 + 0 \times 0 + 0 \times 0 + 0 \times 0 + 0 \times 1 + 0 \times 1 + 0 \times 0 + 0 \times 1 + (-4 \times 2) = -8$. If the obtained value is large, the neuron will be activated and the feature will be found.

Traditional neural networks are not suitable for large images. For example, in a street image, objects such as people, dogs, trees, etc., if the size of the image is $1000 \times 1000 \times 3$, where 3 refers to the 3D RGB color value, then the number of parameters that a neuron needs to train is 6 orders of magnitude. This is just one neuron, and if you count the other neurons, the number of parameters to train would be an astronomical number due to combinatorial explosions. Obviously, this full-join form is time-consuming, and such a large number of parameters can lead to overfitting problems. If people were constructed in advance, dog, trees and other characteristics, then numerical input filter matrix characteristics, such as using a 50×50 matrix to describe the shape of the trees, and then use the filter to identify the object, the matrix can also rotate operation in response to the shooting with a rotation angle of the camera len, and the results of identification through the connection neural network output, will significantly reduce the number of parameters need to be trained.

Although a deep learning approach in large-scale image classification, speech recognition, face recognition and other fields has made amazing progress, but the depth of the network model with the human brain analogy, there are huge limitations, only "feedforward" deep networks, lack of logical reasoning and expressiveness of causal relationship and the lack of short-term memory and unsupervised learning ability, it is difficult to handle with complex spatio-temporal correlation between tasks. These questions prompt us to look for new approaches to artificial intelligence.

Brain-like Computing is a new direction of artificial intelligence, which uses dynamic evolving network systems to break through the limitations of traditional knowledge expression based on symbols and probabilities.

[Case: Deep transfer learning system]
Nicol developed an Deep transfer learning system to detect hard-to-remove bones in chickens. X-rays used in the old system sometimes gave false positives, causing the meat to be thrown away. The company hopes to reduce food waste in chicken processing by 80% within three years. Sagamiya Foods, a tofu maker, uses the remote monitoring image data provided by JMA (Japan medical association) to forecast sales of products affected by temperature. The company said that reducing excess capacity will reduce costs by about $92,500 a year.

3. Behavioral school, or evolutionism, is inspired by the relatively lower organisms in nature and takes biological inspiration and adaptive evolutionary computation as its basic tools. Genetic algorithm, particle swarm optimization, ant colony algorithm, immune algorithm, reinforcement learning are typical examples. Behavioral learning means you can explore unknown space, but it depends on search strategy.

Behaviorist scientists decided to start with simple insects to understand the birth of intelligence, believing that artificial intelligence came from cybernetics. Behavioral school emerged in the face of a new school of artificial intelligence at the end of the twentieth century. The early research work focused on simulating the intelligent behavior and role of humans in the control process, such as the research on the cybernetic system such as self-optimization, self-adaptation, self-stabilization, self-organization and self-learning, and the development of "cybernetic animals".

Genetic algorithm makes bold abstractions of biological evolution in nature, and finally extracts two main links: mutation (including gene recombination and mutation) and selection. On a computer, we can simulate organisms in nature with a bunch of binary strings. The selection function of nature—survival competition, survival of the fittest—is abstract as a simple fitness function. In this way, the evolutionary process of nature can be modeled and simulated on a computer, which is called genetic algorithm.

In 1995, the Boid model was extended and the particle swarm optimization algorithm was proposed, which successfully solved the function optimization problems by simulating the movement of birds. Similarly, there are many examples of using simulated crowd behavior to realize intelligent design, such as ant colony algorithm and immune algorithm, etc. The common feature is that intelligence emerges from the simple interaction rules followed by a large number of individuals from bottom to top and can solve practical problems. Behaviorism is good at simulating the workings of the lower bodies, not the higher intelligence of the human brain.

5.7.3 Intelligent Products

Enterprises do not need to advertise the modernization of production lines, but must explain the value of the products produced. The goal of IM is the high value of the product, rather than IM itself. For example, a machine tool factory production equipment and process are intelligent, but the machine tool produced by it is not intelligent ordinary machine tools, so the machine tool factory itself will not buy this machine tool.

Intelligent products and equipment are an important content of IM. IM should not only focus on promoting the construction and development of intelligent production process, but also constantly improve the intelligent level of terminal products. Germany's "industrial 4.0" report describes intelligent products: "smart products can store them is made in detail and information will be used, so they in interactions with the machine, the staff can answer questions such as "what need to install the parts", "the next step of production steps", "production using what parameters", "where I (the products or tools) should be sent to". Specifically, this data is often stored in ERP, MES and other systems, or stored in the cloud, but by the label on the product bar code, the data can be stored directly into the product embedded RFID chips in a storage capacity, so that you can continue to track the products after the product leave from workshop or when the absence of network conditions.

Cutting tool products, for example, as shown in Fig. 5.22, the cutting tool intelligent management based on RFID solutions, when the handle/tool embedded RFID chips, with the aid of ERP/MES information management, tool became an intelligent product, it can answer such as "how much cutting time is left", "should adopt what kind of processing parameters", provide premise for unattended 24 h of continuous production, extend the service life of cutting tools to improve production efficiency at the same time, to reduce the usage cost products or tools. The description of this system can be seen in the inspiration case in Chap. 3.

As shown in Fig. 5.23, the digitization and networking of products are the primary basis of intelligent products. Installing sensors on products and adding computing functions are the necessary basis of product intelligence. Further, connecting products with fog computing, edge computing and cloud computing is the main way to make products intelligent. Communication is not only between two objects, but also across workshops, factories, and enterprises, across different levels of business systems and supply networks.

[Case: Sensors on construction machinery of Zoomlion]
Zoomlion company has installed sensors on various construction machinery such as excavators it produces and sells all over the country, with an average of 110 sensors per machine. The real-time data collected by the sensor network can be transmitted directly to the control center through the Internet. On the large screen of the control center, the status of construction machinery nationwide can be grasped in real time. The National Bureau of Statistics has set up an economic indicator called the Excavator Index to evaluate China's economic development, and has designed a job to keep tabs on it.

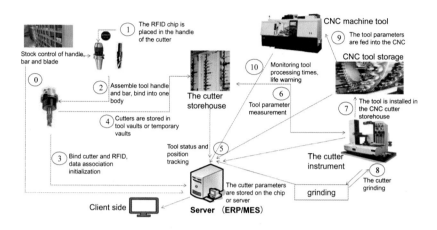

Fig. 5.22 Information exchange between intelligent product and monitoring platform with tool as an example

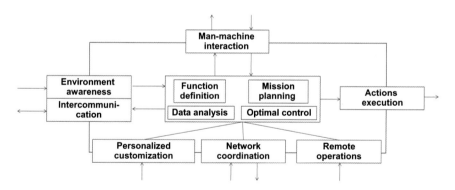

Fig. 5.23 Basic structure of intelligent product

From the level of intelligence, it can be divided into three levels: component level, system level and terminal level. Component-level intelligent products mainly include key basic components and common components, such as car windows that automatically open after falling into the water, car doors that automatically close when strangers approach; System-level intelligent products mainly refer to intelligent equipment, mainly including high-end CNC machine tools and industrial robots, 3D printing equipment, intelligent sensing and control equipment, intelligent detection and assembly equipment, and intelligent warehousing and logistics equipment. Terminal-level smart products refer to all kinds of smart terminals that provide services for users, such as baby carriages that can tell the probability of a cold, refrigerators that can provide personalized health recipes and prompt the purchase demand, as well as various wearable devices, smart home and home appliances.

Referring to the Belief—Desire—Intention (BDI) model of the Agent, intelligent products should have the following intelligent characteristics: on the interactive interface, they should have the characteristics of environment perception, mutual communication, man–machine interaction and action execution; Inside the agent, has a function definition, mission planning, data analysis, the characteristics of the optimal control, through the function defined including environmental characteristics, and its function of "Belief" description of cognition, through the task planning to achieve the plan of the state of "desire", through data analysis and optimization control to realize the objective of "intention" implementation guide; In the emerging business model, it has the characteristics of personalized customization, network cooperation and remote operation and maintenance. The basic structure of intelligent products should be shown in Fig. 5.23.

[Case: Multi-functional wheelchair service robot]
Aiming at the enhanced elderly care needs in the aging society, the developed wheelchair service robot has the following functional characteristics:

1. Multiple postures: can be used as a flat bed for sleep; The hinged bar folding mechanism can shrink the front and rear wheels into a sitting posture for the user's toilet and shower services; If the front and rear wheels continue to contract, they will become a standing posture, which will help the user to lift and assist walking.
2. Remote monitoring and maintenance: the elderly (users) are judged to be out or fall by the sensor or camera by SMS and WeChat notification; By means of infrared sensing and other environmental sensing methods, users can be identified as they move around at night, and then the lights on the chair will be automatically turned on and followed. The camera can be monitored and operated remotely through wireless network and real-time communication.
3. Multi-sensor fusion: sensing and judging human hands and body movements, identifying the user's intention by integrating the pressure changes of head, back, buttocks, feet and other parts to the wheelchair, and providing power and support for the user's sleep, sitting, standing and walking. Electroencephalogram (EEG) electrodes can be set in the posterior parietal occipital region to collect scalp brainwaves at high frequency, and the patterns of brainwaves can be recognized with the help of embedded CPU and software to achieve predefined goals.
4. Task planning and optimization control: the product can carry out machine learning, and can stop some rules, adjust the weight of rules, or supplement new rules through human–computer interaction interface; Can remember the user's living environment, bedroom layout and lifestyle, carry out task planning

according to the functional definition made by personalized customization, carry out path planning for walking and avoid obstacles, evaluate the effect of each task execution, and independently optimize and adapt.

Intelligent products are synonymous with the lifestyle of customers.

Intelligent products are an important part of IM. To develop intelligent product technology and improve the intelligent level of products is a necessary link for the implementation of IM project in China, and will also become an important evaluation index of IM. From smart phones, smart TVS, wearable products, smart cups to smart cars and smart robots, enterprises need to continuously make technological innovations and invest in the intelligence of products. For example, Haier, Gree and Midea are all investing in smart appliances.

5.7.4 Intelligent Services

Intelligent service is to transform the business model of an enterprise from product-driven to data-driven and from selling products to selling services through IoT technology, big data and other IT technologies, focusing on user needs and based on intelligent products, so as to realize the innovation of service model and business model.

Germany's "Industry 4.0" has undergone several upgrades since it was formally proposed in 2011, and its focus has shifted from the production of data-driven smart factories to smart products and services. "Industry 4.0" and "intelligent service" have become the national industrial competition strategy of Germany. Smart services are expected to boost German firms' productivity by 30% by 2025.

Intelligence service implementation is an on-demand and active service, namely by capturing the user's original information, through data analysis, build the demand structure model and the user portrait, conduct data mining and business intelligence analysis, restore the real big scene. In addition to analyzing the user's habits, preferences and other dominant demand, can further mining Implicit demand associated with space–time, behavior, working routine, take the initiative to provide users with personalized, accurate and efficient service. What is needed here is not only the transmission and feedback of data, but also the multi-dimensional and multi-level perception and active and in-depth identification of the system. On the other hand, it is imperative for the service industry to move from low-end to high-end, and the transformation and upgrading of the service industry needs to rely on intelligent services.

[Case: Amazon's sales service]

Amazon uses big data from its 2 billion user accounts to drive sales growth by forecasting and analyzing 10^9 GB of data on 1.4 million servers. Amazon's flexible MapReduce program is built on top of the Hadoop framework. The product catalog of the data on the interface is analyzed and sent back to a different database every 30 min. Every 10 min, Amazon changes the price of items on its site. One of the most popular

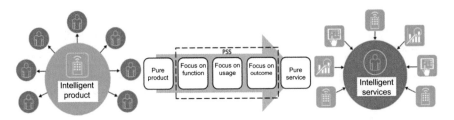

Fig. 5.24 Transition from functional product to service experience

services is Amazon's out-of-stock forecast for sellers, which uses recommendation algorithms to analyze sales and inventory levels for sellers. Amazon also uses graph theory to select the best time arrangement, route and product classification to deliver products to minimize costs. According to a new patented technology called predictive shopping, Amazon delivers the goods that consumers are likely to buy to the nearest express warehouse in advance according to their shopping preferences. Once the buyers place the order, the goods can be delivered to their doorstep immediately. But if the big data algorithms get it wrong, Amazon could face the difficulty of bearing the logistics costs of shipping goods back and forth.

In the future, enterprises will shift their focus from functional products to service experience, as shown in Fig. 5.24. Manufacturers no longer sell pure products, but deliver value to users through product-Services Systems (PSSs). In this process, intangible services play a vital role. Service Dominant Logic (SDL) regards supply chain as a network of value co-creation and resource integration, and provides a useful perspective to explore how PSSs are developed and delivered. In order to gain new competitive advantage, manufacturing enterprises must interact with customers and shift from Product Dominant Logic to Service Dominant Logic.

5.8 Practical Case: Midea Smart Factory

1. Midea group's Strategy

Midea group vision of the future transfers from labor-intensive to technology-intensive, capital-intensive, from low-end manufacturing to high-end manufacturing industry and advanced manufacturing industry, and finally transfers into a technology group with four big businesses of consumer appliances, HVAC, robots and automation systems, intelligent supply chain, driven by science and technology.

The primary axis of Midea's strategy is firstly product leadership, surpassing Japanese home appliances through technological innovation; Secondly, it is driven by efficiency, including manufacturing efficiency, asset utilization efficiency, etc. Midea's order to sales cycle has been compressed from 30 days to 12.5 days. Thirdly, it is a global enterprise. The overseas income of Midea Group accounts for 50% of the operating income. In the past two years, Midea has also put forward the "two-intelligence strategy". First, intelligent products, such as interconnected refrigerators in the kitchen ecosystem, campus Alipay washing machines, and air conditioners that simulate forest air. Second, intelligent manufacturing, including lean and digital transformation, and operation of the whole value chain. This can be reflected in the layout of the acquisition of Kuka.

2.   Addition and Subtraction Tactics

There are two ways to improve profit and employee efficiency:

The first is subtraction. The main subtraction is to reduce the number of product categories. In the past, there were more than 60 categories in Midea, but now the number has been reduced to 32, the number was halved. Secondly, the number of specifications and styles was reduced from 4000 to 2000, which greatly reduced the cost, including the cost of mold, supply chain material, and research and development. Thirdly, it is to remove channel inventory. Midea began to pilot the "T + 3" order system in 2016 from the Little Swan washing machine. From placing an order to satisfying the demand, users can collect customer orders, collect raw materials from the factory, produce and deliver goods in 4 cycles. Each cycle is set to three days, so it is called "T + 3". The lead time on traditional sales orders, which used to be 23 days, has been reduced to 12 days. According to market research, this is the best time for user experience. In the past, many enterprises decided to sell on the basis of production. Every year, they planned to increase stored goods by 10–20% compared with last year, after excessive production to the channel pressure, the channel felt suffering.

The second is to addition, especially in automation and IT investment. In the past, the IT system of each business unit was decentralized and implemented separately. "632" IT project was planned and implemented since 2011, "632" refers to six major operating systems (PLM, ERP, APS, MES, SRM, CRM), three major management platforms (Business intelligence, BI; financial management system, FMS; human resource management system, HRMS), and two major technology platforms (Midea Information Portal, MIP; Midea development platform, MDP), and enterprise process framework (EPF), master data management (MDM), etc., a total of 13 large project group for implementation and promotion. It was finally completed on December 31, 2016, which took four and a half years. In 2017, 7 billion yuan was spent on automation. By 2018 Midea had 1000 IT staff. In the "632" project, more than 2000 interfaces have been set up for the system, eight business divisions have been fully launched with the system, and each business division has about 200 full-time responsible persons. Senior personnel has been hired from Samsung. The huge labor force investment reflects the company's courage and determination to promote enterprise transformation. In the end, Midea realized business connection from order to collection, from purchase to payment, and among internally connected transactions, partner business processes and systems.

Midea in the implementation of IM has formed one whole set of methodology. It includes three stages of work: first, set up the business process architecture, every time of implementation will cover with the architecture department, the department to do the comparison, identify department process framework and overall framework are consistent, create the change point. Business changes first, then systems are integrated. The first requirement of enterprise transformation is lean operation and agile user response. Secondly, IT system integration, including client end to end, product end to end, order end to end. Use the "632" strategy to pull the whole process and realize the unity of IT information management. Finally, it focuses on developing Internet technologies, including big data, mobile, and cloud technologies. At present, it has entered the third stage called Intelligent 3.0, which includes industrial Internet, C2M, customized production, and so on.

3. Big Data Analysis

Big data realizes a transformation process from the original experience-driven to data-driven: it uses big data to cover the R&D side, helping to understand users' ideas and improve products; Support lean manufacturing at the production end; In the market side to help the market research, competitor analysis, until can help the whole channel management and distribution, research on consumers and improve after-sales service. To be specific, Midea made efforts in several directions. One is to connect internal data and internal supply chain, namely, production, research, and development, sales, finance, etc. All the data in these fields are connected and put together to establish a unified data warehouse of Midea, on which they will conduct data operation analysis. Second, it integrates with external data to collect Internet business and public opinion data, including data of e-commerce series, such as Taobao, Tmall, Jingdong and Suning websites; Data from traditional industry websites; Data on the news media; User data, all the data that Midea has contact with users, such as every order, every after-sales phone, every installation record, after cleaning and fusion, user data, namely user portrait, can be obtained; Data on devices, especially data collected using smart home devices.

Several products were developed based on this data. First of all, the IT product for data analysis is called Stargazing Platform. Every software in Midea is a product rather than a system. The product manager, not the project manager, is always in charge of the product from the planning to the final operation. Stargazing products adopt Internet data, make research on the market with data support, such as market capacity calculation, rival discrimination, competitive price, their own company product performance, best-selling or unsalable reasons, sales fluctuations, etc., and operation of online retailers, analysis of why some businesses do well but some is terrible, user feedback, and so on.

Midea's previous product development method was trial and error, leading to a wide variety of products. For example, it could develop 100 rice cookers a year, but only 10% sold well and 90% lost money. Now, based on big data analysis, Midea can produce the most popular products in the market and design products with a higher success rate than before, which is the first successful application of data on the Internet. The second application is using the Internet's unstructured data and Chinese data, perform voice analysis, from the user individual comments gain the viewpoint of value, and these ideas for statistics, such as how many times users have mentioned air conditioning noise, with a specific format that help operating.

Industry information system based on business and public opinion, to tap the market opportunities, and insight into the user's demands, get the user evaluation and product reputation, for the first time to get the negative public opinion, to understand the industry situation, such as air conditioning industry among the top 10 brands, especially in the first three of the competition between Midea, Gree, Aux, including the analysis of their own advantages, industry prices the latest intelligence information, compared with competitors. Public praise is a unified assessment index for the whole group. As long as this product is sold in e-commerce, it will be directly evaluated by negative comments. If the product fails to meet the standard for three months in a row, it may be forced to take the product offline. The market share of various products has changed from 2015 to 2016. For example, the microwave oven accounted for 30% in 2015 and 53% in 2016. the sales volume of electric ovens increased from 20% in 2015 to 29% in 2016. Conduct data analysis every day. The Business Unit will track the data weekly, monthly and quarterly, and then formulate corresponding strategies, including the evaluation of the frequency effect of some single product outbreaks, to drive the marketing activities and policies of products. Take electric ovens as an example, in early 2015, a rival brand was found to have launched hot-style electric ovens. Through big data analysis, Midea finds that its advantages are as follows: its price segment is around 400 yuan, and its capacity is 32 L, which is mainly featured with baking function and culture. Meanwhile, through analyzing user comments, Midea finds that many users cannot bake snacks, cookies and cakes. The Department of Midea planned to change the 38-L oven, whose main function was roasting meat, into a 32-L oven whose main function was baking culture. At the same time, it would also send corresponding small baking accessories and make corresponding tutorials and videos to promote this product. Through three months, Midea has become the first in the electric oven industry. The second is to find out the quality problems of the products and improve them. Take refrigerator as an example, it usually takes half a year for a new product to be launched and put on the market, then the noise problem is fed back to the company in the market, and finally, the improvement is made in the production line. In the next six months, there will be a large number of such substandard and defective products flowing into the market. Now through Internet data analysis, Midea can respond within a single month, push the new batch to the market. If there is still a user complaining about the noise problem, directly locate the user's comment to the corresponding order of the user, and contact the corresponding user in the order to make door-to-door analysis by the engineer. On one occasion, it was found that the hair inside caused noise after being opened, and the noise problem was later solved by improving the air inlet.

Midea has been doing business intelligence (BI) since 2008, but the effect has not been very good because BI is to do the corresponding report to the leadership. Due to the change in leadership needs, about a thousand reports have been made one after another. The utilization rate is less than 10%. The demand response of the report is also very slow. It may take more than one or two months to modify or make a report, which leads to the dissatisfaction of the leaders. The data produced is not only inaccurate but also hard to use. Based on this situation, we made a comprehensive data integration and reconstruction in 2015, sorted out the indicators to be analyzed based on the business process, and built the report system based on the index system. Midea's eight business divisions could check different dimensions to see the report. Then, based on the index system, a complete set of element management systems is constructed. If anything goes wrong with any metric, you can figure out its logical processing steps and determine which business systems are the ultimate data sources through layer upon layer of traceability to ensure scientific indicators to achieve data quality control. For example, once I found the abnormal cost of materials, I traced it to the source through the element management system, found that the problem was caused by mistake in the column of materials in the ERP system, and have been wrong to fill the cell phone number into the price. However, the business system did not check it, so the 11-digit cell phone number became a unit price. After a year of using the system, through closed-loop optimization, almost no department says the data is inaccurate.

Internet big data products also launched mobile terminal applications, core management for business daily, every morning to send business briefing, the management focuses on indicators, such as the total amount, the amount receivable, order delivery remaining days warning, order delay and delay reason, at seven o 'clock every day in the daily push above information, then push e-mail at eight o 'clock, recommend warning at half-past eight. Send in a read-only mode, a simple click to complete the report browse. In this way, the use of the system gradually has been popularized. In order to get the data into the ground and operation more quickly, IT personnel secretly put a big screen in Fang Hongbo's office to show some big data analysis reports. Then, when the operation analysis meeting was held, General Manager Fang could accurately point out the ranking situation and existing problems of the business division, which caused the heads of each business division to be shocked and ask how General Manager Fang knew this data. With this top-down pressure of the data culture, the big data analysis report was applied and popularized.

At present, Midea connects all the data of millions of smart home devices. In the future, Midea will also connect the data of equipment in the production line of industrial Internet to do real-time monitoring and processing and carry out user portraits. Midea's user portraits are based on its own big social data, plus online data, yielding about 188 million user portraits. Marketing campaigns based on user portraits are a feature of Midea. In square or mall when sales promotion, for example, users can be accurately informed around five kilometers within the scope of the user portraits, to different users to send different text messages, to the high consumption crowd pushing high-grade new product information, to the price sensitive crowd pushing preferential product information, blocking spam once complaints, to ensure users to participate in activities, enhance the promotion effect, in the scope of a province Midea can gain a sales scale of 100 million $ per day, which is twice as much as before.

Based on user portraits and user's insight, it can also support the design of the product, such as the development of environmentally friendly products, the environmental protection product corresponding to all the users to find out, and then analyzed the group, including their characteristics, the intention of price, quality, making the designer at the time of design will be better able to make the products to match the user.

4.   The Problem to Be Solved

Every time the IT system is closed-loop optimized, some problems are found and solved. Current issues to be addressed include:

(1)   Production planning and scheduling algorithm model logic is not clear. The Oracle ASCP now used by Midea is an upgrade based on the original MRP, which is more dogmatism. Some logic does not conform to the actual production of Midea, but it is difficult to change. Mapping out the production plan should combine order first, based on MRP demand time, push the start time, as long as the demand is consistent with the time will make the order and material of the merger, Oracle ASCP plan based on the limited resources will have the ability to do things according to the resources, the plan forward or back, logically, has made the merger may not be appropriate, a reasonable practice should be based on resource planning first moved forward or back, and then merge. Midea itself has now developed a production planning system that uses a simple model and only considers bottleneck resources because it takes a lot of time to consider all resources. Midea should further apply next-generation artificial intelligence methods, such as deep learning and visual recognition, in order to better solve the problems in planning and scheduling decisions.

(2)   Slow operation speed. Midea's goal is to complete the planning and production schedule every 2–4 h, but Oracle needs to complete it only once a day, and it often needs to work until night for tomorrow's production. Plans need to be fast and precise, the requirements are higher and higher, but it is difficult to meet. Because Midea's business and products are so complex, some products have as many as ten layers of BOM. The customers of Oracle are mainly the top ten samples of Coca-Cola, IBM and Dell, etc. The software is mainly adapted to those enterprises with very high standardization, but it is difficult to meet the needs of Midea.

(3)   Poor accuracy of marketing prediction. Marketing prediction is not accurate, but the manufacturing side of the order need to deliver quickly, then both ends of production and sales always have such a contradiction. The only way to solve this contradiction is to start from the source, making the prediction relatively accurate, or rolling cycle correction. At present, it is necessary to establish a multi-factor model based on big data analysis to get an accurate result for sales promotion and seasonal demand prediction.

(4)   Low automation degree of man–machine collaboration. It has been a common phenomenon that robots do palletize and carry. However, palletizing, welding, spraying, and other processes have high requirements for equipment. Midea has basically automated equipments, but in the process of final assembly and cabinet loading, there are many model options and tools, and the degree of automation is relatively low because now automation can only be implemented in some degree of standardized mass production. There is still a lack of research on multi-variety and small-batch automation in the complex changing environment. Midea and Kuka have cooperation in this field and set up specialized teams to study. Once these generic technologies make breakthroughs, the

whole Midea or the whole industry can benefit from them. To meet development needs, Midea now plans to build another 1000-mu industrial park in Shunde. It will take on a whole new look by 2025.

Thinking exercise

1. Point out the basic situation of IM of Midea Company.
2. Try to analyze the difficulties of Midea's IM transformation.
3. Try to point out three integrations of Midea's IM.
4. Analyze the steps and stages of Midea IM implementation.

Homework of this chapter

1. Discuss the composition of investment and income for IM. What are the common reasons for upgrading original ERP systems? What factors need to be considered?
2. Please indicate the impact of product modularization on product diversity, manufacturing cost, development cost, types of parts, quality and performance.
3. What benefits did Chin Dynasty's standardized weapons bring?
4. Please point out the importance of semi-finished product packing boxes (material containers) for collaborative efficiency between workshops and enterprises.
5. Why can design solutions be verified easily when using three-dimensional modeling?
6. What factors support the high market value of this software?
7. Please point out the impact of the Industrial Internet on quality management.
8. what is the basic logic of CPS?
9. In this elevator remote monitoring system and the application of big data technology, what are small data and big data respectively? What data analysis algorithms does the company (Hitachi in this chapter) use to support what decisions?
10. What are the hidden dangers of APP technology bullying for users? What data is most important for forming a complete life picture?
11. What are the characteristics of Toyota's knowledge management tool "Tradeoff Curve"?
12. What is the best experience of Siemens in knowledge transmission and sharing channels?
13. What innovation does Comac have in knowledge scenarioization and intelligent push?
14. Please identify the names and contents of Midea's three ecosystems: products ecosystem, production ecosystem, business ecosystem.
15. What changes have taken place in the organizational structure of these enterprises (Handu Clothing co., Ltd. And Huawei)?How should the four elements of organization, process, technology and data work together?

16. Please indicate what results and recommendations will occur when the vibration of the reactor was detected to be 60 dB according to chained reasoning with the rules in this chapter.

17. Please indicate the steps in the transition from a product to a smart product according to the examples of Zoomlion.

18. Please indicate which modules the product "Multi-functional wheelchair service robot" has.

19. Please analyze the advantages of e-commerce companies Amazon's sales service relative to traditional stores.

20. What do you think of the choice path and emphasis of China's IM (Intelligent Manufacturing)?

21. What are the information system planning methods?

22. What are the steps and contents of information system planning?

23. What are the typical problems of information system planning?

24. What is the relationship between lean production and IM?

25. What impact does standardization have on product quality, manufacturing cost, development cycle and cost, diversity, change, parts type and quantity, automation cost, etc.?

26. What is the relationship between the three representative technologies of the Internet, big data and artificial intelligence?

27. What problems should be paid attention to in the implementation of IM?

28. What is the logical relationship between the five phases of IM implementation? What about the tasks in each stage?

29. In what aspects do you think manufacturing is intelligentized?

References

1. Intelligent Manufacturing Capability Maturity Model White Paper (1.0) [EB/OL]. China Institute of Electronic Technology Standardization (2016)
2. Current Standard System of Intelligent Manufacturing System [EB/OL]. National Institute of Standards and Technology of the United States (NIST) (2016). https://www.nist.gov

Chapter 6
Future Research and Application Direction

6.1 Inspiration Case: Apple's Next Generation Industry

Apple is probably one of the most successful technology companies in the world since the release of the iPod in 2001. It has transformed three businesses over the next decade: music, smartphone and tablet. When Steve Jobs died in 2011, his successor, Tim Cook, was left to revolutionize the next generation of industries. In 2015, Cook seems to have three potential product targets: watch, TV, and car. The future of all three targets is highly uncertain. Watch shipments started to rise in the first quarter after the release. Television seems to have come of age, but many companies are trying to change the industry. Cars are Apple's biggest opportunity, but also its biggest risk, in terms of value.

Financially, Apple was the most profitable company in the world in the fourth quarter of 2014, generating $18 billion in net income. But Mr Jobs once famously said: "Apple's success comes from saying no to 1000 things to make sure we don't go wrong or try to do too many things. We're always thinking about new markets we can enter, but only by saying no can we focus on the things that really matter. "Mr Cook and his team asked: Are watch, television and car the right way to go?

1. Apple watches

Cook, along with Apple's design director and the company's Marketing Department, has repeatedly called the Watch the "most personal" product and stressed that the Watch is both technical and stylish. In Cook's words, "It shows your taste and expresses your self-image, just like your clothes and shoes… We recognize that technology alone is not enough, it has to have a stylistic element.

Apple sees the watch leading the way toward "technology meets fashion." The Apple team is trying to change the way people interact with mobile technology by limiting how often and for how long users can interact with the iPhone. At 2013 a study found that the average user turns on their smartphones more than 100 times a day, and analysts estimate that up to two-thirds of interactions can occur on a watch or other wearable device. The necessary information and notifications of the user's mobile phone can be clearly seen on the watch. The watch allows them to focus on other tasks or the world around them. Interactions with the watch must be short, lasting less than five to 10 s. To do this, Apple designers simplified or removed a number of features. As CEO of Apple watches project, Kevin lynch said: "every day people carry their phones, look at the screen frequently, appear to have been used to, but we must be in a more humane way to provide it, it is to keep you away from the iPhone, especially in talks with a key figure, often taking out mobile phone text messages are not polite".

(1)   Smartwatch and wearable technology

In developing the Watch, Jonathan Ive spent months studying the history of time-keeping and invited watch experts to Apple campus to talk about the "philosophy of an instrument that measures time," as one of them is called. "What's interesting is that for centuries, timing technology has found this wrist position and never moved anywhere else," Ive explained. "The historic, crucial position is the wrist."

Several major smartphone vendors released smartwatches by the end of 2014, most of which pair with smartphones and allow users to receive and respond to calls, notifications and run third-party apps. Smartwatch sales are growing, but only 3.6 million units were sold in 2014, or 6.8 million shipped if you include the less features-healthy bracelets, for a total revenue of less than $1.2 billion. Analysts expect the market to grow rapidly after Apple Watch's launch. IHS, a market research firm, estimates that the Smartwatch market will grow from 3.6 million units in 2015 to 34 million, with Apple Watches accounting for more than half of Smartwatch sales. They expect the market to grow to 100 million units by 2020, with the ratio of Smartwatches to Smartphones rising from 1:500 in 2015 to 1:20 in 2020.

In March 2014, Google announced that Android Wear would extend the Android mobile platform to watches and other wearable devices. Samsung, MOTOROLA, SONY, LG and Asus later launched watches based on Android or proprietary platforms. Pebble has also developed a proprietary Smartwatch operating system. As with Smartphones, Android watches are implemented in very different ways, including various specifications, materials, prices, and dimensions (such as flat and circular surfaces). The Android Wear watch works with any Smartphone running the latest version of Android. The watch uses Bluetooth or WiFi to pair with the phone. Android Wear displays text messages, emails, phone calls and other information on a user's phone as small cards that can be refreshed or opened with a tap. It also includes standard health and fitness apps, where music can be stored locally on the watch, and provides time and location sensitive information, such as weather reports, public transport information or traffic conditions. Some have complained that Android's attempts to predict what information a user might want have led to too many notifications being annoying.

Samsung is the market leader, selling 1.2 million units in 2014. Nearly 45% of U.S. consumers who reported owning a smartwatch at the end of 2014 bought a Samsung watch, with LG, MOTOROLA, SONY and Beibao accounting for 8 to 11% respectively. By the end of 2014, Samsung was offering six different watches in its Gear range. Most of the watches run Tizen and require a Samsung Galaxy smartphone to be paired.

The current battery life of Apple Watch is 18 h. In order to enhance the battery life, Apple stepped up the layout of Micro LED related Technology patents after it acquired LuxVue Technology, a Micro LED display Technology company, two years ago. If Apple can really make a breakthrough in Micro LED technology, it will be the first company to bring Micro LED technology to mass production.

(2) Version and pricing

Apple offers users a range of options. There are three versions of the watch, all with the same specifications and different materials for the case and the bracelet. Each version has a 38 mm or 42 mm display. The main line comes with a stainless steel case, sapphire crystal display, and a variety of wristbands called fluoroelastomers, made of leather, stainless steel or high-grade rubber, starting at $549 and going up to $1099, depending on size and choice of wristband. At $349, the cheaper Apple sport watch is more expensive than the most expensive smartwatches from rivals Samsung, LG, MOTOROLA and Beibao. With an 18-karat gold case and sapphire crystal display, it starts at $10,000 and goes up to $17,000.There are 38 different combinations of display size and wristband options available. In addition, customizable digital interfaces mean an almost infinite variety. According to the head of Apple's human–computer interface department, "The only way to get a company's products to be worn by different people is to offer more options. We want millions of variations, and with hardware and software, we can do that." Apple makes typically high profits at these prices, with the material cost of the 38 mm sport watch estimated at less than $84, or 24% of the price.

(3) Lanch to Market

Apple began preorders on April 10, 2015.The initial stock sold out within hours. Apple sold 1.7 million watches in the first two weeks after its launch. Unlike previous product launches, Apple did not initially sell the watch in its retail stores, although customers could book an in-store trial. Most versions of the watch are available online only. Sales reached about $15 million in 2015, more than doubling industry shipments in 2014. But reports that important parts made by one of the two suppliers were defective forced Apple to scrap some finished watches and move all of its manufacturing to a single supplier, which slowed down production. The cooperation between Hermes and apple was reached as early as 2015. Apple said that both Apple and Hermes pursue rigorous manufacturing processes and excellent quality. Apple spends a lot of money on advertising. Between the March 9 launch event and the initial order on April 10, Apple spent $38 million on television ads for the watch. In the past five months, Apple has spent $42 million on TV advertising for the iPhone 6.

(4) Function

With a digital interface to choose from, the Apple Watch is an extremely accurate timer. In addition to timing, the Apple Watch comes with the usual health apps, including a heart rate monitor, an exercise tracker, a calorie counter and regular reminders to stand up and exercise. Like other smartwatches, the Apple Watch transfers some of the phone's functions to the watch when paired with an iPhone 5 or later, such as incoming calls, text messages, emails, calendar events or notifications from third parties. The goal is to make short notifications and messages easy to see, limiting the number of times the wearer can interact with the iPhone. It also allows users to control music playback, store music on the watch, access maps and navigation, and combine the capabilities of third-party apps.

The watch introduces several innovations into its user interface. It includes what Apple calls a "digital crown," which looks like a lever on a traditional watch and allows users to scroll and zoom. The touch screen has a feature called "power touch," which gives users more information when they press hard on the screen. In addition, Apple has developed a new font for the watch, San Francisco, designed for small screens. Another innovation is the Touch engine, which alerts users to different types of notifications by monitoring different patterns on their wrists. Interactions with the watch are aided by Siri, Apple's voice assistant.

Although early assessments were mostly positive, some shortcomings were pointed out. The main problem is that watches usually need to be recharged every day. In addition, the roughly 3500 apps available in earlier versions are just a fraction of the apps in the APP Store, and most are not optimized for the watch, providing too little information or too much action. The bigger question is: Can Apple sustain profits? When competitors respond, will Android have 80% of the market, like it won in smartphones, or will Apple keep its market share? Once the firm is established as a $17,000 watch, it will have to move into other luxury markets, such as cars.

2. Apple TV

"I want to create an integrated TV that's completely easy to use, that will sync seamlessly with all devices, iCloud…It will have the simplest user interface". Still, Apple executives made clear that they thought the TV industry was mature. Cook echoed these sentiments: "The TV interface is terrible…You can't see things until they come, unless you remember to record them…Apple TV will reshape the way you watch TV".

Apple first entered the TV business in 2005, when it began selling television shows to be downloaded through the iTunes store for users to watch on their computers or video iPods. Analysts are waiting for Apple to ship a new high-definition television set following Mr Jobs's comments about "cracking TV". The TV industry sells more than $100 billion a year, but TV makers typically have low margins, averaging 5%, and a technology upgrade cycle takes up to an average of eight years. Apple has been working on televisions for several years, including video calls using ultra high-definition displays, motion sensors and cameras. However, in May 2015, the Wall Street Journal reported that Apple had disbanded the project in 2014.

While Apple has scrapped plans to make a TV, analysts expect the company is planning a subscription-based online video streaming service that could center on a new generation of Apple TV set-top boxes, which haven't been updated since 2012. Apple TV also launched Airplay, which allows consumers to display approved content on their TV from their iPhone, iPad or Mac. Most of the content on Apple TV comes from other providers. In total, Apple sold about 25 million Apple TV boxes by the beginning of 2015. Its relatively small sales have led Apple executives to call its TV products a "hobby." Apple cut the price to $69 in March 2015, possibly to clear inventory. Apple TV boxes have several competitors, including Roku, Amazon and Google, with prices ranging from $35 to $99.

Rising prices for cable services have led to discontent among a small but growing number of consumers, especially younger ones, who tend to cancel cable services in favor of subscriptionable services, which allow them to subscribe to fewer channels for $25 or less a month. Apple is said to be in talks with Disney, 21st Century Fox, CBS Corp, Discovery and other media companies to acquire certain channels.

In 2014, more than 40% of U.S. TV households had subscribed to video on demand (SVOD) access, and 12.5% of households subscribed to multiple services. The number of Americans paying for television services (cable, satellite or fiber) for the first time fell slightly in 2013, by 250,000 to 100 million, and fell further in 2014. The percentage of households choosing cable TV dropped from 71% in 2001 to less than 57% in 2015, while satellite wireless options rose to more than 30 percent from less than 12% over the same period. At the end of 2014, the average American spent more than 141 h a month watching traditional TV, less than 11 h watching video on the Internet and less than two hours watching video on smartphones. However, the amount of time spent watching traditional TV has fallen by nearly 10 h per month (6%) since 2012, but 63% of people still prefer the big screen.

Sales of television advertising reached $78 billion in 2013, up from $64 billion in 2009. Internet advertising revenue was about $42 billion in 2013, according to PWC. Pay-TV operators averaged profit margins of about 40% in 2013. Cable networks are almost as profitable, with margins of about 37%. Satellite TV operators, broadcasters and film and television producers are less profitable at 25%, 17% and 11% respectively.

The average price of a LCD TV in 2017 is less than half of an original LED TV of the same size and function. From LG's first OLED TV, which was released in 2013 and priced at $14,000, the price of a 55-inch 4 K OLED TV dropped to $2000 in 2016. According to 2018 report from IHS Markit, LG, SONY and Skyworth are the top three OLED TV brands in the world. Panel capacity directly determines how much space OLED TVS have for development. In 2018, OLED panel capacity is too low to meet industry demand. When Samsung and LG abandoned the LED business to battle OLED, a new generation of LED-based display technology, Micro-LED, popped up. Micro-LEDs use micro-light-emitting diodes (LED) to reduce pixel distances from millimeters to microns by integrating a high-density, micro-sized LED array on a chip.

3. Apple's car

If Apple wants to launch a game-changing new product, cars are a significant option. Full-year 2014 car sales are estimated at $1.6 trillion, well ahead of $400 billion for smartphones and $266 billion for personal computers. The auto industry is in the midst of two revolutions: electric cars, pioneered by Tesla, and self-driving cars. Driverless, new energy vehicles are becoming the next battleground in global competition.

Apple's idea of building cars was met with considerable skepticism in 2015. Steve Wilhite, Apple's former vice President of global marketing, says Apple will never adapt to the three—to five-year product life cycle of the auto industry. Similarly, Bob Lutz, former head of product development at GENERAL Motors, said, "There are multiple ways Apple can get into the auto business. Do I think they're going to start working on cars? Yes. Are they going to make a whole car? No."

The reasons for scepticism are obvious. The automobile industry is a mature industry, with many well-grounded manufacturers. The engineering complexity of developing a car exceeds that of Apple's previous projects, and cars have safety problems that do not exist in the consumer electronics industry. Even with decades of experience in the auto industry, including giants such as Toyota and General Motors, quality control problems and costly recalls and fines can result in great losses. In addition, the carmaking business is much less profitable than Apple's existing business. In 2014, carmakers' average profit margins averaged just 8%. The US sales and distribution model also creates barriers to entry. Tesla, for example, faces barriers to car sales in many US states because current regulations in several states require cars to be sold through licensed dealers. If an individual wants to buy a Tesla, they have to buy one online and deliver it.

After 12 years in business, Tesla's success is still uncertain. In 2017, Tesla made just 100,000 cars but lost $2 billion. In 2018, Tesla's market value is just $51.3 billion, while Apple's is on the verge of topping $1 trillion—and Apple's cash pile is $285 billion. Tesla CEO Elon Musk admitted that he and Apple had been in merger talks since 2015, but added that Apple had poached 150 employees from Tesla.

While electric vehicles remain the main alternative to conventional gasoline-powered vehicles, vehicles powered by hydrogen fuel cells are emerging as competitors in the zero-emission vehicle category. Some automakers, including Honda and Toyota, are emphasizing hydrogen fuel-cell cars over electric cars.

The combination of electric and autonomous driving means the car industry is undergoing the most rapid changes in a century. In the field of autonomous driving, Alphabet ranks first in the number of patents. But with traditional car companies pouring resources into autonomous driving, Alphabet's edge is shrinking. Apple is currently a laggard when it comes to self-driving cars. Apple's new patents include an 'arbitrary polygon avoidance collision system' and an 'autonomous navigation system'. According to the autonomous driving report card of 2018 released by the California Department of motor vehicles, Apple needs manual intervention every 1.8 kilometers, while Alphabet's Waymo needs manual intervention every 1846.8 kilometers. Apple laid off 200 autopilot team employees in early 2019 and transferred many employees to other "key" projects.

When asked if Apple plans to build a car, Cook declined to comment and stressed that Apple is focused on CarPlay, which integrates users' iPhones with automakers' in-car systems and allows users to access the phone, text, music and navigation app screens built into the car. Cook made clear that Apple recognizes the importance of this car for the future, saying that in February 2015, "We've extended iOS to your car, to your home, to be a vital part of your life, and none of us wants to have different platforms in different parts of our lives." Apple launched CarPlay at the Geneva Motor Show in March 2014, and several automakers launched CarPlay cars in 2015. Google has developed a competing platform called Android Auto. Google also launched the Open Automotive Alliance, a partnership between technology companies and automakers to develop and implement the platform, and several automakers launched vehicles supporting Android Auto in 2015.

With CarPlay or Android Auto, users connect their smartphones to the car's on-board system, and a limited number of programs are displayed on the car's built-in screen for quick access to specific features: phone, text, music and navigation/maps. Like the Apple Watch, the goal is to provide access to phone-related functions without distracting drivers. Importantly, car manufacturers insist on maintaining dominance of the human–computer interface, including how drivers navigate CarPlay or Android Auto. Volvo's CarPlay implementation includes a touch screen similar to an iPhone or iPad, which Mercedes insists is not the ideal human–computer interface. In the 2015 Mercedes C-Class, a wheel was installed on the center console, allowing users to scroll through apps and menu items.

While many automakers have developed their own navigation, phone integration, and audio systems, it seems likely that most will adopt Apple and/or Android for tight integration between phones and cars. In addition, Apple and Google have expertise in developing viable voice recognition technology, which has long been a weak link to in-car systems. Most automakers recognize the need to cater to both Apple and Android users by offering two platforms in their vehicles, rather than forcing buyers to choose one when buying a car. To connect cars and services, in the words of Ford's executive director, "We don't want people to make choices based on their phones. We want to accommodate all our customers and their devices." Similarly, car sound system makers Alpine and Pioneer launched systems that integrate CarPlay and Android Auto. Some analysts see CarPlay as a place for Apple to expand more control over in-car systems, manage the human–computer interface and exert some influence over the interior design of vehicles, even if it does not develop its own cars. Apple appears to have taken the first step, announcing in June 2015 that CarPlay would support various features of the car, such as adjusting microclimate, through the CarPlay interface, with support from automakers. In a 2017 interview with Bloomberg, Cook described self-driving technology, electric cars and ride-hailing services as "the big disruption that's coming." He believes that these three changes will happen almost simultaneously in the same time frame.

Going beyond CarPlay would be a multibillion-dollar bet for Apple, which carries far higher risks than the $100 million to $150 million it spent earlier on Apple Stores and even the iPhone.

Thinking Exercise

1. Evaluate Apple's Watch strategy. What are the ways to increase willingness to buy and reduce costs? What are the pros and cons of doing it or not doing it? Is it worth doing?
2. Evaluate the emerging strategy for Apple's TV business. Would you advise Cook to bet on TV?
3. Should Apple get into the auto industry or focus on providing software for the cars of the future?
4. Consider the above questions from the perspective of the ecosystem.
5. If you had a few minutes with Tim Cook, what advice would you give?

6.2 The Introduction

How to become an industry leader through the transformation of the IM business model? How to increase the added value of products and services through process innovation? How to integrate next-generation artificial intelligence, 5G communication technology, blockchain and 3D printing technology to form a strong core technological competitiveness? How to use the Internet, big data and artificial intelligence to observe and judge market trends? These are crucial for enterprises facing the transformation and upgrading of IM transformation. This chapter interprets these problems and analyzes the challenges, opportunities and direction of change facing the manufacturing industry.

6.3 Future Technology Trends of IM

The focus of enterprises and the scientific and technological support from the government should be placed more on the future point of the timeline. Thinking about future new technologies should not be limited to the actual technical requirements, but should make full use of imagination and walk out of the creative technological development path.

6.3.1 New Generation of Artificial Intelligence

On July 8, 2017, the State Council released the Development Plan for the New Generation of artificial intelligence. The new generation of artificial intelligence is driven by big data. Through a given learning framework, it constantly modifies and updates parameters according to current settings and environmental information, with a high degree of autonomy.

The development of artificial intelligence has entered a new stage. After more than 60 years of development, especially in the mobile Internet, big data, supercomputing, sensor network, a new theory, new technology such as brain science and strong demand under the drive of economic and social development, accelerate the development of artificial intelligence, present a deep learning, cross-border integration, man–machine coordinated, group intelligence, autonomous control. At present, the development of new-generation AI related disciplines, theoretical modeling, technological innovation, software and hardware upgrading and other integrated advances are triggering a chain breakthrough and accelerating the leap from digitalization and networking to intelligentization in all areas of the economy and society. Artificial intelligence has become a new focus of international competition and conflicts, a new engine for the development of the manufacturing industry, bringing new opportunities for social construction and high security risks and uncertainties at the same time. The five technical systems of the new generation of artificial intelligence are explained as follows:

1. Deep learning and big data driven

Deep Learning and big data intelligence are the technological foundations of the latest wave of artificial intelligence. Deep learning is the application of deep neural network technology to solve problems. Due to the multi-hidden layer nonlinear neural network, deep learning has a strong learning ability and the ability to approximate arbitrary functions.

Deep learning can be based on big data or small data. However, when deep learning is associated with big data, better information describing the internal logic of the big data will be obtained, and the effect of $1 + 1 > 2$ will be generated. Deep learning has shown excellent performance in image recognition, machine vision, speech processing and other applications. Moreover, it is suitable to use GPU (graphics processor unit) for parallel computing. In the case of quite a complex model and especially large data volume, it can still achieve an ideal learning speed. Deep learning is the learning foundation of the other four technology systems.

The research focus of big data intelligence theory is to break through difficult problems such as unsupervised learning and comprehensive deep reasoning, establish data-driven cognitive computing model with natural language understanding as the core, and form the ability from big data to knowledge, from knowledge to decision-making.

2. Cross-media integration

Cross-media convergence is also known as cross-media intelligence and cross-media perception computing. The border is the media, including all kinds of sensors, and the cross boundary fusion includes multi-sensor fusion and sensor-side perception. Cross-border fusion is to link all resources such as sound, image, text and natural language together, and enhance intelligence through more complete and higher-dimensional information intake and perception.

Computers can now do a good job of processing image information, processing sound information, processing text information, and beginning to process language information. However, when people use this information to solve a problem, they do not use them separately, but simultaneously use global, sound and language information to jointly support innovative visual and auditory recognition.

The research focus of cross-border fusion theory is to break through such theoretical methods as low-cost, low-energy intelligent perception, active perception of complex scenes, auditory and verbal perception of natural environment, and multimedia autonomous learning, so as to realize superhuman perception and high-dynamic, high-dimensional, multi-mode distributed scene perception.

[Case: Hot-selling product mining based on cross-media integration]
With the popularity of mobile devices, various social media have emerged, such as Twitter, Facebook, Wechat moments, and Weibo. The way people share their

lifestyles and details of their lives is becoming faster and more convenient. Millions of people are happy to share the products they bought and used online. The data shared online, including texts, pictures, and videos, is collected and interpreted by scholars using AI semantic analysis technology. Semantic analysis technology is capable of analyzing text and converting pictures and videos into written descriptions. These technologies can judge the user's preference for specific products. On this basis, we can get the user's taste and preference to recommend dishes and restaurants to them using the vector-space model method on keywords to build the user's preference model. Jalal et al. captured images and auxiliary text information posted by Kenyan residents on Instagram for semantic analysis. They extracted keywords and utilized data sets containing images and text for deep neural network training. The model they came up with has identified 13 popular foods in Kenya and was 81% accurate on converting Instagram images. The results showed that cake and roast meat were the most popular foods in Kenya in March 2019 [1].

With the rapid development of society, the food industry has accumulated many data with vast sources, fast growth rate, low-value density, and high application value. How to use big data and artificial intelligence technology to explore more potential application value of food data to promote the sustainable development of the food industry has become a key research issue. The Internet of Things, blockchain, big data, and artificial intelligence are important ways to solve the problem of how to make innovative ingredients, such as artificial meat, artificial eggs, and personalized food, more palatable and enjoyable to people. Individual data can be collected in real-time through wearable sensors and uploaded to the cloud chain. Big data technology can fuse the collected data and unify the format. Artificial intelligence, natural language processing technology, and image processing technology are applied to analyze the processed data. The trained model can perceive the "delicious" described by natural language or the "delicious" associated with seeing pictures. Finally, the recommendation technology of artificial intelligence is used to customize the unique taste for individuals so that the recommended food most fits the needs of the current user's body.

In conclusion, the data analysis updated in real-time on social media can master people's social eating habits and preferences in a specific region and time to help restaurants or enterprises develop new products and determine corresponding marketing strategies.

3. Human–machine collaborative hybrid intelligence enhancement

A variety of smart wearable devices are now part of this trend. Human–machine collaborative hybrid enhanced intelligence means that people and machines learn from each other, do what they are good at respectively, solve complex problems jointly, and make intelligence more capable and stronger. Human–machine collaborative hybrid enhanced intelligence as an important direction of artificial intelligence (AI) 2.0, through human–computer interaction and collaboration, enhanced the performance of the artificial intelligence system, made artificial intelligence a natural extension and expansion of human intelligence, through the man–machine coordinated more efficiently solve the problem of complex, has profound scientific significance and great prospect of industrialization.

Hybrid intelligence will realize more advanced, robust and enhanced intelligence through man–machine complementation, man–machine collaboration and man–machine fusion. While current AI systems rely on a large number of sample training at different levels to complete "supervised learning," true general intelligence deftly "unsupervised learning" based on experience and knowledge accumulation. It is impossible to get a universal AI by simply using a combination of various AI computing models or algorithms. Therefore, hybrid enhanced intelligence with man–machine collaboration is a typical feature of the new generation of AI. The forms of hybrid intelligence can be divided into two basic realization forms: "Hybrid enhanced intelligence of human in loop" and "Hybrid enhanced intelligence based on cognitive computing". Natural language processing, natural human–computer interaction and other technical characteristics exist in both human–computer collaboration and cross-border fusion.

The theory of hybrid enhanced intelligence focuses on the breakthrough of such theories as human–machine collaborative understanding and decision learning, intuitive reasoning and causal model, memory and knowledge evolution, etc., so as to realize the hybrid enhanced intelligence that is close to or beyond the level of human intelligence in learning and thinking.

4. Group intelligence

Group intelligence is also called swarm intelligence. In the Internet environment, the group composed of humans, machines, computers and robots is changing to cluster and scale applications, relying on the cooperation between neurons and group computing to enhance intelligence. The theory and method of swarm intelligence based on the Internet has become one of the core research fields of the new generation of artificial intelligence.

In practice, swarm intelligence can be divided into three types in application mode from easy to difficult. The first type is crowdsourcing, which divides tasks into different task bearers. The second type is the swarm intelligence of the workflow pattern, which finishes the task alternately many times. The third type is a complex solution problem, which can be solved by using the collective intelligence of multiple models.

In terms of theoretical research, genetic algorithm, ant colony algorithm and particle swarm optimization are three traditional swarm intelligence algorithms. Both of them are inspired by biological evolution or animal group behavior, and they provide a feasible way to solve complex optimization problems without centralized control and global model. Its potential parallelism and distributed characteristics provide a technical guarantee for processing a large number of data in the form of database.

Compared with the simple cooperative behavior of lower animals concerned by the above methods, the intelligent fusion method of expert group is obviously more promising. Qian Xuesen, a famous scientist, proposed the Hall for Workshop of

Comprehensive Integration system in the 1990s, emphasizing that expert groups conduct collaborative discussion in the way of man–machine combination to jointly study the challenging problems of complex giant systems.

In the Internet's big data environment, the research of swarm intelligence is further expanded and deepened. It not only focuses on the expert group, but focuses on large data via the Internet drive system of artificial intelligence and its convergence of mass participants, the participants in competition and cooperation, and other independent cooperative ways to jointly cope with the challenging problem, the open environment of complex system decision making task, emerged in the process of interaction beyond individual intelligence intelligent form. In the Internet environment, massive human intelligence and machine intelligence empower each other, connecting the advantages of human experts in innovation ability, intuition and spirituality with the advantages of computers in computing speed and memory capacity, and forming human–machine integrated ecosystem group intelligence beyond human and machine. This ecosystem swarm intelligence under the Internet big data environment will radiate the whole life cycle of the organization from R&D to business operation, as well as the relationship network formed by the end-to-end interaction of all organizations. Therefore, research on swarm intelligence can not only promote the theoretical and technological innovation of artificial intelligence, but also provide the core driving force for the application, system, management and business innovation of the whole information society.

The research focus of swarm intelligence theory is to break through the theories and methods of swarm intelligence organization, emergence and learning, establish expressible and computable swarm intelligence incentive algorithm and model, and form the theoretical system of swarm intelligence based on the Internet.

5. Autonomous collaborative control and optimization decision-making

   After 60 years of development of artificial intelligence, unmanned systems have developed faster than robots, including unmanned aircraft, unmanned cars and unmanned ships. The key technologies of autonomous collaborative control and optimal decision-making for unmanned systems should include four aspects: situational awareness technology, planning technology, autonomous decision-making technology and task coordination technology.

   (1) Situational awareness technology. Autonomous control and crossover fusion are also inseparable from sensor information acquisition and perception cognition. To realize autonomous control of unmanned systems, it is necessary to constantly develop situational awareness technology and independently model the task environment through sensor network, including 3D environment feature extraction, target recognition, situation assessment and so on.

   (2) Planning technology. The ability of path planning and planning optimization of unmanned system is a must for autonomous control system of unmanned system. That is, the system can plan and optimize the task path of the system in real time or near real time according to the

detected situation changes, and generate the feasible motion trajectory for completing the task autonomously. The typical planning problem of autonomous unmanned systems is how to effectively and economically avoid threats or obstacles, prevent collision and dumping, and accomplish mission objectives.

(3)   Autonomous decision-making technology. For unmanned systems operating in complex environments, it is necessary to have strong autonomous decision-making ability. The main problems to be solved in autonomous decision-making technology include task setting, coordination of different unmanned systems in formation, task decomposition of groups, etc.

(4)   Task coordination technology. The purpose of autonomous control of unmanned systems is to enable them to respond quickly to changes in the environment and tasks. The autonomous control of unmanned systems should have an open platform structure and provide working modes including single-aircraft operation and multi-aircraft formation collaboration. Collaborative control technology mainly includes optimizing the task route of the formation, planning and tracking of the trajectory, coordinating among different unmanned systems in the formation, and realizing reconstruction control and fault management while considering the environmental uncertainty and its own fault and damage.

The research focus of autonomous collaborative control and optimization decision theory is to break through the collaborative perception and interaction for autonomous unmanned systems, autonomous collaborative control and optimization decision, and knowledge-driven three-way human–machine collaborative and interaction, to form the innovative theoretical architecture of autonomous intelligent unmanned systems.

6.3.2   5G Communication Technology

From 1 to 4G, the main solution is interpersonal communication. 2G realized from 1G analog era to digital era, 3G realized from 2G voice era to data era, 4G realized IP-based, data rate greatly improved. 5G will also solve the communication between people and things and between things, namely the interconnection of everything. 5G is a new communication technology with the characteristics of high speed, large capacity and low delay. In terms of speed, 5G can reach up to 10Gps from 100Mbps of 4G, which is 100 times faster than 4G. In terms of capacity and energy consumption, for IoT and smart home applications, 5G networks will be able to accommodate more device connections while maintaining low-power battery life. In terms of low delay, "industry 4.0" smart factories, Internet of vehicles, telemedicine and other applications must have ultra-low delay.

This will enable 5G technology to have new applications in the fields of IoT, remote services, field support, virtual reality, augmented reality and so on, thus becoming an important supporting technology in the field of IM. The application scope of

5G will be very wide, such as unmanned driving, autonomous driving, navigation, manufacturing, circulation, news and so on. 5G is considered as the infrastructure of key networks in the future, and has become the development direction and strategic commanding point of the new generation of information technology. China's R&D level is at the forefront of the world. Currently, the third phase of 5G technology R&D trial has been launched, focusing on pre-commercial product development, verification and industrial collaboration. It is expected that 5G will enter the commercial phase around 2019.

From the perspective of the application of 5G highly reliable wireless communication technology in factories, on the one hand, the wireless of production and manufacturing equipment makes modular production and flexible manufacturing in factories possible. On the other hand, wireless networks can make the construction and renovation of factories and production lines more convenient, and reduce the cost by reducing a lot of maintenance work through wireless. Sensors in IM closed loop control system (such as pressure, temperature, etc.) to obtain the information need to pass through the extremely low latency network, the final data need to be passed to the execution of the system components (such as mechanical arm, electronic valve, heater, etc.) to complete high-precision production operation control, need very high network reliability, and the entire process to ensure the safety of the production process efficiency. In addition, automated control systems and sensing systems in factories can operate from hundreds to tens of thousands of square kilometers. Depending on the production scenario, there may be tens of thousands of sensors and actuators in the production area of a manufacturing plant, which needs to be supported by the massive connectivity capacity of the communication network.

The timeliness of wireless communication under IM is a very big test of IM based on big data. If the timeliness does not meet the requirements, data collection and industrial Internet are of little significance. In the future, data acquisition will be in milliseconds or even microseconds, and real-time analysis and control based on big data will be needed, which will be a difficult generic technical problem. 5G communication technology is one way to solve this problem. When running in 5G mode, 5G's 1 ms ultra-low delay makes the control instructions of the inverted pendulum execute quickly, and it takes only 4 s to start the pendulum to a steady state. By comparison, we can see the great value of 5G low latency (delay) network in automatic control. 5G enables the wireless automation control of the factory, enables the industrial AR application, enables the factory to use cloud robot, and meets the real-time communication requirements between the robot and collaborative facilities.

5G will promote the transformation of business models and the integration of ecosystems. 5G promotes an end-to-end ecosystem that will build a fully mobile and fully connected society. 5G mainly includes three aspects: ecology, customers and business model. It delivers a consistent service experience and creates value for customers and partners through a sustainable business model.

Compared to 4G, 5G has a 100 times higher traffic density. To achieve this goal, two conditions need to be met: (1) network reconstruction, operators need to consider the air interface (refers to the interface protocol between mobile terminals and base stations) and network adaptation, to provide differentiated services, and further open the network capability. (2) Cloud should adapt to local conditions according to different businesses, and it needs to adopt the methods of cluster processing, fog calculation and haze calculation classification processing.

6.3.3 Block Chain

Block chain technology is the use of blockchain to verify the data structure and data storage and the use of distributed node consensus algorithm to generate and update the data, the use of cryptography way to ensure the security of data transmission and access, the use of automated script code intelligent contracts to programming and operating data of a new kind of distributed infrastructure. The blockchain technology adopts a decentralized point-to-point communication mode, which disperses computing and storage requirements among the various devices that make up the IoT and efficiently processes the information exchange between devices.

Blockchain runs on a decentralized peer-to-peer (P2P) network, and we illustrate its working mechanism by the following steps. Generally, a blockchain is activated by a transaction proposed by its users (Step 1), and then the transaction is encrypted and declared legally valid by the network (Step 2). If the transaction is invalid, it should be abandoned; otherwise, it will be broadcasted to the whole network (Step 3). All the validated transactions are collected within a given time interval (Step 4). The gathered transactions are packed into a candidate block with time stamps (Step 5), and the resulted block is spread in the network (Step 6). These above two steps are fulfilled by the users known as mining nodes in the network. The candidate block is verified by the user receiving it (Step 7). Finally, if all validations pass, the valid block will be linked to previous blocks on each network node (Step 8). By repeating steps 1 to 8, the blockchain size extends. The process shows some apparent merits of blockchain, such as immutability, high transparency, disintermediation and decentralization. There are also some hidden merits, such as authenticity, traceability, and anonymity. These advantages are beneficial to overcome the trust and cost dilemma that small and medium-sized manufacture enterprises face. For example, disintermediation is expected to achieve less complexity in manufacture enterprises by reducing intermediaries.

In brief, blockchain is a kind of multi-party participation of encrypted distributed bookkeeping. It has three properties: bookkeeping, encryption, and distributed multi-party participation. First, it's an account book, a record of transactions that users want to make public. Second, encryption means using cryptography to ensure that an account cannot be tampered with. When an account is opened on the blockchain, the system will automatically create a key, with which the account can be operated on the blockchain. Finally, distributed multi-party participation means that the blockchain is distributed in any network node in the world and does not belong to a specific institution or a centralized institution. There is no center here, or everyone acts as a center. Not only is ledger data stored at each node, but each node synchronously shares and copies the entire ledger data. Blockchain is conducive to cost reduction and supply-side reform, and will become a supporting technology for IM, for the following reasons:

First, the blockchain is characterized by decentralization, disintermediation and transparency of information. Blockchain first solves the pain point of trust in the financial field, and can significantly reduce fraud, reduce costs and improve efficiency and security in all aspects of the manufacturing field. Therefore, it has been first applied in the financial field and now also begins to be applied in the manufacturing industry. Take the anti-counterfeiting of wine as an example. If the manufacturer does anti-counterfeiting on the bottle body or bottle cap, the counterfeit dealer will recover the bottle or bottle cap, or even directly crack the manufacturer's coding rules. However, the addition of blockchain technology is expected to solve this problem. For example, when Company A sells a certain commodity to Company B, it will not only write the original commodity information, source information and transaction information into it to ensure the continuity of circulation, but also broadcast the transaction information. Everyone will know that the real product is in Company B, and it is undoubtedly fake if it is obtained through other ways. Blockchain reduces fraud because of its excellent traceability. It eliminates unnecessary work based on mistrust, such as supplier background checks and product quality in-shipment testing, and reduces the intermediate stages of the product life cycle.

Second, blockchain drives the rise of edge computing. Edge Computing is a method to process data in a location close to data generation physically, which is used to solve real-time and interactive computing problems. Edge Computing is relative to cloud Computing. Today, more and more enterprises, organizations, and research institutions are shifting the focus of centralized cloud computing to edge computing and moving the architecture of applications and data from the cloud to the edge. Due to the large number of edge computing devices, the location distribution is relatively dispersed and the environment is very complex. Many of the devices are embedded chip systems with weak computing power, which is difficult to achieve self-security protection. By using blockchain technology, each device can generate its own unique address (hash element value) based on the public key, so as to be able to send and receive encrypted messages with other terminals, enabling edge computation to be implemented securely.

Finally, blockchain will affect the layout of existing industrial cloud enterprises, gradually break the brand differences between industrial cloud platforms, promote the compatibility between platforms and brands, gradually obscure the concepts of public cloud and private cloud, and break through the barriers of different demand environments of public cloud and private cloud. However, due to the transparent characteristics of blockchain information, it is not suitable for transaction records that need to be kept confidential. Moreover, due to the synchronous sharing, replication and redundant storage of information, the data volume increases rapidly, so the general users are not easy to accept such confusion.

Product lifecycle successively contains the following activities: design, manufacturing, distribution, support, and retirement. This section discusses blockchain applications in these activities covered by manufacture enterprises.

1. Blockchain-enabled design

Under the intelligent manufacturing environment, product design is beyond in-house design and means collaborative design (i.e., co-design) fulfilled by a company and its partners. Co-design makes the company approach customer demands at a high level, which reaps enormous strategic benefits, including increased customer satisfaction and loyalty, substantial market shares, stable revenues, social sustainability, etc. Despite these advantages, co-design also faces many challenges, of which low design involvement, intellectual property protection and unreliable co-design environments are the primary issues. Blockchain is a promising solution to these issues.

As for these issues, scholars are devising blockchain-enabled systems as the mainstream solution in existing literature, which is determined by the nature of blockchain. For the involvement issue, blockchain-aided systems allow individuals and organizations to create, discuss, alter and share their feelings, experiences, opinions, and interests about co-creation products in a synergic way. Communication channels are more effective by this distributed means, yielding a higher involvement rate. As to the intellectual property protection and environments issue, blockchain-based systems enable all design members to collect, store and distribute the identical copy of design details in a validated and encrypted manner without intermediaries and plagiarizing, which endows reliable co-design environments marked by immutability, transparency, authenticity, and anonymity. In the co-design of cooperative manufacturing mode, a large number of design ideas and intellectual property rights are transmitted and shared without confidentiality agreement. The blockchain automatically determines the ownership and infringement of intellectual property rights, which improves the efficiency and security of collaborative work.

2. Blockchain-enabled manufacturing

Similarly, manufacturing here means the co-production of artifacts accomplished by a cooperative partner and the related internal plant logistics. Manufacturing locates at the bottom of the smile curve with low added value in the traditional manufacturing model but is expected to jump to the top of the rainbow curve with high added value in intelligent manufacturing mode. As a game-changing technology, blockchain is a solid thrust to this transformation. Blockchain permeates manufacturing in three main aspects: resource scheduling, device control, manufacturing process tracking. In resource scheduling, virtual and physical resources (e.g., software and machines, respectively) are connected to a P2P network. When customer demands or process requirements are input into the network, the available production resources are scheduled by smart contracts, and the corresponding schedule is sent to authorized network nodes for processing. For device control, blockchain can first offer attribute-related records of the more significant number of devices appearing in smart manufacturing. Then, control the devices by enrolling and transacting the attributes as smart or digital. This control method is more flexible in dealing with interference, imbalance and change in the production workshop, compared with centralized equipment control. Concerning manufacturing process tracking and tracing, blockchain is often used to track the evolutionary processes of products, that is, how materials and the relevant actions are converted into final products, and the reverse process of tracing from final product to origin. The intelligent manufacturing ability driven by blockchain can be further elevated when blockchain is integrated with other technologies, such as digital twin, software-defined networking, IoT and SoLoMoCo (Social, Local, Mobile App, Cloud Computing).

3. Blockchain-enabled distribution

Distributing external logistics is exceedingly crucial for the manufacturing industry to ensure the efficient and timely movement of products, services, data, and capital from the origin of raw materials to final users. Logistics in an intelligent manufacturing network are evolving from conventional centralized forms to decentralized and distributed ones. This transformation requires new supporting technologies, mainly information and communication technology (ICT), like artificial intelligence, IoT, and blockchain. The blockchain is one of the most matched and towardly technologies due to its intrinsic features, such as disintermediation and transparency.

The practical and theoretical explorations of blockchain in logistics are varied. Scholars continually propose to improve transportation systems in logistics by blockchain-based methods. Tracking and tracing all actions in physical distribution (e.g., a screwing action was performed by whom, at what time, where the location is, and what is the torque value) using blockchain can improve efficiency, especially when an emergency occurs. For example, during the outbreak of Covid-19, tracing and tracking the movement path of physical products and human virus transmission route using blockchain can identify sources of coronavirus so that remedial and preventive actions can be adopted accurately and quickly. In practice, retail giants Walmart and JD use blockchain to connect drones in last-mile delivery and track beef import respectively. Since blockchain originates in finance, it can also apply to finance in logistics, that is, logistics finance.

4. Blockchain-enabled support

Support refers to using instructions, fault diagnosis, and maintenance for the marketed products and services. It involves many stakeholders, namely manufacturing firms, upstream and downstream firms, and final customers. Blockchain is an newly created alternative for distributed firms and consumers to jointly fulfill supporting activities and is mainly applied in remote monitoring and scheduling, fault diagnosis and maintenance. The process is as follows: (1) A consumer can make a device diagnosis request in blockchain when the device is under abnormal operation or when the routine diagnosis comes. Receiving the request, maintenance workers with different bid prices and credits on the P2P network compete for the order, and the winners are selected. (2) After the competition, the selected workers or vendors analyze product-in-use data based on knowledge base in blockchain to judge if the product has problems. The diagnostic solution is broadcasted and stored on the blockchain. (3) After the fault diagnosis, ex-post or ex-ante maintenance activities are conducted remotely or on-site.

5. Blockchain-enabled retire

Retire is delivering end-of-life products from user-specified sites to companies (i.e., reverse logistics) to be recycled or disposed of. Rational utilization of end-of-life products can conserve valuable resources and reduce environmental damages, not merely as waste discarded. Consumers commonly have a positive feeling towards recycling, but this attitude does not become actual purchasing behavior. Enterprises usually have invested heavily in recycling, such as equipment buying and staff training. The quality uncertainty of sustainable products is one of the main barriers preventing customers from buying products made from recycled materials. Blockchain is a burgeoning technology that can be employed in addition to RFID and QR code technology.

In reverse logistics, instant tracking and transparent tracing of the used products are possible by checking the current state of blocks, which aids reverse logisticians to have better anticipation and assessment in the location and arrival time of the products. Meanwhile, the vitrification of the used products in reverse logistics and former stages is beneficial for enterprises to better know the product quality when making a recycling decision. Blockchain has many applications in recycling, such as credit rating for participants, incentives for recycling, protecting personal information left in the used products, regulating the number of used products in recycling, and tracking recycling processes. Realizing these applications relies on blockchain-based systems or platforms. Financial rewards in cryptocurrency can be granted in exchange for recyclables to motivate customers to participate in recycling. With blockchain, stakeholders may have more confidence in recycling and the related products, and a rise in recycling is possible.

6.3.4 Multi-material 3D Printing

3D printing has become an important Additive Manufacturing technology that keeps abreast of the need to personalize products. This ability to customize large, multi-material composites on demand, without the need for molds, will revolutionize manufacturing. According to relevant standards promulgated by the American society for testing materials, increase material Manufacturing can be divided into seven categories technique, is applied in the "Metal" print, there are four main technology of molten Metal Powder Bed, respectively, Powder Bed Fusion (PBF), Laser Metal Deposition (LMD), gelling agent spraying forming (Binder Jetting), and Laminated Object Manufacturing (LOM).

In China, 3D printing technology has been involved in the manufacturing of titanium alloy components for large aircraft such as Transport 20, C919, J-15, J-31 and other new fighters. The 3D printing technology of high-performance metal materials is the core technical advantage of China's aerospace equipment manufacturing industry. In the United States, GE has long used 3D printing for its aerospace products. Additive manufacturing was included in the 2013 MIT Review's top 10 Breakthrough Technologies. GE plans to start selling 3D metal printers that can be used to produce large parts in 2018, enabling rapid, low-cost printing of metals.

The current trend of 3D metal printing is towards large size, automation and multi-materials. At the most professional exhibition of 3D printing, the 2017 Formnext (International Precision Forming and 3D Printing Manufacturing Exhibition) held in Frankfurt, Germany, GE displayed a printing size of 110.3 m³. The aerospace parts, and claimed that the future 3Dsize can be increased to more than 1 m. In addition, in the automation part, GE also takes the tip of fuel nozzle as an example. Through 3D metal printing, the manufacturing period can be reduced from 15–18 to 3–5 months, and the parts of this jet engine can be integrated into one piece from 20 pieces.

In the future, a 3D printer can print more than three kinds of materials. For example, an airplane wing, like a human arm, is made of a variety of materials, including skin, a large number of organic memorized structural parts, a variety of fillers, hydraulic systems, pipelines and cables, etc., all of which can be printed together on a 3D printer.

6.4 Social Change and New Business Model

IM contributes to the establishment of new business models and at the same time has a comprehensive impact on the development of society.

6.4.1 Social and Job Changes

Thanks to the promotion of artificial intelligence technology, human society is undergoing a comprehensive and profound transformation. Decisions previously made by humans are increasingly made by computers and big data algorithms. Once you have enough data and computing power, software algorithms are better than the human understanding of the operation mechanism and optimization of the manufacturing system paths, and even better than humans understanding people's desires, ideas, and decision-making, such as people wanting to buy a product online, often depend on the intuition, relatives and suggestion of the buyer, and now is increasingly dependent on software algorithms. Amazon's automated recommendation feature, for example, is popular as a substitute for people making decisions. According to the principle of "Use it, or lose it.", human decision-making ability may gradually deteriorate, just as drivers accustomed to using GPS/Beidou navigation gradually lose the ability to find the road and direction. With the increase of amount of data and system complexity, you can grasp the complex custom products production plan and real time optimization of managers, according to Internet big data mining user requirements and find the optimal design scheme of the designer, to understand financial transactions posture and transient respond right stock traders will be less and less, and the algorithm is more and more competent for this job. In the application fields of all walks of life, robots and their algorithms will not only surpass ordinary people physically but also intellectually, while human beings cannot continue to make progress in some abilities and even deteriorate in decision-making ability. Therefore, more and more people will fall behind artificial intelligence and become a class replaced by intelligent robots.

The competition between enterprises will change from competition between people to the competition between software algorithms. Under the pressure of competition, a large number of factory line workers, taxi drivers, couriers and even administrators will lose their jobs. New jobs, such as software engineers, are hard to replace. Countries with high levels of automation have lower production costs, widening the gap between rich and poor.

The impact of AI on society and jobs is wave after wave and constantly changing, with many jobs created but soon disappearing and replaced by new ones. The current assembly-line or simple rule-based work will be largely done by robots and computers, and the jobs that will still be done by people will be those that are highly creative and flexible. Therefore, it should become an important teaching content and research topic to cultivate students' creative and flexible thinking, systematic integration, coordination and balance ability, and research on big data and artificial intelligence new technology.

[Case: Unmanned food factory in China]
According to the 2020 Artificial Intelligence Empowers China's Advanced Manufacturing Industry, China's artificial intelligence technology application in the manufacturing industry is below 2%. In 2025, infiltration application in manufacturing is expected to reach more than 10%, leading the benefits of upgrading manufacturing intelligence to exceed ten billion yuan. For this reason, companies have to increase the AI application in the manufacturing industry. On the one hand, AI technology is applied to all links of production and manufacturing with its advantages of simplifying the production process, improving production efficiency, and improving process technology. For example, in meat processing, pork processing robots, eviscerating robots and other equipment realize intelligent and automatic processing through AI technology. The application of image recognition technology and AI technology in food sorting and quality inspection can realize the real-time monitoring of the whole process of food sorting, packaging, palletizing. Also, it can realize machine learning to continuously improve the accuracy of processing and production. Furthermore, applying artificial intelligence technology in food processing greatly improves efficiency and saves costs. For example, Zhengda "Unmanned food factory" in Hebei using intelligent dumpling production line as shown in Fig. 6.1, from the integrated with noodles, stir paste, knead full-automatic, dumplings output per hour can reach hundreds of thousands, greatly enhance the enterprise's economic benefit. Moreover, now based on artificial intelligence technology of cooking robots, food delivery robots continue to appear and indirectly for the enterprise brand to achieve marketing publicity.

However, there is a shortage of artificial intelligence equipment and interdisciplinary talents in China, and the standard agreement between machinery and equipment is not standardized and unified. So, there is still a long way to go to popularize and apply AI in production workshops.

Fig. 6.1 Unmanned food factory using intelligent dumpling production line

6.4.2   Expansion of Industry Boundaries to New Industries

Expansion to new industries means that enterprises use their technological advantages and infrastructure barriers to enter other industries with common key technologies. The competitive basis of the industry shifts from a single product function to product system performance. For example, in the agricultural machinery industry, the industry boundary extends from tractor manufacturing to agricultural equipment optimization. In the mining machinery industry, from the optimization of individual equipment performance to the optimization of the overall equipment in the mining area, the industry boundary also extends from the individual mining equipment to the entire mining equipment system. For example, with the help of relevant policies such as "Made in China 2025" and "Internet plus" action plans, the State Grid of China makes reference to the major deployment of promoting the integration of telecommunications network, broadcast television network and Internet network, and makes use of power channel resources to realize the deep integration of power grid and communication network infrastructure. The synchronous transmission of energy and information is realized through the demonstration project of power optical fiber to households, and information is regarded as a new content serving the public. Develop charging piles and service outlets for the majority of electric vehicles, and build an open new public service foundation platform.

John Deere and AGCO are working together not only to connect farm machinery and equipment, but also to irrigation, soil and fertilizer systems, so that information such as climate conditions, crop prices and futures prices can be accessed at any time to optimize the company's overall performance. Gree Electric Appliances, a company specializing in air conditioning, tries to develop a mobile phone as the control center of smart home, connecting air conditioning, home energy storage system, lighting system, safety system, family car, etc., in order to evolve into a system integrator, thus occupying a dominant position in the home appliance industry chain. In the future, in the closed loop of network collaborative manufacturing, the roles of users, designers, suppliers and distributors will all change, and the traditional value chain will inevitably be broken and reconstructed.

This expansion requires enterprises to focus on the use of new digital capabilities to transform the original linear traditional value chain corresponding to centralized control into an integrated value rings, so as to realize digital innovation and improvement at each link, and ultimately bring about tight integration, automation and acceleration of internal operations. The digital maturity and reach of different industries is different. To expand into new industries, it is necessary to implement new industry solutions, attract new talents to serve new enterprises, create new partners to serve customers, and adopt new background service processes, such as revenue models, plans, budgets, etc., to meet the needs of new industries.

6.4.3 Enterprise Boundaries Disappear and Full-Channel Customer Experience

Online comprehensive integration of the full-channel customer experience is to connect every contact: face to face contact, retail environment, online behavior, anything through smart phones, and all the other links with the customer, and live with customers in an interactive world. The polymerization of customer behavior into enterprise data supports business decisions. In turn, businesses will be more transparent to their customers than ever before.

The OMO (Online-Merge-Offline) mode is expected to become the next growth point of the Internet.

The innovation dividend of the Internet has brought earth-shaking changes to the traditional industries, and the boundaries of the traditional industrial fields are becoming increasingly blurred, and the industrial and non-industrial sectors will gradually be difficult to distinguish. The OMO (Online-Merge-Offline) mode is expected to become the next growth point of the Internet. OMO means the vanishing of the boundary between online and offline business, which may be offline and integrated in some areas, or offline and integrated online in other mature areas. For example, shirts can be selected from virtual stores online, and then purchased offline products without waiting in line for payment. Stores can even offer product suggestions based on customers' historical preferences, and automatically put personalized small orders into production scheduling. Of course, this business model will require higher standards of product design and automated production methods. OMO will be an opportunity for start-ups in this new wave of innovation.

Manufacturing enterprises will no longer focus on the manufacturing process itself, but will focus on personalized user needs, product design methods, resource integration channels and network collaborative production. Therefore, some information technology enterprises, telecom operators and Internet companies will be closely connected with traditional manufacturing enterprises, and it is very likely that they will become the leaders of traditional manufacturing enterprises and even the secondary industry.

6.4.4 Decentralize End-to-End Products and Services

The Internet has made possible large-scale collaboration of ecosystems across enterprise boundaries. What e-commerce has eliminated is only some intermediate service providers that make use of information asymmetry to survive, but at the same time, it has given birth to a large number of emerging electronic intermediaries with core capabilities, such as Alibaba and Taobao. Another emerging class of intermediary service providers will be different from traditional electronic intermediaries in that they will gradually disappear behind the scenes, using decentralized software to provide end-to-end products and services for direct connections between customers. The decentralized service model has the following advantages: (1) the decentralized software has no central point that can cause the failure of the whole system, so it is more robust; (2) No node will interfere with the work of other nodes. Based on edge computing and other technologies, each node performs parallel computing, which can accelerate the computing speed and reduce the data delay; (3) Each node conducts transactions and interactions directly, which is more efficient and easier to ensure data security. Decentralized social e-commerce is going to be the new hot spot, and the WeChat mini-program has become a decentralized ecosystem.

[Case: Google's Pixel Buds]
The Pixel Buds headphone demonstrates the promise of real-time translation, available in a variety of languages, and easy to use. This is another trend that goes against the trend of "third-party cloud platforms", in which users are separated from the platform and all the software is embedded into the wearable products to achieve direct translation, communication and service between people and end-to-end without the need for a powerful platform or intermediary.

6.4.5 Third-Party Cloud Platforms and Freelancers

IM model could eventually form a digital platform for a particular industry, it has access to the machine, equipment and products throughout the world, on top of it can create many new business models, such as trade surplus capacity of production equipment, manufacturing data, remote monitoring and maintenance of equipment, equipment rental and maintenance outsourcing, a variety of services, etc., can also absorb the social various aspects and industrial chain resources on all aspects of the include large enterprises, small and medium-sized enterprises, self-employed. China is adding 300,000 engineers a year. By 2020, freelancers will account for 43% of the workforce, according to projections from linkedin. The business model innovation that gives full play to the efficiency of various social resources such as freelancers is crucial to the development of the manufacturing industry.

Companies can build strategies around platforms. Platforms are to the new industrial system what value chains are to the old. The platform combines interoperable standards and systems to create a plug-and-play technology basis, on which a large number of vendors, freelancers, and consumers can seamlessly interact through the same set of hardware, software, and services. The most successful platforms should match customers and suppliers, maintain an efficient customer experience, and collect data. In this ecosystem, a group of enterprises exchange products and services, forming a community of Shared future, not just interests.

Germany's Trumpf, for example, has built Axoom, an industrial platform that provides many small companies with laser equipment, welding, metalworking, 3D printing tools and all software access. Individual consumers can propose customized product ideas on the platform, while freelance designers can design products and processes, and achieve cost budgeting, remote manufacturing and logistics distribution on the platform to complete orders. Hospitals, banks and other organizations are also organizing specific supply chain platforms. The emergence of driverless cars also triggers the idea of a smart city platform, which will be based on the operation mode of autonomous driving navigation transportation system.

6.5 Technology Foresight for Industries Related to IM

6.5.1 Research Status of Industrial Technology Foresight at Home and Abroad

Technology foresight is a systematic study of the future directions of science, technology, economy and society. Its goal is to select strategic research fields and analyze potential technological opportunities. Technology foresight has received extensive attention internationally and achieved remarkable results in strategic emerging industries such as information communication and energy technology. How to analyze and predict the development trajectory of strategic emerging industries, find the most potential technology and product categories, and point out the right direction for the development of the industry has become a research hotspot.

Technological foresight began in the United States in the 1930s. More and more countries pay attention to the importance of technology foresight and actively carry out forward-looking research on technology foresight and key technology selection. Japan's ninth Science and Technology Foresight report predicts that robots that can accurately analyze the taste and composition of agricultural products will be developed by 2018, autonomous robots with complex judgment will be developed by 2020, and autonomous deep-sea underground excavators will be developed by 2023. Since 1991, the United States has issued a national Key Technology Report every two years to forecast and select key technology areas for development. In 2017, the United States released National Robotics Program 2.0 to follow up on robot strategy deployment. The program focuses on broader issues: how to interact and collaborate effectively between teams of multiple people and robots; Robots facilitate the completion of tasks in a variety of environments and keep hardware and software changes to a minimum. Using the vast amount of information from the cloud, other robots and people to make robots learn and work more effectively; Let the hardware and software design of the robot ensure reliable operation. The robot technology roadmap pointed out that the key to industrial robot ability include: unstructured environment operation, flexible class people manipulate perception, adaptable and reconfigurable assembly, robot, automatic navigation and human work, assembly line rapid utilization, green manufacturing, based on the model of supply chain integration design, workability, and component technology, nano manufacturing.

At present, the United States, Germany, Japan and South Korea have carried out the technology foresight of the IM equipment industry, and the relevant practice activities are gradually carried out in China. Since 1992, China has carried out technical foresight work. In 2011, based on the technology roadmap method, Shanghai Institute of Science studied the strategy and technology development of 13 emerging industries in Shanghai, including the "Strategic Technology Roadmap for Intelligent Robots". The Strategic Advisory Committee on Building China into a Manufacturing Power published the "Made in China 2025 Key Areas Technology Roadmap" in October 2015, forming a set of relatively sound robot manufacturing technology roadmap with five years as the stage. It is pointed out that the key common technologies of industrial robots include complete machine technology, component technology and integrated application technology. Key components include reducer, controller, servo system, sensor; Key products include domestic welding, handling, spraying, assembly and testing robots; The main technical content of the equipment manufacturing industry includes robot loading and feeding, robot control and positioning system.

Due to the unstructured problem of industrial technology system prediction, traditional technology prediction methods are mainly qualitative and supplemented by statistical methods, including the brainstorm method, Delphi method, affinity graph method, benchmarking analysis and user analysis. The results of qualitative methods are easily interfered by subjective factors. Therefore, more and more researchers pay more attention to quantitative analysis methods based on data analysis. Table 6.1 is a summary of current research methods on technology prediction. The two major trends of current technology prediction methods are the migration from qualitative methods to quantitative methods, and the development from a single method to a comprehensive model of multiple methods. A patent has both technical and market attributes because it needs to satisfy the requirements of originality, technical feasibility and business value assessment. Patent analysis as a quantitative analysis method has been gradually popularized, and data mining has also played a more and more important role. According to empirical statistics, patents contain 90–95% of all scientific and technological knowledge in the world. However, such knowledge resources containing massive technical and commercial knowledge are far from being fully utilized by people. Therefore, patent mining has gradually become an important tool for industrial planning and technology foresight.

Table 6.1 Summary of main research problems and methods of technology foresight research

The research question	Technology life cycle analysis	Technology activity recognition	Technology evolution track and trend analysis	Technical evaluation	Other
	Life cycle assessment; Technology maturity prediction	Key technology identification; Generic technology identification; Emerging technology identification; Blank technique prediction; Disruptive technology identification	Diffusion of technological innovation; Technology transfer; Alternative technology; Technology development path; Trajectory of technological evolution	Technical weight ranking Technical level assessment	Alternative technology ranking; R&D project selection; Knowledge management; Risk management; Technological ecosystem; Technical cooperation and competition
The main method	Patent analysis; Data mining; S curve; Archesuller Patent Mining model; Technology Readiness Level	Patent analysis; Data mining; Social network analysis; Morphological analysis; Bibliometrics; Technology roadmap; Delphi method; brainstorming	S curve; Envelope analysis; Patent analysis; Delphi method; Scenario analysis; Technology roadmap	Delphi method; Expert consultation; Patent analysis; Analytic hierarchy process	Morphological analysis; Analytic hierarchy process; Data mining; S curve; Social network analysis; Delphi method; Scenario analysis

6.5.2 Development Status of Industrial Robot Industry

"Made in China 2025" identifies ten key areas, including the new-generation information technology industry, high-end CNC machine tools and robotics. The development of the new generation of information technology industry focuses on IM basic communication equipment, IM control system, new industrial sensors, manufacturing IoT equipment, instrumentation and testing equipment, IM information security products. The focus of the development of high-end CNC machine tools and robots is to meet the needs of IM, especially the small batch customization, personalized manufacturing, flexible manufacturing, can complete the dynamic, complex mission, and can work in collaboration with humans. The new generation of robots will become advanced high-end manufacturing equipment "brain".

Industrial robots are mechanical devices used in the industrial field that can replace human labor through various automatic operations, usually for handling, welding, spraying and assembly. Early industrial robots copied the mechanics and motion of human arms and had little sense of the environment. In the mid-to-late twentieth century, the development of integrated circuits, digital computers, and miniaturized components made computer-controlled robots possible. In 1954, the world's first programmable industrial robot was born. In the late 1970s, industrial robot has become an important part of flexible manufacturing system automation. After more than 60 years of development, industrial robots have been widely used in automobile, metal products, chemical industry, electronics, food and other industrial fields. Industrial robots can free people from the heavy monotonous and high-risk industrial production activities, and at the same time can improve production efficiency and quality, gradually becoming an indispensable core equipment in the tertiary industry.

Industrial robot system structure as shown in Fig. 6.2, its three core components for the controller, servo motor and reducer, including controller to control system, is the brain of the robot system, servo motor is a kind of drive system of, primarily for actuator section provides the power. Reducer is the deceleration transmission device between the drive system and the perform system.

At present, there are three classification standards for industrial robots, as shown in Fig. 6.3. According to the classification of mechanical structure, industrial robots can be divided into tandem robots and parallel robots. According to the ontology

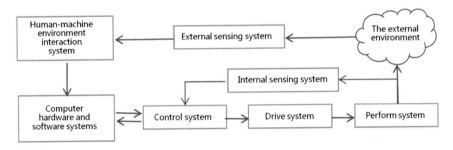

Fig. 6.2 Structure of industrial robot system

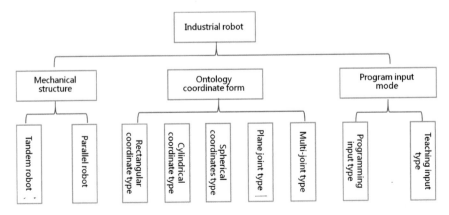

Fig. 6.3 Three classification standards for industrial robots

coordinate form, it can be classified into rectangular coordinate type, cylindrical coordinate type, etc. This classification is the most commonly used classification method. According to program input, it can be divided into programming input type and teaching input type.

Since 2013, China has become the world's largest industrial robot market, and it continues to expand. In 2017, China's output of industrial robots reached 131,079, accounting for about one-third of the global output and an 81% year-on-year increase. According to some data, since the implementation of the "Machine substitution" policy, the product qualification rate in Dongguan city has increased from 86.1 to 90.7% on average, which can reduce the employment of nearly 200,000 people and reduce the unit product cost by 9.43% on average. According to the National Bureau of Statistics of China, there are 50 million manufacturing workers in China, and the potential stock market will reach 18–25 million if two to three workers are replaced by each industrial robot. In 2018 and the next five years, we can see annual sales of 500,000–600,000 units in the domestic market.

The proportion of domestically produced robots is relatively low, only about 30%.Reason is that domestic industrial robot company independent R&D ability is weak, most enterprise is assembled, integrated model, to low-end products, mainly for stacking, loading and handling requirements for accuracy is not high, most of the three axis and four axis robot, used in automobile manufacturing, welding and other high-end industry in the field of six axis or more high-end industrial robot market mainly by Japan and the European and American enterprises, domestic six axis industrial robots new installed capacity of less than 10% of the country's industrial robots. In recent years, although in tunneling, assembly automation, engineering machinery, intelligent robot technology achieved a major breakthrough, the overall condition is still in the abroad developed countries' level of 1990s, particularly in the manufacturing process and equipment is difficult to achieve high accuracy, high reliability, high speed, high efficiency, the key parts and components supporting unit is still in a state of imports. China has such a huge market demand that some advanced robot enterprises in the international market aim at this point and enter China on a large scale, including FANUC, ABB, Kawasaki and other enterprises, which, to some extent, hindering the development of domestic independent brands.

6.5.3 Technology Foresight for Industrial Robot

Innography is a patent information retrieval and analysis platform developed by Dialog company, which includes patent data covering more than 70 countries around the world. And Innography provides a unique patent strength algorithm, can help filter out ultra-low value patents, obtain value concentrated patent data set. Through Innography, the author retrieves and downloads relevant patents in the field of industrial robots. After deweighting, cleaning and screening according to patent strength, 5002 patent data is obtained and stored in the local relational database as the basic data for subsequent analysis.

The author based on SOA method patent similarity matrix, based on similarity matrix formed patent network with the patent map, and then analyzing the knowledge map visualization, it is concluded that industrial robot technology of the correlation between the heat transfer process and the core technology, it is pointed out that the current technology opportunities, and the future direction of technology is forecasted.

1.   Construction of patent knowledge map

(1)   Patentee analysis

Firstly, patent data of industrial robots is analyzed by patentees. In Fig. 6.4, the X-axis represents the technical strength of the patentee, which is calculated by the number of patents, patent classification number and patent cited, the Y-axis represents the economic strength of the patentee, calculated by the enterprise income, litigation situation and the location of the enterprise, and the bubble size represents the number of patents owned. The patentee on the right is listed in descending order by the number of patents owned by the enterprise.

The characteristics of major patentees, such as industry, capacity and scale, etc. are analyzed one by one, and it is found that they come from the following industries and have corresponding characteristics and rules:

1)   Industrial robots and their components. The "four big families" of industrial robots rank among the top in patent ownership. Among them, ABB, Fanuc, KUKA, has the largest number of patent applications. ABB emphasizes the integrity of the robot itself, high precision of multi-axis linkage, and motion control system as its core technology. Fanuc's advantage lies in its high-level digital control system and intelligent reverse clearance compensation technology. Its industrial robot has advantages such as convenient process control, small base size and unique arm design. KUKA's industrial robots are fast

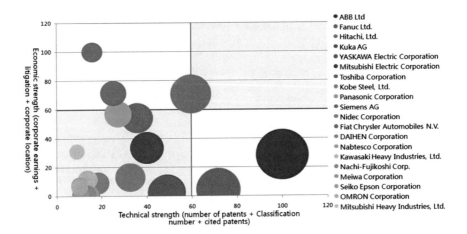

Fig. 6.4 Patentee bubble diagram

and have optimal motion control, but lack capability in decelerators. Yaskawa motor main servo system and motion controller, its technical advantage lies in multi-motor control, vector control and other technologies. Kawasaki company's main controller and transmission arm drive components. A healthy ecosystem promotes specialization.

2) Electronic manufacturing industry. The leading enterprises are mainly Mitsubishi Electronics, Hitachi, Toshiba, Panasonic, Seiko, Omron and other Japanese companies. This indicates that the electronic manufacturing industry has an urgent need for industrial robots, and the electronic manufacturing companies have the motivation to shift their strategy to the upstream of the industrial chain. This also reflects that Japan's electronics manufacturing industry has basically completed the strategic direction of shifting to upstream components and high-end manufacturing equipment. Industry needs to develop industrial robots with high speed, light load, small size and high flexibility to cope with the characteristics of large output, light weight and diversified process types in electronic product production.

3) Automobile manufacturing industry. They include Fiat Chrysler and Mitsubishi Heavy Industries. The automobile manufacturing industry is the industry with the greatest demand for industrial robots. Automobile manufacturers extend to the upstream of the industrial chain, and develop industrial robots with high speed, high precision, heavy load, anti-shaking and large working range, so as to cope with the characteristics of large output, high precision, heavy component and large part size in automobile production.

(2) Patent network analysis

The patent network refers to the idea of the social network analysis method and expresses the technical network through graphs based on graph theory. Based on the semantic similarity of SAO structure, this paper calculates the similarity between two patents and obtains the patent network. The more similar the patent, the smaller the patent distance. Ucinet was selected as the analysis tool for patent network analysis.

In Fig. 6.5, it can be found that there are a total of 6 patent clusters in the shape of islands and a series of related patent clusters. After reading the patents, it is found that these patent clusters appear in different ages, which reflects the transfer of technology hotspots.

1960–1984: the concept of industrial robot was put forward, and the core of the technology in this stage was position control and force (torque) control. Most patents studied whether the robot hand could effectively grasp objects. On the one hand, the improvement of the grabbing ability of the robot hand. In the sensing direction, the TVcamera technology is mainly used to measure the vertical displacement and longitudinal distance with closed circuit TELEVISION camera. At this stage, people have begun to study high degree of freedom robots with six axes and above.

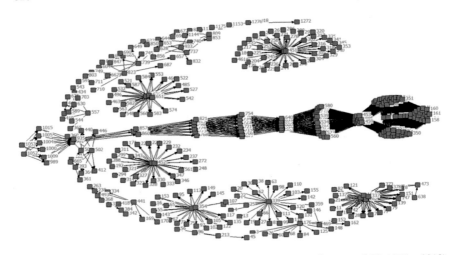

Fig. 6.5 Patent network of industrial robots at various stages (patent distance < 0.08, 1975 to 2018)

From 1985 to 2009, when the industrial robot boom was fading away, the number of patent applications in this stage showed a trend of first declining and then rising, but the technology made great progress. In this stage, early intelligent control has appeared. In terms of control, in addition to improving the traditional torque control, it has also been developed in technical fields such as optimizing path, optimal torque path, compensating static positioning error, tactile feedback, preventing collision, etc., and it is developing towards networked cooperative industrial robots.

Since 2010, industrial robots in this stage have been developing towards high-precision, flexible and reconfigurable, man–machine collaboration and intelligent direction, and the output has also ushered in a burst of growth. In intelligent control, self-diagnosis, robot health management, learning system, neural network, uncontrolled genetic algorithm, multi-objective optimization model, collision free path planning, data fusion, intelligent learning and other technologies are proposed. In terms of flexible cooperation and human–computer interaction interface, technologies such as multi-machine coordination, robustness, multi-axis linkage, graphical operation interface and cloud platform are proposed. In terms of accuracy improvement, the techniques of precise positioning, on-line calibration and error data analysis are put forward.

Middle level of bamboo-like patent embodies the following related technical iterative change: "distance sensor and machine vision" → "collision-free path planning, optimal path planning and multi-sensor fusion, environmental awareness" → "robustness, digital control, fuzzy control, flexible operation" → "smart study, genetic algorithm" → "multi-axis linkage, multi-machine coordination, collaborative robot" → "cloud services, navigation map, off-line programming and autonomous navigation" → "human–computer interaction and security, self-diagnosis and repair". The average technology iteration replacement cycle is 6.1 years based on the average time span from technology emergence to reference.

The process of transfer and replacement of technological hotspots above reflects the evolutionary principles of "system integration and micro-evolution", "increase of controllability and dynamics", "automation and intelligence", "unbalanced development of technology", "increase of ideal degree" and so on.

(3)   Patent map analysis

On the two-dimensional patent map reduced by multi-dimensional scale analysis, as shown in Fig. 6.6, the blank area in the circle is called "patent vacancy". It is the blank area surrounded by existing patents, representing that the circle-indicated technology has not yet appeared. This technique is defined by the set of patents around this blank circle, which are fused together. It is an inevitable trend to fill the patent vacancy, thus, the patent vacancy is the technology opportunity. Clusters of points are patent hotspots, such as the center of the figure, which are productive areas. The point far away from the patent hot spot is the abnormal patent, which contains some budding, non-mainstream technology.

2.   Analysis of technological opportunities

Based on the analysis of 5 patent vacancies in the patent map shown in Fig. 6.5 and the judgment of technology and industrial value, five technological opportunities with high development value are summarized. These are all technical fields worthy of further patent layout:

(1)   Software driving and flexibility. The intelligence of industrial robots is mainly in the software. With intelligence, wisdom, factories, the industrial concept of big data, and so on, people are a higher and higher requirement for the robot's intelligence. Intelligence is mainly manifested in the software of the robot, such as path planning, autonomous learning, genetic algorithms, digital

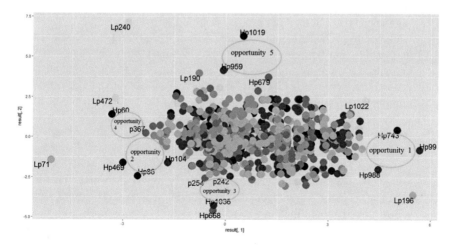

Fig. 6.6 Patent map of industrial robots (1975–2018)

control, fuzzy control, etc. The trajectory planning of the digital workshop, workshop layout and so on all need hardware and software. Combining the development of hardware is not enough. We need to optimize the software system and intelligent algorithm.

Therefore, in order to increase the use of industrial robots, adapt to many varieties of small batch production of small and medium-sized enterprises demand, we need to study to make robots more flexible, mainly through software integration to achieve speed and distance monitoring, power, and power restriction, and other functions. It can use the abstract memory system for adaptive grasping and intelligent product assembly.

(2) Robot with universal interaction, standardized system, communication network, and modular components. The trend of robots in the future is a generalization, standardization, networking, and modularization. On the one hand, it is convenient for robots from different manufacturers to exchange information and realize remote operation monitoring, maintenance, and remote control. On the other hand, it also reduces industry costs. Modular and reconfigurable robots will lead the development of the high-end manufacturing industry in the future. Modular robots have strong adaptability to the environment and tasks, obvious cost advantages, strong versatility, and can adapt to a variety of working environments. Module-based collaborative robot on the technology can be customized and reconfigurable, promote the formation of the ecological system, through the module reuse greatly reduces the cost of late, the production model is given priority to with small batches and customization and not spending too much money on a large scale of production line of small and medium-sized enterprises (SMEs) has a big attraction.

Therefore, the R&D and industrialization of high-performance, modular, and general-purpose controllers should be carried out. Secondary development based on ROS, a universal software platform for robots; An open control system for industrial robots is developed by combining real-time operating system, high-speed bus and modular robot distributed software structure.

(3) Collaborative robot with man–machine fusion. The synergy between machines and people and between machines and machines are a big trend. Synergy emphasizes robot-to-robot cooperation, or robot-to-robot cooperation, to accomplish a given task together. Man–machine collaborative industrial robots combine the intelligence of human beings with the high efficiency of robots to complete the work together, and ordinary workers will be able to operate it just like using electrical appliances. According to the evolutionary path of the "reduce human intervention" technology, the development of man–machine coordinated development industrial robots will be a fleet of coordinated transition of intelligent robots, can at a relatively low cost to make up for any deficiencies in the current industrial robot intelligent level. Therefore, synergy robots are small and medium-sized enterprises, and enterprises with flexible production requirements are effective ways to realize IM. Collaborative robots usually integrate machine vision and have more than six axes of freedom, so their control is difficult. The joint control of multiple collaborative robots

requires more complex control systems and algorithms, which will also be a major challenge for developing collaborative robots.

Therefore, it is necessary to combine machine vision, 3D modeling, independent joint, exoskeleton and tactile sensing array at the human hand level to realize real-time obstacle detection and 3D trajectory optimization, collision detection, and virtual interactive teaching, etc. New materials are used to improve the load to the deadweight ratio of industrial robots. With high-precision touch and force sensors and image analysis algorithms, the production of product parts can be timely detected, and the emotional and physical state of production personnel can be evaluated.

(4) Intelligent sensing and deep learning. The robot is not an isolated individual but an organic system combined with the environment. Robot technology and IoT technology are combined to achieve precise positioning.

The robot that uses artificial intelligence technology for planning and control can perceive environmental information through machine vision, think, recognize and reason independently, and make judgments, planning, and decision. It can automatically complete the target without human intervention and become the main body of the production system, such as the cloud service robot. In the future, by making up for the shortcomings of environmental perception ability and enhancing machine vision and intention judgment ability, flexibility and feedback ability to work environment will be greatly improved.

Deep learning promotes the robot to get rid of the constraints of pre-programming and move towards intelligence. Deep learning enables robots to learn new skills and adapt to unknown work environments just like humans. The application of deep learning in industrial robots is divided into three levels: robots learn new skills through trial and error; Multiple sharing experiences to improve learning efficiency; Robots can prevent and repair faults themselves. If the robot fails, it will be shut down for repair, resulting in loss of production capacity. Therefore, robots should be able to repair themselves in the future.

Therefore, it is necessary to study the comprehensive application of deep learning and artificial intelligence methods, including self-tuning parameters, earthquake suppression algorithm, torque wave compensation, and other integrated control algorithms, to improve motion stability and achieve precision control under high dynamic multi-axis nonlinear conditions.

(5) Industrial robots for subdividing industries. The robot application of the automobile industry is occupied by foreign robot giants, which is related to the earlier development of the automobile industry in developed countries. In China, logistics, high-speed rail, ceramics, 3C, new energy, and other emerging industries are the most competitive fields for Chinese robots, bringing more opportunities to local robot enterprises. Among them, the 3C industry is developing rapidly, and more than 70% of the world's production is in China, which belongs to the blue ocean region where domestic industrial robot companies compete with foreign industrial robot giants. The process of the 3C industry is

complex and diverse. Through patent network analysis, it is found that industrial robots in the 3C industry tend to be more and more miniaturized. Miniaturized industrial robots for the 3C industry will be a subdivision field expected to be broken through. In addition, domestic intelligent warehouse robots and unmanned aerial vehicle (UAV) technology applied in the logistics industry have made breakthroughs. In these new application scenarios, foreign and domestic robots are at the same starting line, and domestic robot companies can take the lead in layout through technological innovation.

Therefore, it is necessary to research and develop special industrial robots for the needs of segmented areas and take the road of differentiation. For example, to develop automatic robotic liquor production lines for the production of various flavoring liquor; Develop sheet metal bending robot with high precision, automatic error compensation, real-time following, and other functions to improve the product quality and production efficiency of hardware factory; The precision gluing robot of electronic manufacturing with track accuracy up to 0.3 mm meets the high-precision assembly requirements of the production line of smartphones. The lightweight desktop arm robot with repeatable positioning accuracy up to ± 0.03 mm can perform flexible operations in a narrow workspace.

6.6 Practical Cases: Jinbaoli Cloud Monitoring Platform for Fine Chemical Industry

Guangdong Jinbaoli chemical technology equipment co., LTD. (hereinafter referred to as JBL) provides chemical process design, automatic control scheme design, chemical equipment type selection design, manufacturing and installation, process pipeline design and construction, electrical engineering design and installation and other services for the five sub industries of coating, ink, resin, adhesive and chemical building materials. It is a supplier of chemical equipment system integration.

The equipment manufacturing industry is the pillar industry of China's national economy, and complex equipment is an important carrier of high-end manufacturing. In recent years, with the development of modern fine chemical production equipment towards the direction of large-scale, continuous, systematic and automation, the production system has become more and more closely linked to each link. Fine chemical production has the following characteristics: (1) more customized demands from final consumers, quick replacement of new products and short production life of products; (2) Multiple varieties, small batches, due to the targeted use of the products, often a type of products have a variety of grades, often using intermittent equipment production; (3) Most fine chemicals are formulation-type products. Formulations and processing technologies have a great impact on product performance. Formulations are usually highly confidential and basically depend on independent R&D of enterprises; (4) Less investment, smaller device scale, high added value and large profit; (5) There are many moving equipment (process equipment with rotating mechanism), which is easy to wear out. The product color change leads to frequent cleaning of the equipment and large energy consumption; (6) The production management is complex, and the equipment has many and complex connections in time and space, and the safety production is under great pressure. Therefore, computer management and decision support are needed.

In the past, when there were abnormalities in chemical equipment, JBL, as an equipment manufacturer, needed to directly send technicians to the customer company to repair and maintain the machinery and equipment. This kind of routinized treatment has fallen behind in the age of information network. The method of artificial periodic stop inspection is no longer suitable for the need for automatic continuous production system. Under the background of globalization, equipment users are distributed in every corner of the world, which brings great difficulties and challenges to the operation and maintenance of equipment. In the context of servitization, equipment manufacturing and services are interpenetrating and integrating with each other, and the traditional one-way manufacturing format of "manufacturing + sales" has begun to transform to the service-oriented manufacturing complex format of "technology + management + service". It has become the general trend of today's manufacturing industry to shift from production-oriented manufacturing to service-oriented manufacturing. Effective health monitoring of key chemical equipment has become a research hotspot in the domestic and foreign scientific community. General manager Li Qiang said: "The industrial development path of an industry will generally follow a law: that is to develop along the road of manual—mechanization—electrification—automation—information—IM." Along the track, the company is committed to paint and resin chemical mode of production automation, clean production and R&D of the process information management, with the raw material conveying, control and automation of production process control, as a way of formula and information management, automatic generation of prevent wrong feeding process control, reference barcode management technology to realize information management of the production process. Since 2016, the company has decided to develop in the direction of IM, planning, design and development of public cloud monitoring platform for chemical equipment, so as to promote the informatization and intelligent development of the industry.

1. Collaborative mechanism between value network and ecosystem

Value network is a perspective of cloud manufacturing monitoring business analysis, as shown in Fig. 6.7. The nodes in the value network represent six roles: manufacturing monitoring platform (provider or owner of the platform), customer (chemical plant), management decision maker (charging and pricing decision maker), manufacturing equipment maintenance personnel, equipment manufacturing plant (or equipment supplier), and network provider. The flow and interaction of tangible or intangible products/services, funds/charges, data/instructions, access/knowledge are among nodes through interaction.

The monitoring platform provides quality of service (QoS) of various types to meet the needs of different customers, and the platform provider's target platform maximizes revenue and improvement. The fee pricing provider first provides pricing

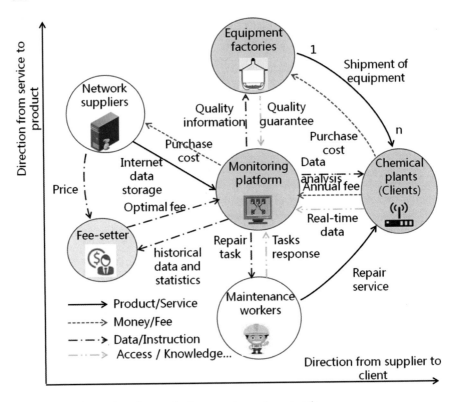

Fig. 6.7 Cloud manufacturing monitoring ecosystem value network

rules and determines the pricing of each customer based on data such as the production and equipment scale of the customer's factory. Then the customer decides whether to participate in the platform and which mode of service to participate in according to its product safety risks, costs and expectations. Customers expect maximum security and minimum security costs. The platform assigns the nearest maintenance personnel from the fault location to carry out repairs to reduce costs. Maintenance workers are divided into two categories: on-site maintenance workers, who are employed by the factory or outsourced to the platform and whose task is to monitor equipment; The other is the third-party personnel, who are closest to the equipment fault location through the response platform task assignment or take the initiative to undertake the maintenance task.

2. Physical network architecture design

Customers are divided into three categories: large customers, medium customers, and small customers. The network system to be built is mainly divided into three parts: one is the sensor network installed on the customer site; The second is the data center, which can be built or leased. Third, the self-built monitoring center.

The equipment information collected by large-scale customer monitoring is relatively complex, including an extensive sensor network, a large amount of video information, a large database server, and a video server, which requires a large amount of data to be transmitted and has its own perfect internal network system. Small customers have a small sensor network and a small database server, or they can directly send the collected data to the cloud network without the need for a server. The internal network system is not formed, so the transmission amount required is small.

Take a large customer as an example. The video data collected by the camera of field equipment information is transmitted to the video server in the form of wireless or wired, while the data collected by some sensors from the equipment or PLC system is also transmitted to the database server in the form of wireless or wired. Then video server and database server access switches, again through the firewall, router, the data transmission on the network, and then the data center to receive information, through a series of processing, then the data information transmitted to Jin Baoli monitoring center, Jin Baoli device status according to the acquired information for diagnosis.

The structure of the cloud platform is shown in Fig. 6.8. The monitoring platform mainly includes data center and monitoring center. The platform gives the company the ability to monitor the operation of globally distributed production equipment for real-time monitoring and early failure prediction.

3. Data flow and service flow design

As shown in Fig. 6.9, the data flow design is divided into information collection, information processing and processing, and information release.

(1) Information collection. Whether it is a system automatic acquisition of equipment operation parameters or the timing of equipment maintenance personnel

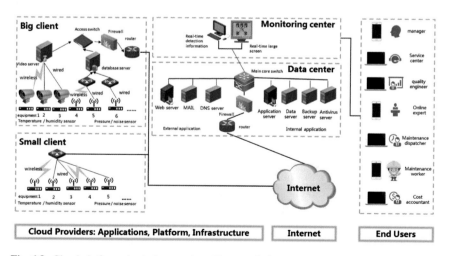

Fig. 6.8 Cloud platform physical network architecture design

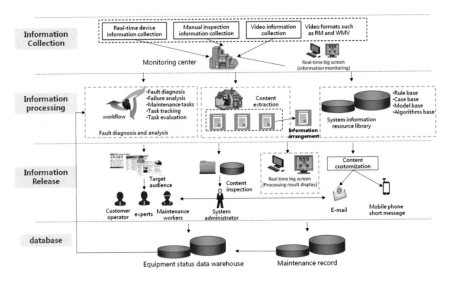

Fig. 6.9 Data flow

inspection equipment operation parameters of input data, and the scene of the client company workshop equipment operation video information, should be in a certain format, through particular physical hardware configuration and software resource support, transfer to the data server, management and monitoring shall be conducted by Jin Baoli monitoring center.

(2) Information processing. Mainly Jinbaoli monitoring center through the information (including manual entry) of the data collected in monitoring, extraction, diagnosis, early warning, analysis, through resource scheduling task generation, maintenance, maintenance task confirmation, assessment of maintenance tasks, and link, the collected data of the remote monitoring system of processing, and makes the corresponding maintenance decisions.

(3) Information release. After information processing, the result of processing and the corresponding maintenance decision will be pushed to the corresponding crowd, such as maintenance workers, personal experts, and customer operators through email, mobile phones, and the system itself. The customer operator can only get the information selected and processed by Jinbaoli, but not all the data and information.

4. System function module design

As shown in Fig. 6.10, the platform mainly consists of five modules, namely, system management module, resource management module, online monitoring module, maintenance task management module, and knowledge management module. Under each function module, it can be further subdivided into each sub-function module.

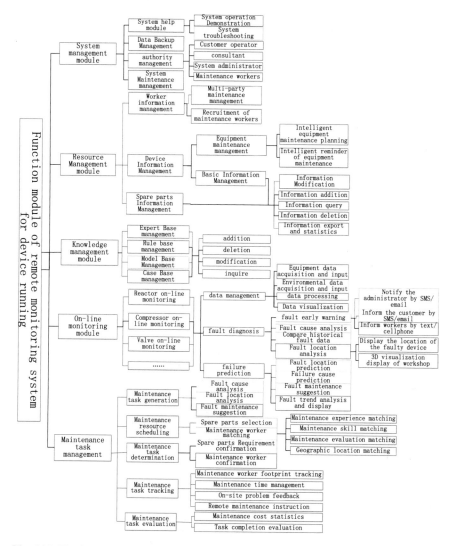

Fig. 6.10 The functional modules of the system

(1)   System management module

The system administrator of the platform management side is responsible for: monitoring device abnormalities, making diagnoses and predictions; Send maintenance task information to maintenance workers after maintenance decision is made; Choose to buy maintenance spare parts on the platform and to send them to the maintenance location; Evaluate the maintenance task according to customer feedback after maintenance.

Maintenance workers, some of whom are the platform management's own main-tenance workers in the preparation, are not many. They implement the salary model of "fixed salary + maintenance salary". The other part is that in the long-term cooper-ation between the platform management side and the maintenance workers recruited from the customers, their maintenance skills are generally recognized. In addition, some maintenance workers of the customer themselves are transferred to the platform management party in terms of affiliation after the platform runs, and the customer himself can no longer configure equipment maintenance personnel. Maintenance workers are mainly responsible for timely logging in the system and receiving and confirming maintenance tasks timely according to their own reality (such as the idle state); After receiving the tasks, according to the maintenance task list and within the prescribed time limit, take transportation to the site of the customer company, use the delivered spare parts to carry out maintenance according to the information sent by the platform and system administrator, and consciously accept the supervision and inspection of the customer; Responsible for uploading invoices and cost information to the system administrator for review; Responsible for writing the summary of this maintenance and uploading it to the system.

The consulting expert is mainly responsible for the enrichment, improvement, and updating of the knowledge base of diagnosis, calculation, model, and maintenance in the system, to fully ensure that these knowledge bases can effectively solve practical problems.

(2)   Resource management module

For maintenance spare parts, the spare parts inventory of the platform manager, the spare parts inventory of the customer, and the spare parts supplier shall be integrated into the spare parts warehouse of the platform, and the maintenance shall be purchased on the platform. The maintenance spare parts are published on the platform by the parts supplier, whose ownership belongs to the VMI model. Through the integration of spare parts resources, the cost of spare parts will be reduced, and the quality will be improved. The platform also enables maintenance spare parts suppliers to be integrated into the industrial chain advocated by Jinbaoli.

(3)   Online monitoring module

The online detection module is the key functional module of the system, which has fulfilled the functional requirements of data collection, diagnosis, and prediction. Given the variety and complexity of equipment, it is not easy to apply the same set of algorithms and models to all the equipment. Therefore, it is necessary to monitor equipment operation in different categories and in a targeted way. Now the equipment is basically divided into reaction kettle, compressors, valves, and other categories, respectively, for the collection and management of operation data and equipment fault diagnosis and prediction. Run the data management module in the equipment, solve the problem of automatic data collection or entry, and solve the problem of visualization and panel display after data collection or entry.

The visual display includes parameter displays and comparisons of different periods, different factories, different workshops, and different equipment. Firstly, regularities and anomalies are found through comparison, and secondly, fault diagnosis and prediction are made through the algorithm.

In the fault diagnosis and analysis module, diagnose the current fault, predict the future direction, analyze the cause and location of the fault, and generate the preliminary maintenance task. When there is a fault warning, the platform management company, customer company, and maintenance personnel can be automatically notified through SMS or email, and the real-time geographical location of the fault equipment can be visualized to facilitate maintenance personnel to quickly arrive at the designated place for maintenance tasks.

(4)   Maintenance task management module

The Maintenance task management module is the core and main functional module of this system, which has completed the generation of maintenance tasks, maintenance resource scheduling, maintenance task determination, maintenance task tracking, maintenance task evaluation, and other functional requirements.

In the maintenance task generation module, there is a preliminary maintenance task list based on the results of fault location and cause analysis. The list includes fault location, fault cause, requirements for maintenance mode, requirements for spare parts, requirements for maintenance time, and geographical location of equipment. Maintain the resource dispatching module, select spare parts according to the preliminary maintenance task list and combine with maintenance resource management. At the same time, the maintenance task should be matched with the maintenance workers according to their previous equipment maintenance conditions. There are four ways: maintenance experience matching, maintenance skills matching, maintenance evaluation matching, and maintenance workers' geographical location matching. Maintenance experience matching, that is, to investigate whether the name and type of equipment that maintenance workers have repaired in the past match with this maintenance task list; Maintenance skill matching, that is, whether the qualification certificate and professional skills of maintenance workers are matched with the maintenance task list; Maintenance evaluation matching, that is, inspect the evaluation scores obtained by maintenance workers in previous maintenance, and give priority to maintenance workers with high evaluation scores; Geographical location matching means that the distance between the maintenance workers and the customer company in need of equipment maintenance is investigated, and the maintenance workers with relatively close geographical location are given higher priority, which requires real-time and visual sharing of the maintenance workers' location through the mobile APP.

Maintenance task determination confirms maintenance workers and required accessories after maintenance resources are dispatched. The system automatically matches the appropriate number of workers and pushes them to the corresponding worker's mobile phone through the system. Workers in the specified time, according to their idle state and the actual situation, to receive and confirm. The system administrator can set the total number of validations. If the number of confirmed workers exceeds this number, no other workers will be allowed to accept this repair task. In general, when a maintenance worker confirms the acceptance of the task, the push of the maintenance task will be automatically terminated. Once a maintenance worker accepts a maintenance assignment, it must be performed.

In the maintenance task tracking stage, the track of workers' footprints (movement track), the time-consuming tracking of the maintenance task, the problem feedback of the field maintenance task, and the guidance of the platform management for remote maintenance. Similarly, using a mobile phone APP to track the movement track of maintenance workers is also easy to supervise and check the movement path and track of workers, as one of the bases for checking the fare invoice. Tracking maintenance task time is also to better monitor the efficiency of maintenance workers and reduce maintenance costs. The purpose of providing problem feedback to the field maintenance task and providing remote maintenance guidance to the platform manager is to fully integrate the maintenance resources and improve the efficiency and effect of maintenance. After the maintenance workers finish the maintenance, the customer company will check and check the invoice information (fare invoice, spare parts invoice, etc.), especially the information of amount and time, to ensure that the invoice information uploaded to the system is complete and accurate.

In the maintenance task evaluation stage, the customer company comprehensively evaluates the maintenance attitude, maintenance quality, use of spare parts and maintenance effect of maintenance workers. The platform assists the system administrator in managing invoices, calculate and calculate maintenance costs. The evaluation scores of maintenance workers will be permanently archived and stored in the system and serve as an important reference for the system to select maintenance workers in the future.

(5)   Knowledge management module

The knowledge management module is the necessary supplement of system function, the guarantee of automatic and intelligent operation of the system, and the primary basis of fault diagnosis, analysis and maintenance decision. It is divided into expert library management, rule library management, model library, and case library management. All four blocks have the functions of creating and importing, updating and refining, export and statistics. At the same time, in addition to the three functions of case base management, there are two major functions of maintenance experience sharing and maintenance experience exchange, which are mainly for maintenance workers and consulting experts to communicate, sort out and improve.

In the organization structure, Jinbaoli added IM division, plan through continuous IM change, providing customers with automation, clean, information, intellectual production device, and system solutions, make customers realize the low pollution, low labor input, low labor intensity but an efficient model of "three low" high production.

Jinbaoli's goal was to ensure sales growth in 2018, based on no less than 30% in predicting maintenance and maintenance systems, autonomic computing and big data cloud platform, etc. Research direction further achievements make predicting maintenance, accordingly promote the business model transformation, and training equipment maintenance system for the company's future business growth momentum.

Thinking Exercise

1. What kinds of cloud architecture are there?
2. How many options are there for the charging mode of the cloud monitoring platform?
3. How many services can cloud monitoring provide? What are the innovative business models?
4. What steps should be taken before cloud monitoring?
5. What derivative work can be done to further create value after the implementation of cloud monitoring?

Homework of This Chapter

1. Please point out the benefits of decentralized products and services such as Google's Pixel Buds.
2. How does 3D printing improve the efficiency of R&D and manufacturing? Is 3D printing a digital technology for design or manufacturing?
3. What is the relationship between the new technologies of IM?
4. What is the relationship between new IM technologies and new business models?
5. What are the driving factors for the transformation of IM?
6. Why is 5G communication technology, blockchain and 3D printing important for IM?
7. What are the directions of social change driven by intelligent technology?
8. Please state the basic idea of conceiving and designing a new type of industrial robot or service robot.
9. What are the emerging technology developments that could bring about huge growth in intelligent manufacturing?

Reference

1. Jalal M, Wang K, Jefferson S, et al. Scraping social media photos posted in Kenya and elsewhere to detect and analyze food types. In: 5th international workshop on multimedia assisted dietary management (MADiMa'19), October 21, 2019, Nice, France. New York, NY, USA: ACM; 2019. 10 p. https://doi.org/10.1145/3347448.3357170

Correction to: Background, Basic Concepts and Methods

Correction to:
Chapter 1 in: C. Lai *Intelligent Manufacturing*
https://doi.org/10.1007/978-981-19-0167-6_1

In the original version of the chapter, the author provided corrections have been incorporated in Fig. 1.13. The book and the chapter have been updated with the changes.

The updated version of this chapter can be found at
https://doi.org/10.1007/978-981-19-0167-6_1

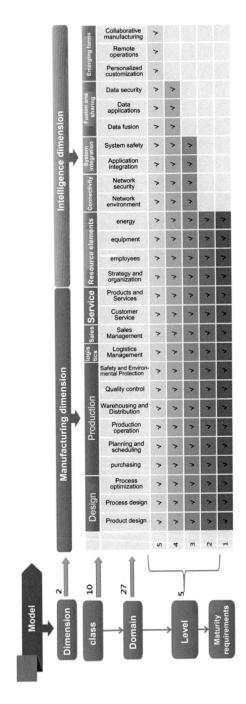

Fig. 1.13 Relationship diagram between model architecture and capability maturity Matrix [4]

Appendix
Research and Evaluation Outline
of Intelligent Manufacturing Level

China Institute of Standardization of Electronic Technology has put forward the White Paper on Intelligent Manufacturing Capability Maturity Model (1.0).The purpose of this white paper is to provide guidance for enterprises to implement intelligent manufacturing, help enterprises understand their own development stage, self-assessment and diagnosis according to the Capability Maturity Model, and achieve the purpose of targeted enhancement and improvement of intelligent manufacturing capabilities.

Intelligent manufacturing capability maturity model (CMM) is on the basis of the research on domestic and foreign relevant maturity model, combined with the characteristics of intelligent manufacturing in China and enterprise's practical experience summed up a set of methodology, it gives to organize the implementation of intelligent manufacturing to achieve goals and the steps of the evolution path, put forward to realize the core elements of intelligent manufacturing, characteristics and requirements, It provides a framework for internal and external stakeholders to understand the current state of intelligent manufacturing, establish strategic objectives and implement planning of intelligent manufacturing. The model reference the intelligent manufacturing standard system construction of the national guide (2015 edition) "put forward by the architecture of intelligent manufacturing system life cycle, system hierarchy and intelligent features three dimensions, to refine the core characteristics and components of intelligent manufacturing summary, summarized as" Intelligence + manufacturing in two dimensions, the last show for one dimensional form, that is, design, production, logistics, sales, service, resource elements, interconnection, system integration, information fusion, emerging business 10 categories of core competencies and refined 27 domains. In the model, related domains are classified and required from low to high levels (planning level, standard level, integration level, optimization level and leading level).According to the different needs of users, it can be divided into the overall maturity model and the single capability model.

Table A.1 Corresponding relationship between scores and grades

Level etc.	Corresponding scoring interval
Level 5 leading level	$4.8 \leq X \leq 5$
Level 4 optimization level	$3.8 \leq X < 4.8$
Level 3 integration level	$2.8 \leq X < 3.8$
Level 2 specification level	$1.8 \leq X < 2.8$
Level 1 planning level	$0.8 \leq X < 1.8$

The relationship between model architecture and Capability Maturity Matrix is shown in Fig. 1.13.

First of all, the enterprise chooses the appropriate model (whole or single item) in combination with its own development strategy and goals, chooses the evaluation domain (process or discrete) according to the characteristics of the industry, judges whether the maturity requirements are met through the form of "problem" investigation, and calculates the score according to the degree of satisfaction, and gives the result. The evaluation process is as follows:

1. Different problems will be set for each capability maturity requirement, and the satisfaction degree of the "problems" will be evaluated as the input of intelligent manufacturing evaluation. The evaluation of the problem requires experts to collect evidence on the spot, compare the evidence with the problem, and get the score of the problem, which is also the score of the maturity requirements. According to the degree of satisfaction to the problem, set 0, 0.5, 0.8, 1 four grades of scoring principle. If the question has a score of 0, the rating is not passed.

2. After scoring the maturity requirements, the weighted average will form the score of the domain, and then the score of the category will be calculated. Finally, the total score value of the organization will be obtained and the grade will be given.

3. The average principle is adopted for setting the domain weight. When an organization applies for the evaluation of a certain level, the average score of all the classes involved in the level must reach 0.8 to be considered as meeting the requirements of the level, and it can only apply for the evaluation of a higher level after meeting the requirements of a lower level. (Note: If the score of any question at the same level does not equal 0, or the score of any field < 0.5, it will be deemed that it does not meet the ability requirements of this level).

The corresponding relationship between the final results and grades is shown in Table A.1.

A Manufacturing Dimension

1 Design

Design is to form the realization of design requirements through the process of product and process planning, design, reasoning verification and simulation optimization. The improvement of design capability maturity is from experience-based design and reasoning verification, to parameterization/modularization and modeling design and simulation optimization based on knowledge base, and then to the collaboration of design, process, manufacturing, inspection, operation and maintenance of the whole life cycle of products, reflecting the rapid satisfaction of personalized needs.

1.1 Product design:

Level 1	Execute 2D CAD based on design experience, and develop product design related standards and specifications
Level 2	To realize 3D CAD and the internal coordination of product design
Level 3	Construct the 3D model of integrated product design information, carry out the design simulation optimization of key links, and realize the parallel collaboration of product design and process design
Level 4	full-dimensional simulation and optimization of design process manufacturing are realized based on knowledge base, and the cooperation of model-based design, manufacturing, inspection, operation and maintenance is realized
Level 5	Realize product design cloud service based on big data and knowledge base, realize product personalized design and collaborative design

1.2 Process design (discrete manufacturing):

Level 1	Realizing computer aided process planning and process design
Level 2	The simulation of key links of process design and the internal coordination of process design are realized
Level 3	Computer-aided 3D process design and simulation optimization are realized, and information interaction and parallel collaboration between process design and product design are realized
Level 4	The process design and simulation based on process knowledge base are realized, and the coordination between process design and manufacturing is realized
Level 5	Based on knowledge base assisted process innovation reasoning and online independent optimization, it realizes multi-field, multi-region and cross-platform comprehensive collaboration and provides real-time process design services

1.3   Process optimization (process-based production)

Level 1	Possess process model and parameters conforming to national/industry/enterprise standards
Level 2	The process model is applied to the site and can meet the requirements of site, safety, environment and quality
Level 3	Be able to use off-line optimization platform to establish unit level process optimization model
Level 4	The whole process optimization is realized based on the process optimization model and knowledge base
Level 5	Establish a complete 3D digital simulation model of the process, complete the digital simulation of the whole production process, and realize the real-time online optimization of the process based on the knowledge base

2 *Production*

Production is to control the five production factors of human, machine, material, law and environment through the integration of IT and OT, so as to realize the intelligent scheduling and adjustment and optimization of the whole production process from front-end procurement, production planning management to back-end warehousing and logistics, so as to achieve flexible production. The improvement of production capacity maturity starts from the information management with production tasks as the core, to the centralized control of various elements and processes, and finally to the closed-loop and self-adaptive process of the whole process from procurement, production planning and production scheduling, production operations, warehousing and logistics, and completion feedback.

2.1   Purchasing

Level 1	Have a certain information basis to assist the procurement business
Level 2	Enterprise level procurement information management, including supplier management, price comparison procurement, contract management, etc., to achieve internal procurement data sharing
Level 3	To realize the integration of purchasing management system with production and warehouse management system, and to realize the synchronization of planning, flow, inventory and documents
Level 4	Realize the combination of procurement, supply, sales and other processes, realize partial data sharing with important suppliers, and be able to predict replenishment
Level 5	Inventory can be perceived in real time, and real-time purchasing plan can be formed by analyzing and making decisions on sales forecast and inventory. Realize data sharing with supply chain partner enterprises

2.2   Planning and scheduling:

Level 1	Realize master production plan management, can generate master production plan and scheduling scheme from sales order and market forecast information
Level 2	To realize the operation of material demand plan, the production plan and purchase plan generated by the operation results are still unlimited capacity plan, which requires manual participation in adjustment and scheduling
Level 3	Based on safety stock, purchase lead time, production lead time and other factors to realize material demand calculation, and automatically generate production plan and purchase plan
Level 4	To realize the operation of production resource planning, balance the capacity load analysis and detailed capacity planning comprehensively, and make the production plan granularity to the day
Level 5	Based on the production scheduling algorithm and the standard time database established based on constraints (process order, processing resources, working time, etc.), advanced production scheduling and scheduling can be realized

2.3   Production operation:

Level 1	Equipped with automatic and digital equipment and production line, with on-site control system
Level 2	Can use information technology means to send electronic documents such as all kinds of processes and work instructions to production units to achieve data collection of personnel, machines, materials and other resources
Level 3	It can realize the business integration of resource management, process routes, production operations, warehousing and distribution, collect and store the real-time data information of the production process, provide the real-time updated analysis results of the manufacturing process and visualize them
Level 4	Be able to optimize production process through production process data, output, quality and other data;
Level 5	Improve production efficiency and quality by monitoring the whole production process, automatically warning or correcting abnormalities in production

2.4   Warehousing and distribution:

Level 1	Be able to realize inventory and inventory management of raw materials, middleware, finished products based on information management system
Level 2	Be able to use RFID/ QR code/label and other technologies to realize digital identification of raw materials, middleware, finished products, etc., and be able to realize automatic or semi-automatic inbound and outbound management based on identification technology
Level 3	It can realize the integration of warehousing and distribution, production planning, manufacturing execution and enterprise resource management

(continued)

(continued)

Level 4	Able to pull material distribution based on actual production situation of production line and adjust target inventory level based on customer and product demand
Level 5	Be able to achieve optimal inventory and immediate delivery

2.5   Quality control:

Level 1	Establish quality inspection specifications, which can be inspected by measuring instruments that meet the requirements and form inspection data
Level 2	Establish quality control system, use information technology means to assist quality inspection, through the analysis and statistics of inspection data to achieve quality control chart
Level 3	Realize the online inspection of key process quality, and automatically judge and warn the inspection results through the integration of inspection procedure and digital inspection equipment\system
Level 4	Establish a knowledge base for the disposal of product quality problems, predict the possible abnormality of product quality in the future based on the online test results of product quality, and automatically provide the corrective measures for the production process based on the knowledge base
Level 5	Through online monitoring of quality data analysis and prediction based on data model, automatic repair and adjustment of relevant production parameters to ensure the continuous stability of product quality

2.6   Safety and environmental protection:

Level 1	Information technology has been used for risk, hidden danger, emergency safety management and environmental data monitoring and statistics
Level 2	Can realize the whole process information management from the clean production to the end management
Level 3	Through the establishment of safety training, typical hidden danger management, emergency management knowledge base to assist safety management; Real-time online monitoring of all environmental pollution points, integration of monitoring data with production and equipment data, timely warning of pollution sources exceeding standards
Level 4	Support multi-source information fusion on site, establish emergency command center and carry out emergency response through expert database; Establish environmental protection management model and real-time optimization, online generation of environmental protection optimization scheme
Level 5	Based on the knowledge base, it supports the analysis and decision of safe operation and realizes the integrated management of safe operation and risk control. Use big data to automatically predict the overall environmental conditions of all pollution sources, and automatically formulate and implement governance plans based on real-time data of governance facilities, production, equipment, etc.

3 Logistics

3.1 Logistics management

Logistics management is to transport product downstream enterprises in the process or user, using bar codes, RFID, sensor as well as the global positioning system (GPS) and other advanced Internet technologies, through the information processing and network communication technology platform to realize the automatic transportation of operation, visual monitoring and management on the optimization of vehicle, path, in order to improve the transportation efficiency, reduce energy consumption. The improvement of logistics capability maturity starts from the information management of orders, planning and scheduling, and information tracking, to the management with a variety of strategies, and finally to the realization of lean management and intelligent logistics. It focuses on order management, transportation planning and scheduling management, logistics information tracking and feedback, transportation path optimization, etc. The levels and their characteristics are as follows.

Level 1	Manage the logistics process by means of informationization and make simple tracking and feedback to the information
Level 2	To achieve order management, scheduling, information tracking and capacity resource management through information system
Level 3	Realize the integration of outbound and transportation process, realize multimodal transportation, and push logistics information to customers
Level 4	The knowledge model is applied to realize lean order management, path optimization and real-time location tracking
Level 5	Realize UAV (Unmanned Aerial Vehicle) transport, IoT tracking etc.

4 Sales

4.1 Sales management

Sales management is based on customer demand as the core, using big data, cloud computing and other technologies to analyze and predict sales data and behavior, driving the optimization and adjustment of production planning, storage, procurement, supplier management and other businesses. Sales capability maturity evolves from the sales plan, sales orders, price, distribution plan, information management of customer relationship, to the customer demand forecast/customer actual demand production, procurement and logistics planning, finally realizes through more accurate sales forecast for the enterprise customer management, supply chain management and production management optimization, as well as the personalized marketing and so on. It focuses on sales data mining, sales forecasting and sales planning, integration of sales business and related businesses, and new models of sales. The levels and their characteristics are as follows.

Level 1	Simple management of sales business through information system
Level 2	Through the information system to achieve the whole process of sales management, strengthen customer relationship management
Level 3	Integrate sales, production, warehousing and other businesses to achieve product demand forecasting/actual demand driving production, procurement and logistics planning
Level 4	Apply knowledge model to optimize sales forecast and make more accurate sales plan. Integrate all sales methods through e-commerce platform to realize the automatic adjustment of purchasing, production and logistics plans according to customer demand changes
Level 5	Can realize the e-commerce platform of big data analysis and personalized marketing and other functions

5 Service

Service is based on customer satisfaction survey and usage tracking, statistical analysis of product operation and maintenance, feedback to relevant departments, maintenance of customer relations, improvement of product process, to achieve the vertical digging of customer requirements for product functions and performance, and then horizontally expand customer base. The improvement of service capability maturity is the transformation of service mode from offline, online, cloud platform and mobile client, customer service robot/field, online and offline remote guidance, remote tools, remote platform, AR/VR, and finally can provide personalized customer service and innovative products and services based on knowledge mining.

5.1 Customer service

Level 1	Set up customer service department, manage customer service information through information means, and feed back customer service information to relevant departments to maintain customer relations
Level 2	With a standard service system and customer service system, customer service management through the information system, and customer service information feedback to the relevant departments, maintain customer relations
Level 3	Build customer service knowledge base, provide customer service through cloud platform, and integrate with customer relationship management system to improve service quality and customer relationship

(continued)

(continued)

Level 4	To realize the fine knowledge management for customers and provide mobile customer service
Level 5	Through the intelligent customer service robot, to provide intelligent services, personalized services

5.2   Products and services

Level 1	Set up product service department, manage product operation and maintenance information through information means, and feed back customer service information to relevant departments to guide product process improvement
Level 2	It has a standardized product and service system, conducts product and service management through information system, and feeds back product and service information to relevant departments to guide product process improvement
Level 3	The product has the functions of storage, network communication and so on. The product fault knowledge base can be established to provide product services through the network and remote tools, and the product fault analysis results can be fed back to the relevant departments, so as to continuously improve the design and production of old products and provide the basis for the design and production of new products
Level 4	Product has the function such as data acquisition, communication and remote control, remote operations services platform, providing online detection and fault early warning, predictive maintenance and operation optimization, remote upgrade services, through integration with other systems, the information feedback to the relevant departments, the design of the continuous improvement of old products production, and provide the foundation for new product design and production
Level 5	Through Internet of Things technology, augmented reality/virtual reality technology, cloud computing and big data analysis technology, intelligent operation and innovative application services are realized

B   Intelligence Dimension

6   Resource Elements

Resource elements are the planning, management and optimization of the organization's strategy, organizational structure, personnel, equipment and energy and other elements, which provide the basis for the implementation of intelligent manufacturing. The improvement of the capability maturity of resource elements reflects the transformation from the planning of management vision to the application of information management to the intelligent decision-making, and reflects the improvement of the intelligent management level of the organization.

6.1 Strategy and organization

Level 1	The organization has a vision to develop intelligent manufacturing and a commitment to include financial input
Level 2	The organization has formed a strategic plan to develop intelligent manufacturing and established a clear fund management system
Level 3	The organization has implemented intelligent manufacturing in accordance with the development plan, and the fund has been invested. The development strategy of intelligent manufacturing is promoting the transformation of the organization and optimizing the organizational structure
Level 4	Intelligent manufacturing has become the core competitiveness of the organization. The strategic adjustment of the organization is based on the development of intelligent manufacturing
Level 5	The intelligent manufacturing development strategy of the organization has created higher economic benefits for the organization, and the innovative management strategy has brought new business opportunities and new business models for the organization

6.2 Employees

Level 1	Ability to determine the personnel capabilities required to build an intelligent manufacturing environment
Level 2	Can provide employees with access to the corresponding capabilities
Level 3	Ability to provide continuous education or training to employees based on their intellectual development needs
Level 4	Can analyze the ability level of existing employees through the information system, so that the skills of employees and the development level of intelligent manufacturing to keep pace with the improvement
Level 5	It can motivate employees to acquire the skills needed for intelligent manufacturing in more areas and continuously improve their abilities

6.3 Equipment

Level 1	Can use information means to realize the daily management of some equipment, start to consider the digital transformation of equipment
Level 2	Continuously carry out digital transformation of equipment, and be able to use information means to realize the state management of equipment
Level 3	It can use the equipment management system to realize the life cycle management of the equipment and remotely monitor the key equipment in real time

(continued)

(continued)

Level 4	The digital transformation of equipment has been basically completed, enabling experts to diagnose the equipment online remotely, and the operation model of key equipment has been established
Level 5	Be able to carry out predictive maintenance of equipment based on knowledge base and big data analysis

6.4   Energy

Level 1	To start the informationization of energy management and realize the collection and monitoring of some energy data
Level 2	Be able to collect and statistic the main energy data through the information management system
Level 3	It can monitor the production, storage, conversion, transmission and consumption of energy, and integrate the energy plan with the production plan
Level 4	It can realize dynamic energy monitoring and fine management, and analyze the weak links of energy production, transportation and consumption
Level 5	Based on the collection and storage of energy data information, it can provide optimization strategies and schemes for energy consumption and production scheduling, and optimize the energy operation mode

7   *Connectivity*

Interconnection is the deployment and application of fieldbus, industrial Ethernet and wireless network in the factory, so that the factory has the environment of organically connecting people, machines and things. The improvement of connectivity maturity is the interconnection between equipment, workshop, factory and enterprise upstream and downstream systems, which reflects the support for system integration and collaborative manufacturing.

7.1   The network environment

Level 3	It can realize the interconnection and communication between manufacturing equipment and information collection and transmission
Level 4	To realize the interconnection between production management and enterprise management system
Level 5	It can realize the interconnection between the upstream and downstream systems and realize the seamless integration of production and operation

7.2   Network security

Level 3	With redundancy capability of network key equipment, carry out subnet management, with intrusion detection, user identification, access control, integrity detection and other security functions
Level 4	Ensure the security of data transmission and important subnet, and have self-recovery ability, with network protocol information filtering and data traffic control functions, can check the integrity of the network boundary
Level 5	To ensure the security of cloud data center access, to provide dedicated communication protocol or secure communication protocol services, to resist the attack of communication protocol damage

8   *System Integration*

The purpose of system integration is to realize the interconnection and interoperation of various businesses and information in the enterprise, and finally achieve the state of complete information physical integration. The improvement of the maturity of system integration is from the single application and interconnection and interoperation between systems within the enterprise to the integration of all systems within the enterprise and upstream and downstream among enterprises, which reflects the full sharing of resources.

8.1   Application integration

Level 3	Can focus on the core production process, part of the implementation of production, resource scheduling, supply chain, R&D design and other different systems interoperability
Level 4	Fully realize the interoperation of different systems such as production, resource scheduling, supply chain, R&D and design
Level 5	The integration of business between enterprises can be realized based on cloud platform

8.2   System safety

Level 3	Safety management requirements, incident management and corresponding systems should be formulated for industrial control systems, and safety risk assessment of major systems should be carried out regularly
Level 4	Be able to monitor non-local processes, conduct safety tests before the system goes into production, and conduct regular training, testing and drills according to emergency plans
Level 5	It can realize active defense and vulnerability scanning security protection for industrial control system

9 Information Fusion

The core of information fusion lies in the development and utilization of data. Through data standardization and application of data model, optimization of design, production, service and other processes can be realized, and the ability of prediction and early warning and independent decision-making can be improved. The improvement of information fusion maturity is a process from data analysis, data modeling to decision optimization.

9.1 Data fusion

Level 4	The enterprise builds a unified data model to realize data integration and transmission between databases and R&D systems
Level 5	Enterprises realize the network integration and application of database (cloud database), and can build multi-function data model according to the self-adaptive transmission of data, and realize the real-time floating transmission of data

9.2 Data applications

Level 4	Be able to analyze and model various business data such as R&D, manufacturing, product and service, and output relevant strategies of the enterprise
Level 5	Be able to optimize business processes online using models

9.3 Data security

Level 4	To ensure the confidentiality of stored information and realize the availability of data and system
Level 5	The establishment of remote disaster preparedness center and special communication channel ensures data security, integrity and confidentiality, which can provide integrity verification and recovery functions for system management data, identification information and important business

10 Emerging Forms

Emerging business forms are new business models that are formed by enterprises rethinking and constructing the production mode and organization mode of the manufacturing industry by adopting information methods and intelligent management measures under the promotion of the Internet. The capability maturity of emerging business forms is mainly reflected in the advanced stage of intelligent manufacturing, which realizes the purpose of meeting the personalized needs of users quickly and cheaply, remote control of equipment, interactive sharing of information resources, and collaborative optimization of various links among enterprises and departments.

10.1   Personalized customization

Level 5	Through the personalized customization platform to realize the user's personalized needs docking; Be able to use industrial cloud and big data technology to mine and analyze users' personalized demand characteristics, and feed back to the design link for product optimization; Personalized customization platform can realize collaboration and integration with enterprise RESEARCH and development design, production planning, flexible manufacturing, marketing management, supply chain management, after-sales service and other information systems

10.2   Remote operations

Level 5	It can realize remote data acquisition, on-line monitoring and so on, and realize early warning and optimization through data mining and modeling

10.3   Collaborative manufacturing

Level 5	It can realize the sharing of innovation resources, design capacity and production capacity among enterprises and departments, and the parallel organization and collaborative optimization of design, supply, manufacturing and service among upstream and downstream enterprises

Printed in the United States
by Baker & Taylor Publisher Services